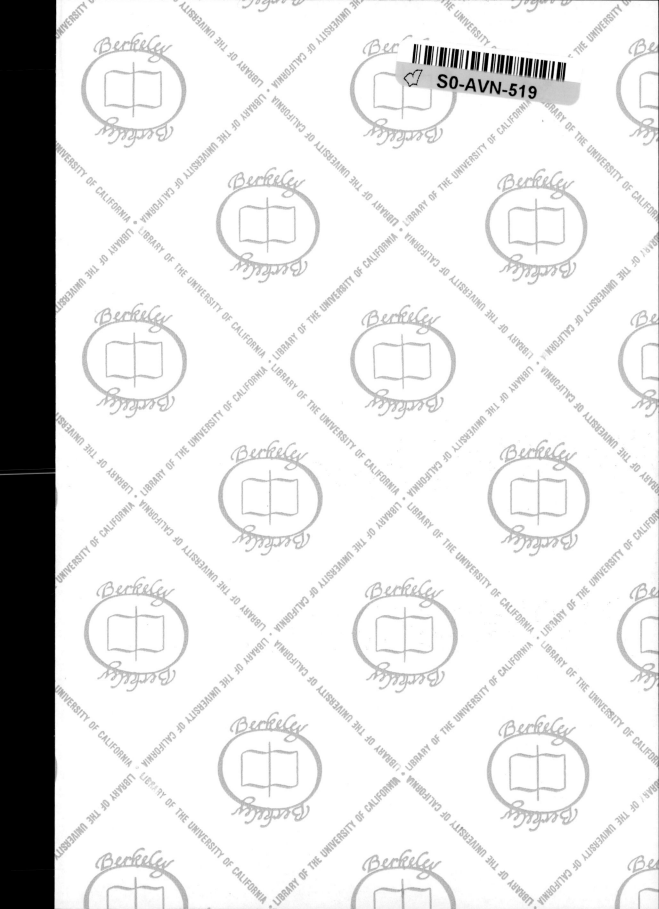

FIRST ASM

HEAT TREATMENT

AND

SURFACE ENGINEERING

CONFERENCE

IN EUROPE

Pt. 1

FIRST ASM HEAT TREATMENT AND SURFACE ENGINEERING CONFERENCE IN EUROPE

Amsterdam, The Netherlands, May 22nd - 24th, 1991

Conference Location :   RAI International Congress Centre

Conference Chairman :   R. Speri, Vide et Traitement (F)

Organizing Committee : W. Moerdijk (Chair), Houghton Benelux (NL)
                        R. von Bergen, Edgar Vaughan & Co Ltd. (UK)
                        R. Gijswijt, Staps Industries (NL)
                        T. Khan, ONERA (F)
                        A. Klosterman, Eindhoven University (NL)
                        B. Lineberg, Lineberg Consult AB (S)
                        B. Vandewiele, Surface Treatment Company (B)
                        H. Veltrop, Hauzer Techno Coating (NL)

Advisory Committee :   E.J. Mittemeijer (Chair), Delft University of Technology (NL)
                        T. Bell, University of Birmingham (UK)
                        G. Bodeen, Lindberg Corporation (USA)
                        O. Ericson, Brukens Nordic (S)
                        M. Gantois, Ecole des Mines (F)
                        B.M. Korevaar, Delft University of Technology (NL)
                        G. Krauss, Colorado School of Mines (USA)
                        H. Kunst, Degussa AG (D)
                        P. Mayr, Institut für Werkstofftechnik (D)
                        W. Münz, Hauzer Techno Coating (NL)
                        R. Roberti, Politecnico di Milano (I)
                        J. Roos, University of Leuven (B)
                        J. Sundgren, Linköping University (S)

List of Sponsors :   The Benelux Heat Treatment Society (VWT)
                        The Heat Treating Division Council of ASM
                        Industrial Heating Equipment Association (IHEA)
                        The International Federation of Heat Treatment (IFHT)
                        Metal Treating Institute (MTI)
                        Netherlands Society for Materials Science (BvM)
                        The Wolfson Heat Treatment Centre

Proceedings of the First

ASM HEAT TREATMENT

AND

SURFACE ENGINEERING CONFERENCE

in Europe

Pt.1

held May 22nd - 24th, 1991, in Amsterdam, The Netherlands

Editor

E.J. Mittemeijer

Laboratory of Materials Science
Delft University of Technology
Delft, The Netherlands

TRANS TECH PUBLICATIONS
Switzerland - Germany - UK - USA

Cover: Microstructure of TIG-cladded Stellite No. 6. For further details, please see the paper by C.F. Magnusson et al., pp. 443-458

Copyright © 1992 Trans Tech Publications Ltd

ISBN 0-87849-642-4

Materials Science Forum Vols. 102-104, Pt. 1 & Pt. 2

Distributed *in the Americas by*

Trans Tech Publications
c/o Ashgate Publishing Company
Old Post Road
Brookfield VT 05036
USA
          Fax: (802) 276 3837

and worldwide by

Trans Tech Publications Ltd
P.O. Box 254
CH-8049 Zürich
Switzerland
          Fax: (++41) 62 74 10 58

Preface

HEAT TREATMENT AND SURFACE ENGINEERING AS A BASIC DISCIPLINE OF METALLURGY

Heat treatment of metals dates back to ancient times. The apparently advanced ways employed for the making of Damascener and Samurai swords serve as examples. In this sense the innovation of the blast furnace by Darby, in the 18th century, and the discovery of the "martensite" structure named after Martens, a century ago, can be considered as (only) modern developments. These technological advances all pertain to steel. The breakthrough for aluminium as a major material was realized also "recently", in the beginning of this century, with the development of age-hardened aluminium alloys as "duraluminium".

The dynamics of the solid state, on the basis of thermal activation, never fail to fascinate the scientist, physical metallurgist, in his struggle to unravel the operating basic principles. Our society enormously stimulates such researchers by providing the necessary funds. Obviously, a principal cause for this generosity is the technological crop expected by this irrigation, as suggested by the above paragraph. Here it should be realized that the great developments in the field of heat treatment of metals are invariably due to a more or less direct interaction of scientist and engineer. This does not imply that their actual cooperation was and is usual.

In the past, in many cases practical development was clearly ahead of scientific understanding, although sometimes the reverse occurred. Nowadays it appears that more intensive interaction of laboratory research and practical application is in order, particularly because new methods of structural investigation allow understanding, in unprecedentedly detailed way, of basic principles involved. Thus tuning the properties of the products by optimizing the process parameters becomes feasible. So, a meeting ground for scientists and engineers working in this discipline is in order.

To meet the ever increasing demands, design of materials of increasing quality appears required. However, the development of new *bulk* materials can be an extremely costly enterprise. In many cases the *surface* of a workpiece is subjected most severely to external loads (e.g. corrosive agents, stress). Hence, surface treatments bringing about an upgrading of the quality of the surface adjacent material of workpieces can be very attractive. This led to the emergence of **Surface Engineering** as a separate field. Prominent and relatively old branches thereof are thermochemical processes as carburizing and nitriding, while recent and becoming increasingly important facets are processes based on physical and chemical vapour deposition.

The fragment of a historical sketch given here illustrates hopefully the interwoveness of (the older) Heat Treatment and (the newer) Surface Engineering. For both fields it holds: atomic mobility leading to phase transformations, as induced by supply of heat, is essential;

the properties generated should allow the practical application sought for and the same allied persons, i.e. scientists and engineers, are active in the combined fields. Against this background the occurrence of a conference on Heat Treatment and Surface Engineering is appropriate.

Without any intention to ignore any significant contribution, I may perhaps state that a principal activity in the area concerned occurs in Europe. Therefore it is fitting that a European Conference on Heat Treatment and Surface Engineering has been organized. The large attendance and participation in Amsterdam (22nd - 24th May, 1991) illustrates that Europe provides a focus for the field.

Considering the contents of the proceedings one may say that the really new scientific and technological developments involve: (i) the substitution of phenomenological knowledge, often inherited from a long past, by fundamental insight on the structure - property relations (example: nitriding and nitrocarburizing); (ii) the modelling and control of various processes as, in particular, the classical quenching and tempering and the thermochemical processes; (iii) the emergence of dedicated surface treatments, as those based on ion beams and the numerous ways for physical vapour deposition and (iv) the rise to maturity of vacuum and plasma technology including the application of (atmosphere) sensors of various kinds.

I have been active in this discipline since 1978, coming from a "world" dominated by "pure" science. It has been one of the most rewarding, fruitful experiences of my life until now, to perform and conduct research at the interface of science and technology, thereby communicating with both scientists and engineers. Others have felt similarly. I hope these proceedings give a reader the similar flavour as well as a useful survey of ongoing activities.

E.J. Mittemeijer

January, 1992
Delft, The Netherlands

Editorial Notes

The papers have been arranged in each chapter in an order primarily on the basis of logic of the sequence.

Although I fully appreciate the symbiosis of science and technology in the discipline Heat Treatment and Surface Engineering (see "PREFACE"), to my disappointment I had to refuse a number of manuscripts which lacked sufficient quality. In particular it should be realized that papers in these proceedings can not be merely commercial pamphlets (advertisements); instead they should contain a nucleus of new, scientific and/or technical information.

The papers marked with * at the page number in the Table of Contents did not conform, as manuscripts, to the "Instructions for Authors" in some respect:

1. because of a misunderstanding many authors of non-invited papers thought it was allowed to submit papers longer than 6 pages. A considerable number of these papers have been shortened by me by using the space available more efficiently, e.g. by reducing the size of figures and tables and by rearranging these (in particular I tried to limit the size of a manuscript to an *even* number of pages). However, often this was not possible. I apologize to the authors who did confine the length of their paper to 6 pages and now might have the feeling that they have been treated unfairly;

2. page length was not 24.5 cm (14.5 cm for title page) an/or page width was not 17 cm;

3. figures and tables were not of correct size and/or not fixed at the appropriate places in the manuscript;

4. coloured pictures were not allowed (although this was not explicitly stated).

With a few exceptions, I did not correct the English of those manuscripts which, even on the basis of my limited command of the English, exhibited a considerable amount of grammatical and orthographical errors.

All modifications have been performed as carefully as possible.

E.J. Mittemeijer

Table of Contents

i   **Preface**: Heat treatment and surface engineering as a basic discipline of metallurgy

ii   **Editorial Notes**

Part 1

I. HEAT TREATMENT OF FERROUS ALLOYS

T. Araki, M. Enomoto and K. Shibata
Microstructural identification and control of very low carbon modern high strength low alloy steels   3

S.J. Mikalac, A.V. Babdemarte, L.M. Brown, M.G. Vassilaros and M.E. Natishan
Thermal cycling in the austenite temperature region for grain-size control in steels   * 11

A. Kulmburg and B. Hribernik
The heat treatment of P/M tool steels   * 31

S.K.E. Forghany
Surface chemical properties of heat-treated industrial steel   43

J.C. Kelly
Heat resistant alloy performance in the heat-treatment industry   53

T.L. Elliot
Commercial heat treatment to 2000 AD   * 63

II. HEAT TREATMENT OF NON-FERROUS ALLOYS

C.W. Meyers, K.H. Hinton and J.-S. Chou
Towards the optimization of heat-treatment in aluminium alloys   * 75

M.J. Starink, V. Jooris and P. van Mourik
Hardness, melting reactions and heat treatment of Al-Si-Cu-Mg alloys reinforced with aluminiumoxide particles   85

B. Cina and F. Zeidess
Advances in the heat treatment of 7000 type aluminium alloys by retrogression and reaging   99

V.S. Zolotorevsky, N.A. Belov, A.A. Axenov and Das Goutam
Effect of high temperature treatment on the structure and mechanical properties of aluminium foundry alloys   109

---

* See 'Editorial Notes' just before 'Table of Contents'

E.W.J. van Hunnik, J. Colijn and J.A.F.M. Schade van Westrum
Heat treatment and phase interrelationships of the spray   115
cast Cu15Ni8Sn alloy

G. Moulin and F. Charpentier
Influence of heat treatment on the microstructural and   125
chemical evolution of $Au_{80}Sn_{20}$ alloys for welding applications

III. CARBURIZING

P. Casadesus and M. Gantois
Gas flow regulation in a carburizing heat treatment of steels   149

J.P. Souchard, P. Jaquot and M. Buvron
Plasma overcarburizing of chromium steels for hot working and 155
wear applications

B. Vandewiele
Influence of the base material hardenability on effective case   169
depth and core hardness

K.A. Erven, D.K. Matlock and G. Krauss
Bending fatigue and microstructure of gas-carburized alloy steels 183

A. Melander and S. Preston
Influence of surface microstructure on bending fatigue strength 199
of carburized steels

F. Kühn and B. Lineberg
Case hardening in the next century   211

IV. NITRIDING AND NITROCARBURIZING

M.A.J. Somers and E.J. Mittemeijer
Model description of iron-carbonitride compound-layer   223
formation during gaseous and salt-bath nitrocarburizing

L. Sproge and J. Slycke
Control of the compound-layer structure in gaseous   * 229
nitrocarburizing

Hong Du and J. Ågren
Kinetics of compound-layer growth during nitrocarburizing   * 243

F. Cavalleri, G. Tosi and P. Cavallotti
Nitriding of sintered steels and nitrocarburizing of steels in   * 249
fluidized bed furnace

M.F. Danke and F.J. Worzala
The structure and properties of ion nitrided 410 stainless   259

R. Roberti, G.M. La Vecchia and G. Colombo
A fracture mechanics approach to toughness characterization   271
of nitrided layers

J. Straus and D. Hitczenko
Influence of thermochemical treatment on corrosion   279
resistance of steels

G. Wahl
Modified process parameters in nitrocarburizing improve   285
component properties and give greater economy

P. Jacquot, M. Buvron and J.P. Souchard
Plasma nitriding and plasma carburizing of pure titanium   301
and Ti_6Al_4V alloy

V.   INDUCTION HEAT TREATMENT

P.K. Braisch
The influence of tempering and surface conditions on the   319
fatigue behaviour of surface induction hardened parts

P. Archambault, M. Pierronnet, F. Moreaux and Y. Pourprix   335
Induction heat treatment of case hardening steels

G.D. Pfaffmann
Selective surface treatment of gears by induction profile   *   345
hardening

C.C. Lal
Induction hardening of cast iron cylinder head valve seats   365

R.C. Gibson, W.B.R. Moore and R.A. Walker
TFX; an induction heating process for the ultra rapid heat   373
treatment of metal strip

R.S. Ruffini and R.J. Madeira
Production and concentration of magnetic flux for eddy   *383
current heating applications

VI.   LASER-BEAM TREATMENT

J.Th.M. de Hosson, J. Noordhuis and B.A. van Brussel
Reduction of the tensile stress state in laser treated materials   393

V.M. Weerasinghe, D.R.F. West and M. Czajlik
Laser surface nitriding of titanium and a titanium alloy   401

R. Lin and T. Ericsson
Fatigue properties and residual stresses in laser hardened   409
steels

S.M. Copley, E.Y. Yankov, J.A. Todd and M.I. Yankova
Coupled growth in eutectic systems   417

W. Cerri, G.P. Mor, M. Balbi and T. Zavanella
CO_2 laser subcritical annealing of 2Cr-1Mo steel laser   433
welded joint

C.F. Magnusson, G. Wiklund, E. Vuorinen, H. Engström and
T.F. Pedersen
Creating tailor-made surfaces with high power CO_2-lasers   443

---

*   See 'Editorial Notes' just before 'Table of Contents'

Part 2

VII.   ELECTRON-BEAM TREATMENT

R. Zenker
Electron-beam surface modification; state of the art (review)   459

D.G. Rickerby
Electron-beam mixing in copper-aluminium surface layers   * 477

VIII. ION-BEAM TREATMENT

J.P. Rivière and J. Delafond
Surface modification of materials produced by dynamic ion   485
mixing

M.A. El Khakani, G. Marest, N. Moncoffre, J. Tousset and
M. Robelet
Surface characterization and tribological improvements   * 495
of Ti or (Ti+C) implanted AISI M2 steel

J. Rieu, A. Pichaty, L.M. Rabbe and M. Robelet   505
Ion implantation for biomaterials

IX.   SHOT PEENING

C.P. Diepart
Controlled shot peening today   * 517

X.   CHEMICAL VAPOUR DEPOSITION

A.B. Smith and Z. Naeem
Chemically vapour deposited alumina coating on high speed   533
steel cutting tools

B. Eastwood, S. Harmer, J. Smith and A. Kempster
Characteristics of chromium-carbide layers produced on   543
martensitic stainless steel using different chemical vapour
deposition (CVD) techniques

B. Formanek. L. Swadzba, A. Maciejny and D. Rozlach
Multicomponent hard coatings for WC-Co sintered carbides   553

XI.   PHYSICAL VAPOUR DEPOSITION

S.L. Rohde
Plasma characteristics related to thin-film microstructures in   563
unbalanced magnetron sputtering processes

J. Vyskocil
Possibilities of suppression of droplets in arc evaporation   573
process

P. Robinson and A. Matthews
    Further developments of the multiple-mode ionized vapour   581
    source system

O. Knotek, H.-G. Prengel and J. Brand
    Deposition of Ti(C,N) coatings by arc evaporation   591

J.P. Celis, J.R. Roos, E. Vancoille, S. Boelens and J. Ebberink
    The development of (Ti,Al)N and (Ti,Nb)N ceramic coatings   599
    produced by steered arc ion-plating

A. Leyland, P.R. Stevenson and A. Matthews
    Tribological performance of plasma treated and coated   * 615
    surfaces

W. König and D. Kammermeier
    Performance of TiN-, Ti(C,N)- and (Ti,Al)N-coated cutting tools 623

A.S. James and A. Matthews
    Deposition of stabilized zirconia onto metallic substrates   633
    using a DC thermionic triode ion plating technique

XII. COATINGS; MISCELLANEOUS

P. Holiday, A. Debhbi-Alaoui and A. Matthews
    A comparison of techniques for producing diamond-like   643
    carbon coatings

M. Zlatanovic and T. Gredic
    Combined plasma nitriding/PVC processes   655

H.-D. Steffens, J. Lebküchner-Neugebauer and M. Dvorak
    Diffusion treatment of surface coatings by hot isostatic   667
    pressing

P. Gimondo, B. Grifoni and F. Sintoni
    Depositions of welded coatings to improve hot wear and   679
    thermal fatigue resistance of hot die steel

L. Sánchez, F. Gutiérrez, J.J. González and J.A. Alvarez
    A study of the adherence and cracking by deformation in a   689
    multilayer metallic coating of plain carbon steels

A. Koutsomichalis, H. Badekas and C. Panagopoulos
    A study of a laser irradiated molybdenum plasma sprayed   * 699
    coating on steel

F. Alonso, I. Fagoaga and P. Oregui
    Plasma-spray coatings on carbon-epoxy substrates   709

L. Swadzba, B. Formanek, A. Maciejny and J. Biedron
    Microstructure and resistance to cracking of modified   721
    protective diffusion coatings on nickel-base superalloys

B.A. Bellamy
    The use of analysis in surface treatments   729

E. Schenuit
    Reliable hardness testing with a fast ultrasonic method (UCI)   733

---

* See 'Editorial Notes' just before 'Table of Contents'

XIII. MODELLING AND CONTROL

H.M. Tensi and A. Stich
Martens hardening of steel; prediction of temperature   * 741
distribution and surface hardness

T. Réti and M. Gergely
Computerized process planning in heat treating practice   755
using personal computers

A. Thuvander and A. Melander
Calculation of distortion during quenching of a low carbon steel   767

R. Arola, H. Martikainen and J. Virta
Computer aided simulation of the temperature and the   783
distortion of components during heating up in heat-treatment
furnaces

M. Isac, D. Pascota and S. Ciuca
Mathematical modelling of the heat treatments (quenching   * 791
and tempering) of a Cr-Mo-B-V, high strength low alloy steel

N. Kanetake
Integrated computer control of induction hardening   799

F. Kühn
Computer-aided optimization of heat-treatment processes for   * 809
the automobile industry

W. Stein
A new generation of process computers for the heat-treatment   821
processes

H.W. Bond
Oxygen sensors; a review of their impact on the heat treating   831
industry

J.E. Powalisz
Electronic flow meter and valve integration with computer   * 839
controlled process equipment

XIV. FURNACE TECHNOLOGY

B. Edenhofer, F. Bless and W. Peter
Progress in vacuum- and plasma-technology   849

Altena
Universal vacuum hardening plant; application and results   * 859

B. Lhote and O. Delcourt
Gas quenching with helium in vacuum furnaces   * 867

A. Kay
Multiflow pressure quenching in vacuum furnaces   * 885

D. Pye
Trends and developments in vacuum furnace technology   * 897

---

* See 'Editorial Notes' just before 'Table of Contents'

I. HEAT TREATMENT OF FERROUS ALLOYS

Materials Science Forum Vols. 102 - 104 (1992) pp. 3-10

MICROSTRUCTURAL IDENTIFICATION AND CONTROL OF VERY LOW CARBON MODERN HIGH STRENGTH LOW ALLOY STEELS

T. Araki (a), M. Enomoto (b) and K. Shibata (c)

(a) Kobe Steel, Ltd., R&D Bureau, 1-8 Marunouchi, Tokyo 100, Japan
(b) National Research Inst. for Metals, 2-3- Nakameguro, Meguroku, Tokyo 153, Japan
(c) University of Tokyo, 7-chome Hongo, Tokyo 113, Japan

ABSTRACT

In order to obtain an optimized balance between higher strength and required properties such as ductility, toughness and workability etc. in the most modern very low carbon HSLA steels, selection and control of the so-called intermediate stage microstructures are studied in conjuntion with the alloy designing and the factors of the thermo-mechanical control process (TMCP). Difficult identification of the main intermediate phases (Zw-α) constiuent in these steels is the bottleneck of planning systematic design with principles.   Such nonclassical intermediate microstructures consisting in these continuously cooled steels are classified into three major categories ,i.e. α q, α w, α°_B.   The morphology, characteristic natures and transformation mode of each Zw-α phase are discussed in respect with influencing metallurgical factors. The microstructural strengthening mechanisms are reconsidered from a basical viewpoit in correlation with the mechanical properties.

INTRODUCTION

The most modern non-heat-treated type HSLA steels can exert a good cost performance by improved balance between higher strength and other mechanical properties, such as ductility, toughness and workability (weldability, shapablity and machinability).   The energy for the heat treatment, e.g. quenching and tempering, can be also conserved by using this kind of steels, that is also favourable for solving global pollution problems.

As one of the non-heat-treated type high strength steels, the already developed vanadium microalloyed medium carbon steels are not good enough for the variety of heavy duty automotive parts and also for the higher level weldable construction steels.

The newly developing HSLA steels which have excellent properties and cost performance are utilizing the advanced process technology such as the thermomechanical control process (TMCP) and applying appropriate alloy design plan in parallel with introducing extensively low carbon content from 0.1% down to about 0.02%.[1,2]   Their consisting microstructures are much different from the conventional heat treated type HSLA steels and generally very complicated due to the variety of thermomechanical processed history of each steel, so that the classical knowlede on the microstructural constitution of such kind of steel is not enough to identify and classify the microconstitutions.   Sometimes there are serious confusion problems in the terminology concerning the observed microstructures.

In this paper the basical aspects of these TMCP type very low carbon HSLA steels are analyzed and reconsidered in reviewing the present status of knowledge on the metallographical investigations and the controlling technics for developing these modern very low carbon HSLA steels.

CONTROL OF MICROSTRUCTURES

In order to attain a fairly high strength level well balanced with enough ductility, toughness and workability, the microstructure of the steel composed of simple ferrite and pearlite is empirically known as inadequate for some heavy duty usages.   Meanwhile the microstructures which are evolved by the controlled rolling and accelerated cooling of (very) low carbon microalloyed steels are one of the most hopeful candidates for this purpose. The functions of microalloying elements, in particular of Nb and B, are utilized to obtain high performance structure in the said balance between higher strength and other mechanical properties.

For simplified discussion from now, the subject steels in this chapter are focused on the TMCP type steels roughly in the following category: i.e.

1) C = less than 0.08% down to 0.01%, Mn+Cr+Ni etc. around 1.5~2.5%, with some microalloying elements such as Ti, Nb, V, and B.
2) Tensile strength level: 700~1000 MPa with sufficient toughness and workability etc.
3) Desirable microstructures: other than the conventional ferritic-pearlitic microstructures; actually the so-called bainitic and bainite-like intermediate structures (Zw) and also martensite.

INTERMEDIATE STRUCTURE: Zw-α

In order to enhance the strength and toughness of the above said ferritic HSLA steels without quench-and-temper heat treatment, the controlled accelerated cooling technics are applied after rolling or forging process; that will usually give rise to a kind of Zw-ferritic phases including the bainitic ferrite as the matrices, except the cubic martensite in a case of extremely high cooling rate.

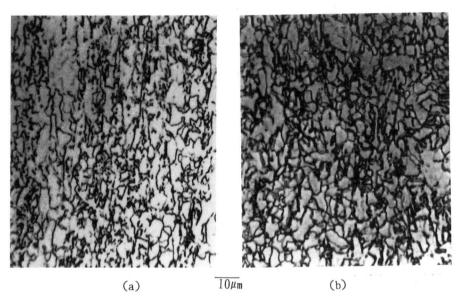

(a) $\overline{10\mu m}$ (b)

Photo. 1. Micrographs of acicular ferritic structures in 0.03~4% -C thermo-mechanical control processed microalloy HSLA steels: a) 780℃ roll finished, cooling rate 6℃/s [3], b) 800℃ roll finished, cooling rate 10 ℃/s [4].

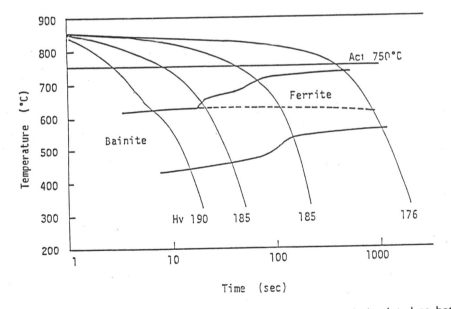

Fig. 1. CCT diagram of 0.02%-C microalloyed HSLA steel simulated as hot-roll-finished at 850℃.

For convenience, abbreviation α will be used for ferritic phases; e.g. α p stands for the polygonal ferrite (representative diffusional product), α w for Widmanstätten ferrite, $\alpha°_B$ for bainitic ferrite and α 'm denotes cubic martensite with (very) low carbon content. So far as the above classification is concerned, $\alpha°_B$ and α w are the said Zw-α phases, although the α w is not common in the (very) low carbon TMCP steels.

An example of microstructure of 0.03 % carbon TMCP type HSLA steel is shown in Photo. 1. Such a structure is often called "acicular ferrite" of line pipe steels[4,5,6]. The microstructure was ever regarded as consisting of a mixture of "acicular shaped" ferritic phases i.e. fine degenerated α p and bainitic ferrite accompanying a little part of tiny secondary particles. However by carefully investigating these constitutions, the ferritic phases have a many variety in morphology. Each ferritic grain seems to be formed by an individual formation mechanism respectively in accordance with the successive temperature change by the continuous cooling process.

At enough low temperatures around 700°C that is below the thermodynamical T_0 or T_0'temperature, the above α p-like phase tends to take much irregular shape surrounded by high angle/low angle mixed boundaries and sometimes in closely associated with the succeeding $\alpha°_B$.

This kind of ferrite is in fact off-polygonal or off-equiaxed in nature and its formation kinetics often seem to be different from the normal polygonal α p.

Therefore it should be rather regarded as a Zw-α phase, and will tentatively be denoted as α q: quasi-polygonal ferrite in contrast to the pure diffusional (polygonal)- α p, which is fully recrystallized.

In CCT curves by TMCP-simulated experiment of (very) low carbon HSLA steels as shown in Fig. 1 [7], each ferrite start curve has an inflexion point which suggests a transition from α p to α q in nucleation and growth kinetics.

Thus the Zw-α matrix phases of very low carbon steels shall be divided into following three wide categories, namely:

α q, α w, and $\alpha°_B$. For instance the microstructure in Photo. 1 will be analyzed as a combined mixture of principal α q plus a few part of $\alpha°_B$ accompanying few dispersion of C-enriched microconstituents: it can be described as:

α q + $\alpha°_B$ - (MA)   where (MA) means martensite/austenite connstituent
                                       which was enriched by carbon partitioning equilibrium.

In the case of somewhat higher carbon content, the α w will become more noticeable in the matrices of microstructures as discussed later.

INFLUENCE OF CARBON CONTENT

Among the chemical composition, the carbon content is a very critical factor influencing the microstructures and mechanical properties. The solubility of carbon in ferritic phase at the intermediate stage temperatures was computed by using Hillert Staffanson's model in the metastable equilibrium state with austenite [8]. Its result indicated that the solubility of carbon in α of Fe-

2%Mn system is about <u>0.02%</u> at about 700~ 500°C.

If the carbon content of TMCP steel does not exceed this 0.02%, the Zw-α phase can be formed directly from γ in a composition-invariant manner, therfore the microstructure virtually accompanys no C-enriched miner constituents. Namely the pure intermediate stage structures may be analyzed only for the ferritic matrix-phases:i.e. α q, α w and $\alpha^{\circ}{}_B$.

Photo. 2 shows two TEM micrographs of major phases in comparing an extra-low carbon with another very-low carbon TMCP type HSLA steel. It can be seen that the substructure ($\alpha^{\circ}{}_B$) of the 0.00x%-C steel is coarse and deformed (or distorted) and furthermore the recovery process seems much proceeded. In such level of C -content the role of carbon may be very significant for the formation mechnism particularly interacting with other cluster forming alloying elements.

In the case of extra-low carbon content, $\underline{\alpha\,q}$ often occurs clearly as a kind of the massive transformation product [9,10,11], of which microstructural aspect shows also typical massive type e.g. in extra-low carbon 1.8% Mn steel [12], 3%Ni-3%Mo steel [13] and also boron added HSLA steel [14].

When the carbon content exceeds the said 0.02%, both $\underline{\alpha\,q}$ and $\underline{\alpha^{\circ}{}_B}$ will be approaching to the general feature usually found in the said acicular ferrite steels. The Widmanstätten ferrite $\underline{\alpha\,w}$ is not usually found in extra-low carbon TMCP steels but can be observed as individual plate-like ferrite grains in very low carbon TMCP steels.

In actual TMCP type HSLA steels, the carbon contnt, alloying /microalloying elements and applied cooling rate will be selected in accordance with the required strength level, and by continuous cooling process the nucleation and growth of successively forming ferritic phases will give rise to a mixture of various choice of combination: fine α p, $\underline{\alpha\,q}$,(α w), $\underline{\alpha^{\circ}{}_B}$ and in some case α 'm.

MICROSTRUCTURAL STRENGTHENING MECHANISMS:

To analyze the physico-metallurgical factors influencing the strength properties, the microstructural strengthening mechanism operating in those steels will be first examined.

As described in Table 1, the strengthening mechanisms in the microstructure of matrix phases are largely divided into four categories according to the existing lattice imperfections. In the present TMCPed very low carbon steels the significant factors are the strengthening by both dislocation [B1] and grain boundry [C1] and subboundary [C2] which are evolved by the transformation in the range of the above stated intermediate stage temperatures.

The influence of each strengthening mechanism on the fracture resisting properties such as ductility, toughness and fatigue endurance was qualitatively surveyed in detail elswhere [l5|. The low temperature toughness will be favoured by grain- as well as subgrain- refining of α -phase resulted from the hot-deformed and unrecrystallized austenite followed by a sufficient cooling rate. The quasi-polygonal ferrite $\underline{\alpha\,q}$ has generally low or medium dislocation density

Table 1 Microstructural Strengthening Mechanisms and Characteristics

Defects of lattice	Symbol	Strengthening Mechanism	(Characteristics) ↑Ductility	↑Toughness	Strengthen. $\Delta\sigma\infty$
no	[O]	Theoretical strength: ~ 10 GPa	↑~0-flow	↑very low	
Point defects	[A]	Solution strengthening			
	[A1]	Substitutional SS hardening.	↑mild	↑fairly good	$(C_s)^n$
	[A2]	Interstitial SS hardening	↑ bad	↑ low	$(C_i)^m$
One dimensional defects	[B]	Dislocation strengthening	(ρ :dislocation density)		
	[B1]	Even dislocation density	↑ mild	↑moderate	ρ^n
	[B2]	Uneven distribution	*	* not good	depending on distribution mode
Two dimensional defects	[C]	Grain boundary strengthening	(d= grain diameter)		
	[C1]	by high angle boundaries	↑good	↑excellent	$d^{1/2}$
	[C2]	low angle sub-boundaries	↑ "	↑very good	d_s^n
Three dimensional defects	[D]	Particle/dispersion strengthen.	(λ = particle spacing)		
	[D1]	coherent particles< 5 nm	↑bad	↑not good	ε^n
	[D2]$_f$	[fine Incoherent " <50 nm	↑not good	↑not bad	λ^1
	[D2]$_c$	[coarse " > 50 nm	↑not bad	↑conditional	n

and very fine grain size such around 2~4 μ m with mixed grain boundaries of high and low angle in nature. Subsequently thus obtained strengthening through the transformation is generally compatible with the fracture resistant mechanical properties of the matrix.

The bainitic ferrite $\alpha^{\circ}{}_{B}$ usually has fairly high dislocation density and fine substructure, that means favourable strengthening by [B1] and [C2].

However the intergranular fracture mode via the prior austenite grain boundary is unfavourable for the bainitic ferrite $\alpha^{\circ}{}_{B}$ which is not capable of crossing over the prior γ -boundary in growth. An example micrograph of $\alpha^{\circ}{}_{B}$ structure of TMCPed very low carbon seel, in Photo. 2, shows the traces of prior γ -grain boundaries . The toughness, ductile-brittle transition, of the higher strength bainitic-ferritic steel can be improved by introducing some amount of fine α q phase which likely nucleates at such inherited high angle boundaries and is able to grow by crossing over this prior γ -grain boundaries. Thus it is necessary to control the microstructures to obtain the desirable combination of constituent phases by adapting the cooling rate and alloy design in accordance with the strength level.

For higher strength level e.g. over 850 MPa, beside the precipitation strengthening by [D1] by means of Nb, V, Cu etc. which mechanism is not good for the toughness, some portion of low carbon martensite, auto-tempered martensite and/or lower-bainitic phase shall be included as a shared major constituent strengthened by [B1] and [D2] in the microstructures. The carbon content of the steel shall then be enhanced e.g. up to 0.08% or a little more, therefore the resulted low temperature transformation products stated above are those of somewhat higher carbon content.

In contrast the higher temperature transformation products such as pearlite, degenerated pearlite and/or upper-bainitic phases with cementite, accompanying the Zw-α matrices, tend to act as unfavourable factors when the influences of these carbon enriched secondary phases to the ductility and toughness of the steels are considered.

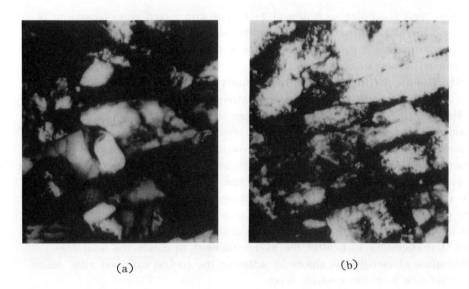

(a)   (b)

Photo. 2. Examples of substructures of very low carbon microalloyed HSLA steels by transmission electron microscopy X 10000.

REFERENCES

1] T.Araki and M.Enomoto: Procdgs. MatTech-90 Sympo., Helsinki Univ.Technology, (1990), Intern. Journ. Mater.& Prod. Technology (1991) to be pressed

2] T.Araki, K.Shibata, M.Enomoto; Procdgs. ICOMAT-89, Sydney, Mater. Scie. Forum, Vol.56, 1990, p.275

3] I.Kozasu, M.Suga et al.(NKK): Ref.2-5 Bainite Committee/ISIJpn (1990)

4] K.Abe,M.Shimizu, S.Takashima, H.Kaji; Proccdgs. THERMEC-88, Tokyo, ISIJ, (1988), p.332

5]Y.E.Smith, A.P.Colderon, R.L.Cryderman; Toward Improv. Ductility Toughness, Intern. Sympo.in Kyoto,(Amax), (1971), p.119

6]T.Tanaka; Internat. Metals Review, $\underline{4}$ (1981), p.185

7]H.Ohtani, Okaguchi; Ref.3-14 Bainite Committee/ISIJpn, 1990

8] M.Enomoto, T.Araki; Preprint 107th General Ass., Jpn. Inst. Metals, in Sendai,(1990), p.76

9] E.A.Wilson: Scripta Metallurg. $\underline{12}$(1978), p. 961

10] D.A.Mirzayev, V.M.Schastlivtsev; Procdgs. ICOMAT-86, Nara, JIM, (1986), p.282

11] E.O.Räsänen; Doctor thesis, Helsinki Univ. Tech.(1969) pp.1-86

12] M.J.Roberts: Metall. Trans. $\underline{1}$ (1970), p.3287

13] A.J.McEvily, R.G.Davies, C.L.Magee, T.L.Johnston; Transform. Hardenability of Steel,(Climax Molyb. Co.), Univ. Michigan, (1967), p.69

14] S.Sanagi, H.Yada; unpublished work (NSC), to JIM Bainite Committee (1986)

15] T.Araki; Procdgs. HSLA Steels-85 in Beijing, (CSM-ASM), (1985), p.259

Materials Science Forum Vols. 102 - 104 (1992) pp. 11-30
Copyright Trans Tech Publications, Switzerland

THERMAL CYCLING IN THE AUSTENITE TEMPERATURE REGION FOR GRAIN SIZE CONTROL IN STEELS

S.J. Mikalac, A.V. Brandemarte, L.M. Brown, M.G. Vassilaros and M.E. Natishan

David Taylor Research Center, Annapolis, MD, USA

ABSTRACT

An improvement in the strength and toughness of steel alloys when the average grain size is reduced has been well-documented; however, the grain size distribution may also be an important factor in controlling fracture toughness. For example, cleavage crack propagation in some steels has been shown to be controlled by the largest grains in a population and not necessarily by the average grain size. Repeated austenitization heat treatments are commonly used to achieve grain refinement although the effects on the grain size distribution have not been measured. This study determined the effects of repeated thermal cycling in the austenite temperature region on the average grain size and the grain size distribution of several high strength, low alloy (HSLA) steels. Repeated cycling generally refined the grain size and resulted in a more homogenous distribution of grain sizes. A model to explain this behavior is discussed. Experimental results demonstrated that the amount of grain refinement and homogenization due to thermal cycling was strongly dependent on the chemistry of the steel.

ACKNOWLEDGMENTS

The success of this metallurgical study can be contributed largely to the efforts of two summer aides, Karen Moore and Marc Raphael, whose enthusiasm and diligence were greatly appreciated. Thanks also go to C. Roe and J. Nasso. The authors gratefully acknowledge the sponsorship of I.L. Caplan, David Taylor Research Center (DTRC) Materials Technology Manager.

INTRODUCTION

The Navy high strength, low alloy (HSLA) steel system has taken advantage of modern steelmaking capabilities and thus is comprised of "clean" steels with very low carbon content, less than 0.06 weight percent. These steels exhibit high strength and toughness without the welding fabrication problems caused by higher carbon contents [1]. Strengthening is obtained primarily from fine copper and carbide precipitates. Tight control over the grain size, via chemistry and thermo-mechanical processing, is also used to enhance both the strength and toughness [2].

Grain size reduction is one of the few metallurgical means available for improving simultaneously both strength and toughness [3-5]. A simple and economical method to reduce the grain size in quenched and tempered structural steel is to repeat the austenitization heat treatment performed after the final rolling [6]. However, the relative efficiency of this method and the limits in strength and toughness improvements due to the grain size reduction are not known. The distribution of grain sizes may also change with repeated austenitization and significantly affect the toughness of the steel.

BACKGROUND

THE TRANSFORMATION TO AUSTENITE

The nucleation of new, strain-free austenite grains begins when a steel is heated above the lower critical temperature, Ac_1. For an austenite grain in a matrix of ferrite/bainite/martensite to occur and remain stable, a critical amount of energy needs to be obtained [7]. This energy is obtained thermally, and nucleation is also aided by occurring at areas within the microstructure which are sources of stored energy. These favorable heterogeneous nucleation sites include prior austenite grain boundaries and ferrite/bainite/martensite packet boundaries, as well as carbides and other precipitates [8]. The nucleation of austenite grains at favorable locations continues while in the dual-phase field.

If the steel is heated above the upper critical temperature, Ac_3, it becomes energetically favorable for the steel to become austenitic. A strong driving force now exists, the energy barrier is small in comparison, and thus nucleation of new austenite grains occurs with relative ease. The transformation is completed when the new grains begin to impinge upon one another within seconds of attaining the upper critical temperature [8]. As the number of austenite nuclei increases, the amount of growth of the newly-formed grains that can occur before impingement is decreased and thus the smaller the size of the grains at the time of impingement [4,5,7,8].

Remaining above the upper critical temperature after the transformation to austenite is complete will result in post-impingement grain growth with larger grains growing at the expense of smaller grains. Grain growth is energetically favorable because grain boundaries are high-energy sources; grain growth results in a reduction of grain boundary area and thus reduces the energy of the system. Grain growth is a thermally-activated process so the higher the temperature, the faster the growth rate [4,5,7,8].

OBJECTIVE

The objective of this study was to investigate and model the effect of repeated thermal cycling in the austenite temperature region on the prior austenite grain size and grain size distribution of several 690 MPa yield strength level Navy HSLA steels.

MATERIAL

Three HSLA steels, DTRC codes GRP (high carbon HSLA steel), GLB (high copper HSLA steel), and GLG (low carbon HSLA steel), were used in this study.

The chemistries of these steels are shown in Table 1.

The steels were received after rolling and austenitization at 900°C for one hour per inch of thickness followed by a water quench. Optically the microstructure appeared martensitic; however, the low carbon content of the steels implied a bainitic or ferritic structure. Detailed transmission electron microscopy has shown the microstructure of these HSLA steels to be composed of packets of intergranularly-nucleated ferrite or bainite packets with some retained austenite [2,9].

Table 1:   Chemical Compositions of HSLA Steels (Weight Percent)

Material Code:	GRP High Carbon HSLA Steel	GLB High Copper HSLA Steel	GLG Low Carbon HSLA Steel
Element			
Carbon	0.06	0.047	0.04
Manganese	0.83	0.85	0.86
Phosphorus	0.010	0.010	0.004
Sulfur	0.002	0.005	0.002
Silicon	0.37	0.22	0.27
Nickel	3.48	3.59	3.55
Chromium	0.58	0.57	0.57
Molybdenum	0.59	0.60	0.60
Copper	1.66	2.00	1.58
Niobium	0.028	0.025	0.030
Aluminum	0.028	0.021	0.032

EXPERIMENTAL PROCEDURE

HEAT TREATMENT

The temperatures used for the thermal cycling were: 845, 900, 955, and 1010°C. For each of the steels studied, 2.5 x 2.5 x 2.5-cm cubes were cut, held for one hour at temperature and subsequently water quenched. This thermal treatment was repeated up to three times. After the thermal cycling was completed, the heat-treated specimens were prepared for metallographic examination with the orientation transverse to the rolling direction.

MICROSTRUCTURAL EVALUATION

After receiving the austenitization heat treatment, the specimens were subjected to a temper embrittling heat treatment consisting of heating to 595°C, holding for 8 hours, and air cooling, in order to improve the etching response of the grain boundaries. Etchants based on a saturated solution of picric acid have proven to be the most effective at etching prior austenite grain boundaries [10-13]. These etchants have been shown to attack grain boundaries where the interstitials segregate during austenitization and/or tempering and are most successful with steels that have high phosphorus content [14-15]. However, due to their relatively low phosphorus content, less than 0.015 weight percent, HSLA steels do not readily respond to these

saturated picric etchants. The etching response of the steels was improved by
the temper embrittling heat treatment which allowed for the diffusion of
interstitial elements to the grain boundaries without an effect on the
austenite grain size.

After tempering, the specimens were sectioned with a water cooled cut-off
wheel and wet ground with silicon carbide abrasive discs in five steps
beginning with 60 grit paper and finishing with 600 grit paper. Polishing was
done in two steps using medium nap cloths charged with 6 and 1 micron diamond
spray, respectively.

A consistent etching technique was developed which yielded sufficient
contrast between the grain boundary and the grain interior for reliable grain
size measurements. Etching was accomplished using modified Winsteard's
reagent: 4g of picric acid was dissolved in 20 ml ethanol, 400 ml distilled
water was added followed by 2g of sodium dodecylbenzenesulfonate which acted
as a wetting agent. 100 ml of the modified Winsteard's reagent was poured
into a glass beaker, 3 drops (about 0.1 ml) of hydrochloric acid added, and
then the beaker was placed in a sonic cleaner. Freshly polished specimens
were immersed in the vibrating reagent. A dark deposit soon began to form on
the polished surface and this was swabbed off while keeping the specimen
submerged, every 30 seconds. The total etching time, not less than 4 minutes,
varied slightly for each specimen. The specimen was then swabbed, rinsed in
running water, ethanol, and blown dry. At this time the specimen was examined
with an optical microscope for grain boundary delineation and completeness.
If more etching was required, it was done in 1 or 2 minute intervals.
Frequently, a compromise had to be struck between grain boundary completeness
and loss of contrast between the boundaries and the grain interior.

Due to the etching response of this alloy, direct automatic image
analysis could not be employed. Even sophisticated gray image processing and
binary image editing did not alleviate problems such as incomplete grain
boundaries and grain interior "noise." In order to obtain high-contrast grain
boundaries, the following procedure was used: Photomicrographs were taken of
each specimen at either 500x or 1000x and placed in a photographic mounting
page with a clear acetate cover. The grain boundaries were then traced on the
acetate using a fine point, indelible marker. Upon removal of the
photomicrograph, a high contrast image of the grain boundaries was all that
remained. The tracing was then placed on a copy stand fitted with a charge
couple device video camera and the image inputted to an image analyzer. The
clean, high contrast image required only a minimum amount of binary image
editing to insure complete separation of the individual grains. Only those
grains completely within the image frame were kept and measured. A minimum
population of approximately 200 individual grains were measured on each
specimen. The raw data, which included measured parameters such as grain size
(feret), grain area, and grain perimeter were recorded.

STATISTICAL ANALYSES

For each material and thermal cycle, the variance of the grain size
distribution was calculated as an indication of the homogeneity of the grain
sizes. Average grain size (feret) was also measured, and to statistically
verify how the composition of grain sizes within a distribution were affected
by the thermal cycling, the following test was used:

The Wilcoxon-Mann-Whitney (WMW) test is designed to statistically test the equality of two distributions [16-17]. The WMW test is a nonparametric test and is ideal for comparing two non-Gaussian distributions, which was the case for the grain size distributions measured in this study. Two assumptions are required for validity of the WMW test and were met in this investigation: 1) observations from the two distributions are independent, and 2) each observation within a population is an independent random sample.

The hypotheses for the one-sided WMW test is generally stated as:

$$H_o: \quad F_x = F_y$$
$$H_1: \quad F_x < F_y$$

where F_x is the cumulative distribution for the random variable X, and F_y is the cumulative distribution for the random variable Y. The null hypothesis, H_o, states that the distribution for the random variable X is equal to the distribution of Y. The alternative hypothesis, H_1, states that the distribution of X is smaller than the distribution of Y. Under H_1, the random variable X is stochastically smaller than the random variable Y, thus an observation from the population X will tend to be smaller than an observation from Y.

Acceptance of the null hypothesis is based on the following test statistic U_x:

$$U_x = n_x(n_x+1)/2 - W_x$$

where n_x is the number of X observations and W_x is the sum of the ranks for X. To determine W_x the observations from X and Y must first be grouped as one set of data and then ranked in ascending order. The sum of the rank assignments associated with each X variable is defined as W_x.

The null hypothesis, H_o, is rejected if $U_x < U_a$, where U_a is the critical value for the random variable U with parameter a. The parameter a is preselected to guarantee a desired confidence level for the statistical test. Typically, a is preselected at 0.05 to guarantee a confidence level or accuracy of 95% for the test. For sample sizes greater than 20, an asymptotic normal approximation for U_a can be implemented. An excellent source for the critical values, U_a, is Beyer [18].

Alternatively, the probability that U_a is less than U_x, $P[U_a < U_x]$, can be examined to determine whether H_o should be rejected. If $P[U_a < U_x] < a$ the null hypothesis must be rejected, concluding that the random variable X is stochastically smaller than Y.

In summary, for a given steel chemistry and temperature, the results from the WMW test indicated how the grain sizes comprising a given distribution were changing due to thermal cycling. This test is a statistically significant verification (or contradiction) of the measured average grain size.

RESULTS

High Carbon HSLA Steel (GRP)

After rolling, this steel was austenitized at 900°C. In the as-received condition (AR), the average prior austenite grain size was 14 microns with a distribution of grain sizes as shown in figure 1 and a measured grain size variance of 38 (the "units" of variance are microns$_2$). One, two, and three cycles were performed at 845, 900, 955, and 1010°C. The effects of these temperatures and cycles on the average grain size, grain size distribution, and variance are documented in table 2 and shown in figures 1-6.

The first set of austenitization cycles was performed at 845°C. After the first cycle at 845°C, the average grain size was reduced from 14 microns to 13 microns. The second cycle resulted in a decrease to 9 microns. The third cycle at this temperature resulted in an increase in the average grain size to 12 microns. The results of the Wilcoxon-Mann-Whitney (WMW) test verified that the grain sizes comprising the distribution shifted downward after the first cycle, again after the second cycle, and increased after the third cycle. The variance of the grain size distribution decreased after the first cycle, decreased further after the second cycle, and increased after the third cycle.

The second series of austenitization cycles was performed at 900°C. After the first cycle, the average grain size dropped from the as-received value of 14 microns to 10 microns. The second cycle did not further reduce the grain size. The third cycle caused an increase in the average grain size to 13 microns. The results of the WMW test verified that the grain sizes comprising the distribution shifted downward after the first cycle, did not change with the second cycle, and then increased after the third cycle. The variance of the grain size distribution decreased after the first cycle, decreased further after the second cycle, and increased after the third cycle.

Table 2:   Change in Average Grain Size and Variance due to Thermal Cycling in the Austenite Temperature Region of High Carbon HSLA Steel

Austenitization Temperature (°C)	Cycle #	Average Grain Size (microns)	Variance
As-Received	-	14	38
845	1	13	33
845	2	9	22
845	3	12	33
900	1	10	24
900	2	10	21
900	3	13	29
955	1	14	35
955	2	15	35
955	3	14	49
1010	1	18	57
1010	2	24	138
1010	3	25	175

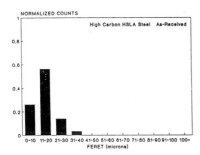
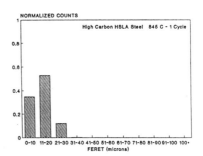

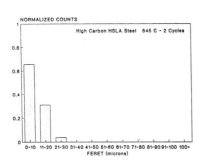
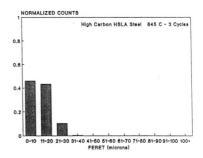

Figure 1.   Grain size distributions of the high carbon HSLA steel in the as-received condition and after austenitizing at 845°C.

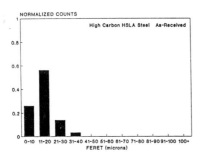
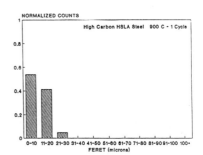

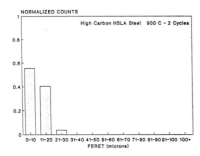
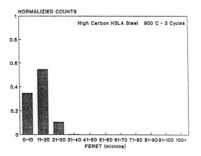

Figure 2:   Grain size distributions of the high carbon HSLA steel in the as-received condition and after austenitizing at 900°C.

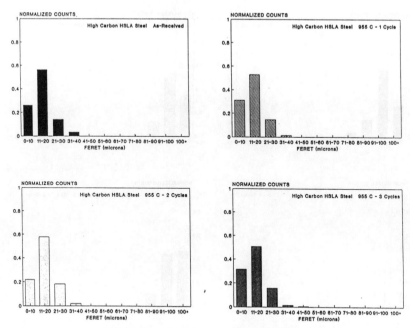

Figure 3: Grain size distributions of the high carbon HSLA steel in the as-received condition and after austenitizing at 955°C.

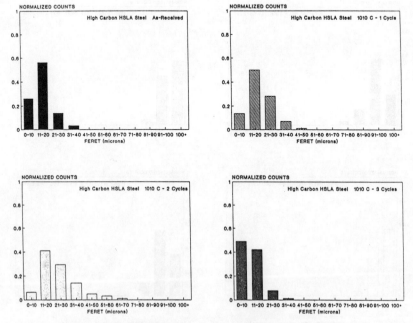

Figure 4: Grain size distributions of the high carbon HSLA steel in the as-received condition and after austenitizing at 1010°C.

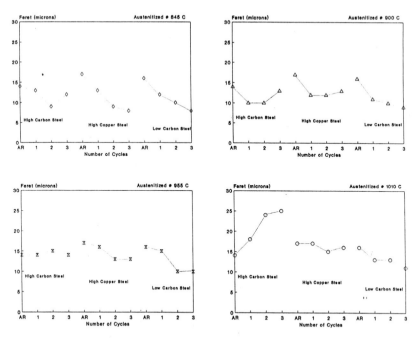

Figure 5: Change in average austenite grain size due to thermal cycling.

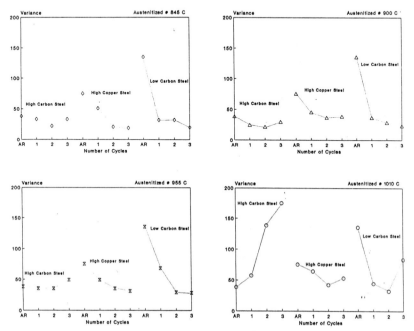

Figure 6: Change in the variance of the grain size distributions due to
thermal cycling.

The third series of austenitization cycles was performed at 955°C. The first cycle did not affect the average grain size. The second cycle saw an increase in the average grain size from 14 microns to 15 microns. After the third cycle, the average grain size dropped back down to 14 microns. These results were confirmed by the WMW test. The variance of the grain size distribution decreased after the first cycle, did not change with a second cycle, and then increased after the third cycle.

The fourth series of austenitization cycles was performed at 1010°C. After the first cycle, the average grain size increased from the as-received condition of 14 microns to 18 microns. The second cycle resulted in an increase in the average grain size to 24 microns. After the third cycle, the measured average grain size increased to 25 microns. The WMW test showed that the grain sizes of the distribution increased after the first cycle, increased further after the second cycle, and did not change further after the third cycle. The variance in the grain size distribution increased continually with each cycle.

High Copper HSLA Steel (GLB)

After rolling, this steel was austenitized at 900°C. In the as-received condition (AR), the average prior austenite grain size was 17 microns with a distribution of grain sizes as shown in figure 7 and a measured variance of 75. One, two, and three austenitization cycles were performed at 845, 900, 955, and 1010°C. The effects of these temperatures and cycles on the grain sizes and variance are documented in table 3 and shown in figures 5-10.

The first series of austenitization cycles was performed at 845°C. After the first cycle, the average grain size dropped from 17 microns to 13 microns. After the second cycle, the average grain size dropped to 9 microns. The third cycle at this temperature caused another drop in the average grain size to 8 microns. These results were statistically confirmed by the WMW test. The variance in the grain sizes also continually decreased with each cycle.

The second series of austenitization cycles was performed at 900°C. After the first cycle, the average grain size dropped from 17 microns to 12 microns. The second cycle did not further affect the average grain size. The last cycle increased the average grain size to 13 microns. The WMW test indicated that the grain sizes of the population decreased after the first cycle but were not affected after the second and third cycles. The variance in the grain sizes decreased after the first cycle, decreased again after the second cycle, and increased after the third cycle.

The third series of austenitization cycles was performed at 955°C. After the first cycle, the average grain size dropped from 17 microns to 16 microns. The second cycle reduced the prior austenite grain size to 13 microns. The third cycle did not further affect the average grain size. The WMW test indicated that the grain sizes comprising the distribution did not change significantly after the first cycle. The second cycle decreased the grain sizes while the third cycle did not further affect the grain sizes. The variance continually decreased with each cycle at this temperature.

The fourth series of austenitization cycles was performed at 1010°C. The measured average grain size remained at 17 microns after the first cycle, dropped to 15 microns after the second cycle, and then increased to 16 microns

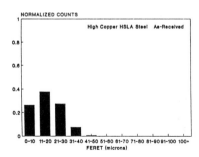
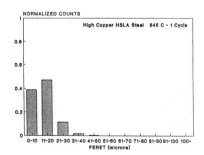

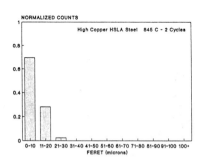
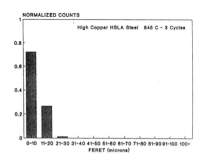

Figure 7: Grain size distributions of the high copper HSLA steel in the as
received condition and after austenitizing at 845°C.

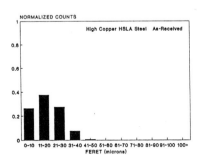
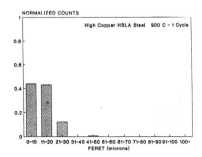

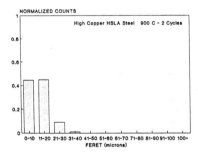
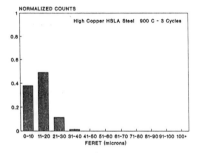

Figure 8: Grain size distributions of the high copper HSLA steel in the as-
received condition and after austenitizing at 900°C.

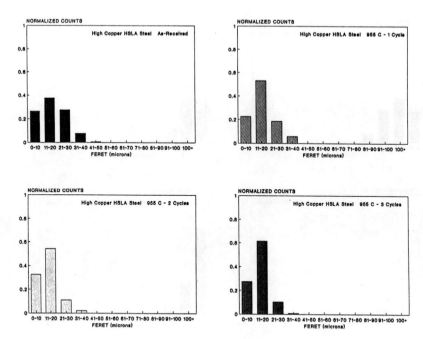

Figure 9: Grain size distributions of the high copper HSLA steel in the as-received condition and after austenitizing at 955°C.

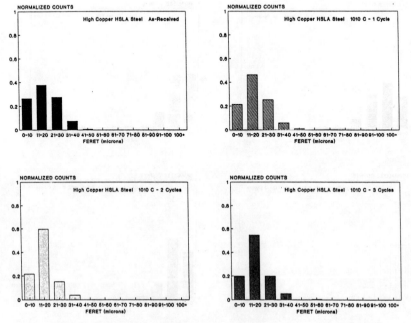

Figure 10: Grain size distributions of the high copper HSLA steel in the as-received condition and after austenitizing at 1010°C.

after the third cycle. The WMW test indicated that the grain sizes of the
distributions were not significantly changing due to the cycling. The
variance in the grain sizes decreased after the first cycle, decreased further
after the second cycle, and increased after the third cycle.

Table 3: Change in Average Grain Size and Variance due to Thermal
Cycling in the Austenite Temperature Region of High Copper
HSLA Steel

Austenitization Temperature (^{o}C)	Cycle #	Average Grain Size (microns)	Variance
As-Received	-	17	75
845	1	13	51
845	2	9	21
845	3	8	19
900	1	12	45
900	2	12	36
900	3	13	38
955	1	16	49
955	2	13	35
955	3	13	31
1010	1	17	64
1010	2	15	42
1010	3	16	53

Low Carbon HSLA Steel (GLG)

After rolling, this steel was austenitized at 900°C. In the as-received
condition (AR), the average prior austenite grain size was 16 microns with a
distribution of grain sizes as shown in figure 11 and a measured variance of
135. One, two, and three austenitization cycles were performed at 845, 900,
955, and 1010°C. The effects of these temperatures and cycles on the grain
sizes and variance are documented in table 4 and shown in figures 5-6, 11-14.

The first series of austenitization cycles was performed at 845°C. After
the first cycle, the average grain size decreased from 16 microns to 12
microns. After the second cycle, the average grain size dropped to 10
microns. The third cycle at this temperature resulted in another decrease in
the average grain size to 8 microns. These results were confirmed by the
WMW test. The variance of the grain size distribution decreased after the
first cycle, remained the same after the second cycle, and decreased again
after the third cycle.

The second series of austenitization cycles was performed at 900°C.
After the first cycle, the average grain size dropped from 16 microns to 11
microns. A second cycle resulted in a decrease to 10 microns. The last cycle
resulted in another decrease in the grain size to 9 microns. The WMW test

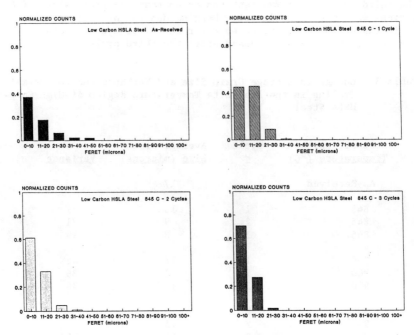

Figure 11: Grain size distributions of the low carbon HSLA steel in the as-received condition and after austenitizing at 845°C.

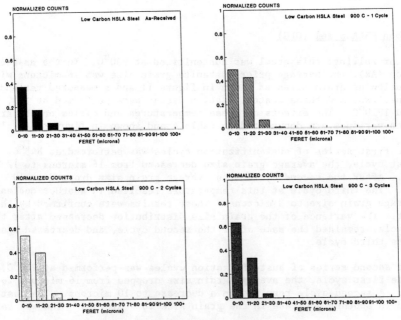

Figure 12: Grain size distributions of the low carbon HSLA steel in the as-received condition and after austenitizing at 900°C.

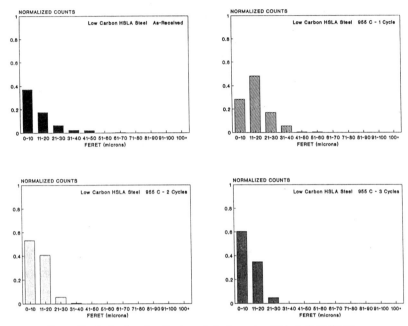

Figure 13: Grain size distributions of the low carbon HSLA steel in the as-received condition and after austenitizing at 955°C.

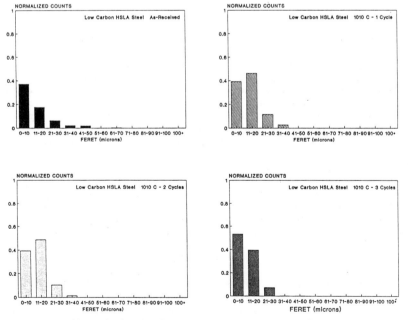

Figure 14: Grain size distributions of the low carbon HSLA steel in the as-received condition and after austenitizing at 1010°C.

indicated that the first cycle did decrease the grain size, but the second
cycle was not statistically different from the first. The third cycle caused
a decrease in the grain sizes of the distribution. The variance continually
decreased with each cycle at this temperature.

Table 4:   Change in Average Grain Size and Variance due to Thermal
           Cycling in the Austenite Temperature Region of Low Carbon HSLA
           Steel

Austenitization Temperature ($^{\circ}$C)	Cycle #	Average Grain Size (microns)	Variance
As Received	-	16	135
845	1	12	32
845	2	10	32
845	3	8	20
900	1	11	36
900	2	10	28
900	3	9	23
955	1	15	68
955	2	10	29
955	3	10	28
1010	1	13	44
1010	2	13	32
1010	3	11	83

The third series of austenitization cycles was performed at 955°C. The
first cycle saw a decrease in the average grain size from 16 microns to 15
microns. A second cycle decreased the grain size to 10 microns while the
third cycle did not change the grain size further. The WMW test indicated
that the first cycle did not cause a statistically significant change in the
grain sizes of the distribution. The second cycle decreased the grain sizes
as did the third cycle although the measured average austenite grain size did
not change from the second to the third cycle. The spread of grain sizes in
the distributions continually decreased with each cycle at this temperature.

The fourth series of austenitization cycles was performed at 1010°C. The
grain size decreased from 16 microns to 13 microns after the first cycle. The
second cycle had no effect on the average grain size. The third cycle
resulted in another decrease in the grain size to 11 microns. These results
were verified with the WMW test. The variance decreased after the first
cycle, decreased after the second cycle, and increased after the third cycle.

DISCUSSION OF RESULTS

When reaustenitizing and then water quenching a ferritic/bainitic steel,
there are three possible changes in the prior austenite grain size: 1) the
grain size is reduced, 2) the grain size remains unchanged, and 3) the grain

size is increased. Physical models for these will be described and applied to the results discussed above.

When HSLA steel is water-quenched to room temperature from above the upper critical temperature, where the structure is austenitic, ferrite/bainite packets are nucleated at prior austenite grain boundaries and grow towards the center of prior austenite grains until another packet or a prior austenite grain boundary is met.

GRAIN REFINEMENT DUE TO RE-AUSTENITIZATION

Austenite grain refinement can occur when the steel is quickly reheated from room temperature into the austenite phase field. Once the lower critical temperature is reached, new and strain-free austenite grains begin to nucleate at areas of high stored energy such as prior austenite grain boundaries and packet grain boundaries. Since within a single prior austenite grain there may be multiple ferrite/bainite packets, several new austenite grains may be formed. Nucleation of austenite grains continues until they begin to impinge upon one another and the transformation to austenite is completed just above the Ac_3 temperature. Because the number of new austenite grains has been increased, the growth of these new grains will be more limited and, at the time of impingement, are much smaller than the austenite grains prior to quenching.

Further time spent at temperatures above the Ac_3 results in post-impingement grain growth, with smaller grains generally being sacrificed to larger grains. The lower the austenitization temperature, the slower the post-impingement grain growth rate, so that when the heat treatment is ended with water quenching, the new austenite grains remain refined.

Upon quenching, ferrite/bainite packets begin to form. Because the prior austenite grain size has been reduced, the number of potential ferrite/bainite nucleation sites has been increased. Also, the growth of a packet before encountering a prior austenite grain boundary or another packet will be reduced, thus the size of the ferrite/bainite packets will be smaller. With the smaller packets, the steel has increased the number of potential austenite nucleation sites so that when the material is reaustenitized, further grain refinement is possible as described above. However, a lower limit probably exists for the obtainable grain size reduction. If the prior austenite grains were so small that they contained only a single ferrite/bainite packet, there would no longer be an increase in the number of potential austenite nucleation sites and reaustenitization would not further refine the structure.

STABLE GRAIN SIZE

Grain size can be maintained despite reaustenitization. The steel is quickly reheated and the size of these new grains at the completion of transformation is much smaller than the prior austenite grains as discussed above. The scenario thus far is identical to that described above for grain refinement. The difference lies with the austenitization temperature at which post-impingement grain growth occurs. The temperature is higher so grain growth occurs more quickly and at the time of quenching, the steel has achieved a grain size similar to the old austenite grains. Since the austenite grain size has not changed, there will not be a refinement of the ferrite/bainite packet size upon cooling. Therefore the number of potential austenitization nucleation sites has not been changed, and if the steel **was** reaustenitized at the same temperature for the same amount of time, the **grain** size would remain the same.

GRAIN GROWTH DUE TO RE-AUSTENITIZATION

Reaustenitization heat treatments can also result in grain growth. The ferritic/bainitic steel is quickly reheated into the austenite phase field. Once the Ac_1 temperature is reached, newly-formed austenite grains begin to nucleate and the size of these new grains at the completion of the transformation, somewhere above the upper critical temperature, are much smaller than the prior austenite grain boundaries. The scenario thus far is identical to those described for grain refinement and for stable grain size. The difference again lies with the even higher austenitization temperature at which post-impingement grain growth occurs. Grain growth occurs more quickly and at the time of quenching, has extended beyond the old austenite grain boundaries. Since the new austenite grain size is larger, the ferrite/bainite packet size will generally be larger. This is due to a reduction in the number of potential packet nucleation sites and also because the packets that do form will be able to grow further before encountering a grain boundary or another packet. The room temperature microstructure will then have fewer favorable austenite nucleation sites, and if the steel is reaustenitized, the grain size will again increase. This cycle is expected to continue until a stable grain size is achieved.

EFFECTS OF CHEMISTRY

In this study, grain refinement was achieved for all three steels at 845°C, but the relative amount of refinement with each cycle varied between the steels. The grain size of the high carbon HSLA steel was reduced after two cycles but increased after the third cycle. Thermal cycling at this temperature caused the grain sizes of the high copper and the low carbon HSLA steels to continually decrease with each cycle although the relative amount of grain refinement with each successive cycle decreased.

At the higher temperature of 900°C, all of the steels again experienced a reduction in the average grain size from the as-received condition. Because the austenitization temperature had been increased and thus grain growth rates were faster, the amount of grain refinement was not as great as that achieved at 845°C. The high carbon HSLA steel had the least stable response to thermal cycling. The high copper HSLA steel reached a stable grain size after a single cycle. The grain size of the low carbon HSLA steel continued to be reduced with each cycle, although the relative amount of reduction decreased.

At the still higher austenitization temperature of 955°C, no substantial grain size reduction or enlargement was achieved for the high carbon HSLA steel. The high copper HSLA steel achieved a stable grain size of 13 microns after the second cycle was completed. The low carbon steel achieved a stable grain size of 10 microns after 2 cycles.

The highest austenitization temperature was at 1010°C. The high carbon HSLA steel experience grain growth immediately and appeared to be approaching a stable grain size of 24-25 microns. The austenitization temperature of 1010°C did not affect the grain size of the high copper steel at any cycle. The low carbon HSLA steel experienced grain size reduction even after 3 cycles.

EFFECTS OF RE-AUSTENITIZATION ON THE GRAIN SIZE DISTRIBUTION

Within the larger austenite grains of a distribution, many ferrite packets may form upon quenching while within the smallest grains only a single ferrite packet may form. Upon reaustenitization, the larger grains may have more potential austenite nucleation sites than smaller grains leading to reduction in their size while the smaller grains remain stable. Therefore, if the grain size is generally being reduced with repeated austenitization cycles, the distribution of grain sizes will also become more homogenous.

It would also be expected that if the austenite grain size is not being changed with repeated austenitization, the distribution of the grain sizes should also remain stable. If the austenite grain size is generally increasing with larger grains growing at the expense of smaller grains, it may be expected that the distribution of grain sizes will also increase.

In this study, the changes in the grain size distribution (as measured by the variance) generally followed the changes in the grain size as discussed above; that is, if the average grain size decreased, the variance also decreased. If a stable grain size was achieved, the variance did not change. If the grain size increased, the variance increased. Experimental results suggested that as the austenitization temperature was increased, the relative amount of homogenization achieved decreased.

RESPONSE TO RE-AUSTENITIZATION

Why do the three HSLA steels respond so differently to repeated austenitization? Some of the possible factors are: One, the starting grain sizes and grain size distributions were different. Second, the upper critical temperature (Ac_3) of the steels was different due to chemistry. Third, the amount, size and distribution of grain-pinning carbides may be different. These factors are discussed below.

The as-received average grain size was smallest in the high carbon HSLA steel, at 14 microns. The high copper HSLA steel and the low carbon HSLA steel had similar average grain sizes of 17 and 16 microns, respectively. If this factor was important, one would expect the average grain sizes of the high copper and the low carbon HSLA steel obtained due to thermal cycling to be similar, and larger than the high carbon HSLA steel. This was not the case.

The as-received distributions of grain sizes were very different, see figures 1, 7, and 11. The high carbon and high copper HSLA steel have much tighter distributions than the low carbon HSLA steel. If this factor was important, one would expect this advantage would be maintained throughout thermal cycling. This was not the case.

The relative amount of heating above the upper critical temperature of the steel has been shown to affect the response to thermal cycling [Grange, 1971]. An estimation of the upper critical temperatures of these steels using equations in Leslie [5] indicated no statistically significant difference.

Niobium carbo-nitrides have been shown to effectively pin ·grain boundaries and are used to prevent grain growth during high temperature thermo-mechanical processing [2]. This fact suggests that the steel with the highest level of carbon would form the largest niobium carbo-nitrides which would be least effective in grain boundary pinning. However, the size and distribution of such precipitates also depend strongly on precipitation temperature and prior thermo-mechanical processing. Since the high carbon HSLA steel responded the least positively to reaustenitization for grain size refinement and homogenization, it would appear precipitates have a profound effect on reaustenitization response. The data presented here support this hypothesis; however, additional microstructural analysis is required for confirmation.

SUMMARY AND CONCLUSIONS

In summary, this study determined the effects of repeated thermal cycling in the austenitic region of low-carbon steels on the average grain size and grain size distribution. Several plates of Navy HSLA steel were subjected to one, two, or three cycles at one of several temperatures in the austenite phase field.

Repeated thermal cycling of steel at temperatures in the austenite region generally resulted in grain refinement until a stable grain size was reached, beyond which no further changes were noted. As the austenitization temperature was increased, the relative amount of grain refinement was decreased.

Grain size distributions generally followed the same trends established by the average grain size; that is, when the grain size was refined, the distribution of grain sizes narrowed (became more homogenous). When the grain size was enlarged, the distribution of grain sizes widened. When the grain size stabilized, the distribution generally also remained stable. The lower the austenitization temperature used, the better the response of the steel in terms of both grain size reduction and homogenization.

REFERENCES

[1] Czyryca, E.J., "Development of Low-Carbon, Copper-Strengthened HSLA Steel Plate for Naval Ship Construction," DTRC-SME-90/21.

[2] Mattes, V.R., "Microstructure and Mechanical Properties of HSLA-100 Steel," Naval Postgraduate School (December 1990).

[3] Pickering, F.B., Physical Metallurgy and the Design of Steels, London Applied Science Publishers Ltd., London (1978).

[4] Samuels, L.E., Optical Microscopy of Carbon Steels, American Society for Metals (1980).

[5] Leslie, W.C., The Physical Metallurgy of Steels, Hemisphere Publishing Corp., New York, N.Y. (1981).

[6] Wilson, A.D., Hamburg, E.G., Colvin, D.J., Thompson, S.W., and G. Krauss, "Properties and Microstructures of Copper Precipitation Aged Plate Steels," presented at Microalloying '88, Chicago, IL (September 1988).

[7] Porter, D.A. and K.E. Easterling, Phase Transformations in Metals and Alloys, Van Nostrand Reinhold Co. Ltd., Berkshire (1981).

[8] Grange, R.A., "The Rapid Heat Treatment of Steel," Metallurgical Transactions, Vol. 2 (January 1971).

[9] Thompson, S.W., Colvin, D.J., and G. Krauss, "Continuous Cooling Transformations and Microstructures in a Low-Carbon, High-Strength Low-Alloy Plate Steel," Metallurgical Transactions, Vol. 21A, pp. 1493-1507 (June 1990).

[10] Bechet, S. and L. Beaujard, "New Reagent for the Micrographical Demonstration of the Austenite Grain of Hardened or Hardened-Tempered Steels," Rev. Met., Vol. 52, pp. 830-836 (1955).

[11] Bodnar, R.L., V.E. McGraw, and A.V. Brandemarte, "Technique for Revealing Prior Austenite Grain Boundaries in Cr-Mo-V Turbine Rotor Steel," Metallography, Vol. 17, pp. 109-114 (1984).

[12] Dreyer, G.A., D.A. Austin, and W.D. Smith, "New Etchent Brings Out Grain Boundaries in Martensitic Steels," Metal Progress, Vol. 86, pp. 116-117 (1964).

[13] Krahe, P.R. and M. Desnoues, "Revealing the Former Austenite Grain Boundaries of High-Purity Iron-Carbon Alloys," Metallography, Vol. 4, pp. 171-175 (1971).

[14] Preese, A. and R.D. Carter, "Temper-Brittleness in High-Purity Iron-Base Alloys," Journal of the Iron and Steel Institute, Vol. 173, pp. 387-398 (1953).

[15] Uciski, A.H., McMahon, C.J., and H.C. Feng, "The Influence of Intercritical Heat Treatment on the Temper Embrittlement Susceptibility of a P-Doped Ni-Cr Steel," Metallurgical Transactions, Vol. 9A, pp. 321-329 (1978).

[16] Bain, L.J. and M. Engelhardt, Introduction to Probability and Mathematical Statistics, PWS-Kent Publishing Co., Boston, Mass. (1989).

[17] Hogg, R.V. and A.T. Craig, Introduction to Mathematical Statistics, 4th edition, Macmillan Publishing Co. Inc., New York, N.Y. (1978).

[18] Beyer, W.H., Handbook of Tables for Probability and Statistics, 2nd edition, CRC Press Inc., Boca Raton, Fla. (1981).

Materials Science Forum Vols. 102 - 104 (1992) pp. 31-42

THE HEAT TREATMENT OF P/M TOOL STEELS

A. Kulmburg and B. Hribernik

Böhler GesmbH., Austria

ABSTRACT

Ledeburitic 12% Chromium steels are used to produce various
types of cold work tools.
12% Cr tool steels with 1.5/2.0% C are charactcrised by a high
proportion of coarse primary carbides ($M_7 C_3$) which lead to
inhomogeneous material properties. Grades with even higher
carbide content have a greatly reduced hot forging capability.
These problems can be solved by using the powder-metallurgical
(PM) manufacturing route. The PM technique offers the
possibility of producing new, highly alloyed grades which
cannot be manufactured by ingot metallurgy.
In the transformation of plastics the mechanical stresses to
which the individual tool parts are subjected are rather low as
compared to the previously discussed cold work tools. The
predominant stresses are wear and corrosion.
Materials with good corrosion properties, such as chromium
steels or austenitic Cr-Ni steels, possess poor or insufficient
wear resistance.

This paper describes the characteristic properties and the heat
treatment of the cold work tool steel X230CrVMo13 4 PM and of a
newly developed steel X190CrVMo21 4 PM with high wear and
corrosion resistance for the plastic technology.
The hardenability of these steels is discussed using CCT-,
cooling-time and quantitative phase diagrams. The heat
treatment of these steels in vacuum furnace is dealt with
especially.

The continuously increasing demands on tool performance, tool life and thus on tool steels led to an intensive search for new processes for the production of tool materials featuring improved properties.

As early as at the end of the sixties the R & D department of BÖHLER carried out feasibility studies to examine the possibility of producing tool steels by the powder metallurgy process.

Last decade BÖHLER was the first producer to develop a P/M ledeburitic chromium steel for heavy-duty cold work tools.

Another outcome of this development work has been a wear and corrosion resistant steel for plastic processing machines, knives, which is currently tested in practical application by some of our important customers (1-3).

X230CrVMo13 4 PM - a P/M LEDEBURITIC CHROMIUM STEEL OF COMPLEX ALLOY CONTENT FOR COLD WORKING TOOLS

Cold work tool steels for blanking and cutting, cold forming and wear protection are exposed in operation to complex stresses which can be devided into mechanical and tribologic ones. In all cold work steel applications the active tool parts are usually those which are subjected to highest stresses, so that only limited tool life is to be expected.

Maximum wear resistence is achieved with ledeburitic chromium steels with a carbon content of 1.5 - 3.5% and a chromium content of 12 - 13%.

The shortcomings of conventional production of ledeburitic steels of complex alloy content and high carbide content include difficulties encountered in production as well as very poor machinability and brittleness of the material. This is especially due to the coarse carbide structure originating from the solidification process.

Only at the beginning of the eighties the capacities in powder production and consolidation were sufficient to allow economical production of the cold work tool steel containing. 2,3% C, 12% Cr, 1% Mo and 4% V by the powder metallurgy process (Böhler K190).

The homogenous structure of this steel is shown in Fig.1 compared to a conventionally produced ledeburitic steel.

The excellent wear resistance as determined by measurement has proved true in practical application.
Fig.2 compares the wear of punches and dies as a function of the carbide content of various blanking tool material hardened to suitable hardness levels. From this it can be seen that the wear resistance of K190 PM is increased by a factor of 3 as compared to other ledeburitic steels.

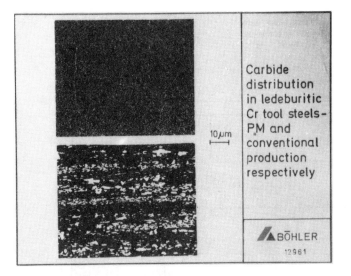

Carbide distribution in ledeburitic Cr tool steels- P.M and conventional production respectively

.Fig.1

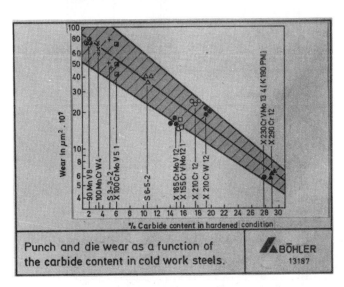

Punch and die wear as a function of the carbide content in cold work steels.

Fig.2

Fig.3 shows the influence of austenitizing temperature on the hardness in the oil hardened condition. The maximum secondary hardness depends on the hardening temperature as it is shown in fig.4. With increasing hardening temperature the maximum secondary hardness rises to higher tempering temperature due to the higher chromium,-molybdenum-, and vanadium content in the matrix.

Influence of the austenizing temperature on the hardness after hardening prior to tempering
(hardening from neutral atmosphere)
quenchant: oil
specimen size: ø 20 mm
holding time: 30 minutes

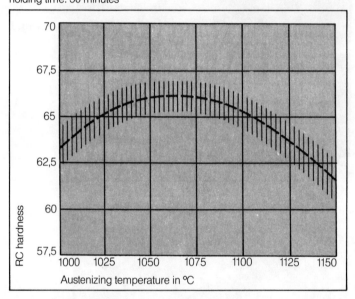

Fig.3

Tempering chart
(hardening from neutral atmosphere)
quenchant: oil
specimen size: ø 20 mm
holding time: 2 hours, min.

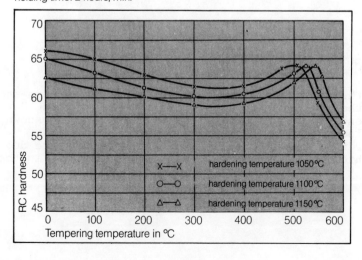

Fig.4

The transformation behaviour is described by the CCT-diagram (fig.5) and the quantitative-phase-diagram (4) (fig.6).

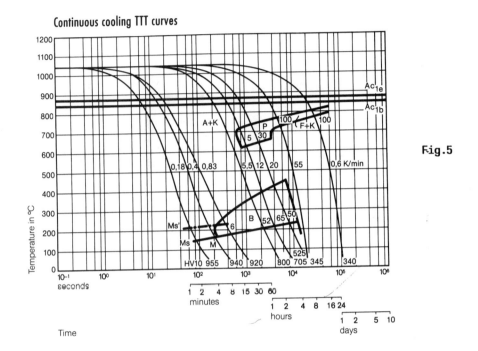

Fig.5

Chemical composition in %
C   Si   Mn   S   Cr   Mo   V
2.25 0.29 0.32 0.011 12.34 0.98 3.94

Austenizing temperature: 1050 °C
Holding time: 10 minutes
6 ... 100 phase percentages
0.18 ... 55 cooling parameter, i. e. duration of cooling from 800 to 500 °C in s x 10⁻²
0.6 K/min ... cooling rate in the 800– 500 °C range

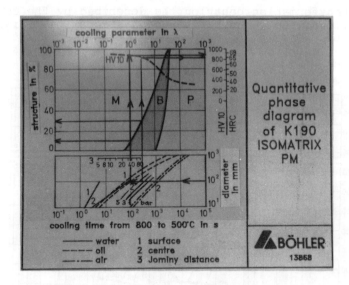

Fig.6

In this diagram the structure in phase vol % is plotted versus the cooling time from 800 to 500°C in s. Every cooling curve in the CCT-diagram corresponds to a vertical line in the quantitative phase diagram.

The lower part of this diagram shows the relationship between the cooling time from 800 to 500°C and the diameter of the workpiece (10^1 to 10^3 mm) depending on the cooling medium (water, oil, air) and the measuring point (surface, center). Furthermore one can find curves for cooling with nitrogen using a pressure of 1,3 or 5 bar. The arrows in fig. 6 show that a workpiece with 100 mm diameter of the steel K190 has 10% bainite and 90% martensite in the structure after oil hardening. The hardness will be about 66 HRC. After cooling in nitrogen with 3 bars pressure the structure will have about 30% bainite and the hardness will decrease to less than 65 HRC.

The compressive strength of the ledeburitic 12% Cr steel
produced by the powder metallurgy process comes up to that of
high speed steels and is superior to that of conventionally
produced ledeburitic 12% Cr steel. Fig. 7 compares the results
obtained from specimens with a hardness of 63 Rockwell C.

Due to the fine carbides the toughness of K190 is much better
than other ledeburitic steels conventionally produced (fig. 8).

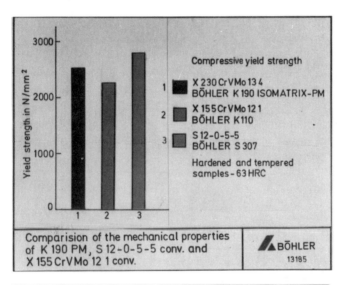

Fig.7

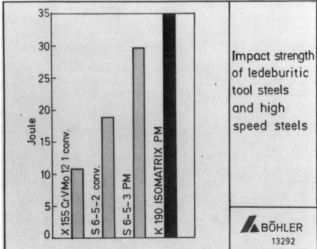

Fig.8

The dimensional stability during heat treatment governs the determination of tolerances and grinding allowances. Minor size changes mean lower repair expenditure and increased economic efficiency.

The maximum size changes in longitudinal direction after hardening and tempering are indicated in fig. 9 for the usual heat treatment cycles. A comparison reveals that the steel X230CrVMo13 4 is best suited for the manufacture of precision tools.

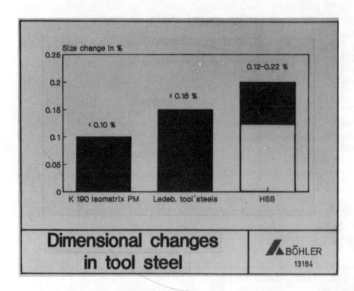

Fig.9

The best heat treatment cycles of this steel is summarized in table 1.

TABLE 1

Best heat treatment of K190

For supreme wear resistance

Hardening: 1150°C/ ≳ 20 min. holding time/oil or warm bath,
               ≳3 bar N₂

Tempering: 2x 540°C/2ʰ/air

Hardness: 61 - 63 HRC

For high toughness:

Hardening: 1080 - 1120°C/≳ 20 min. holding time/oil or
               warm bath, ≳ 3,5 bar N₂

Tempering: 535°C/2ʰ/ air + 515°C/2ʰ/air

X230CrVMo13 4 is suited for all applications where
conventionally produced ledeburitic 12% Cr steels are used,
e.g. high performance cutting tools, cold forming tools,
drawing, extrusion and engraving tools, thread rolling tools,
sintering tools for ceramics and metals, pressing tools for the
pharmaceutical and plastics industries.

P/M CHROMIUM STEEL FOR PLASTICS PROCESSING

In the transformation of plastics the mechanical stresses to
which the individual tool parts are subjected are rather low as
compared to the previously discussed cold work tools. The
predominant stresses are wear and corrosion. The ever
increasing use of asbestos, wood and, more recently, glass-
reinforced moulding materials results in an analogous increase
of wear in both plastics processing machines (on screws,
cylinders, backflow valves and dies) and moulds.

Ledeburitic chromium steels, such as X210Cr 12 (Böhler K100) or
X155CrVMo 12 1 (Böhler K110), exhibit good resistance to
abrasive wear. They corrode however very rapidly when plastics
with corrosive additives are processed. Materials with good
corrosion properties, such as chromium steels or austenitic Cr-
Ni Steels, possess poor or insufficient wear resistance.

The objective therefore was to develop and produce a material
for plastics processing machines and tools, which combines
maximum wear resistance, such as X230CrV13 4, and excellent
corrosion resistance.

These considerations led to the production of the following
alloy:
1,9% carbon, 21% chromium, 1% molybdenum, 1% tungsten,
4% vanadium (Böhler M390M).

Fig.10 shows the structure with the very fine carbides.

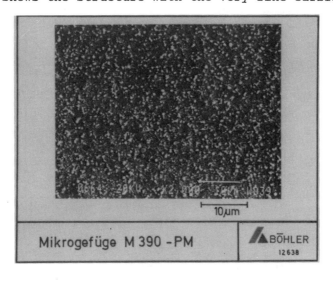

Fig.10

10μm

Mikrogefüge M 390 -PM   BÖHLER
                               12638

The transformation behaviour of this steel is described by the
cooling-time-diagram (4) (fig.11).
In the upper part of the diagram the regions of austenite
transformation into pearlite and martensite are shown, the lower
part of this diagram is identical to the quantitative phase
diagram (see fig.6). One can see that the steel M390 is pearlite
free hardenable in oil up to a diameter of about
120 mm, but only up to 60 mm diameter when cooled in pressurized
nitrogen with more than 3 bars.
The depencence of the hardness on the hardening and tempering
temperature is similar to K190 (fig.12).

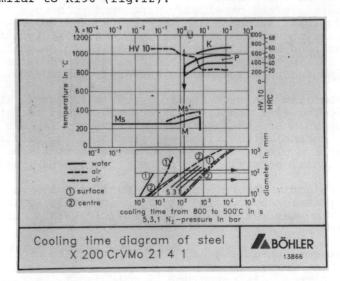

Cooling time diagram of steel
X 200 CrVMo 21 4 1

BÖHLER
13866

Fig.

Tempering chart (salt bath hardened condition)

RC hardness

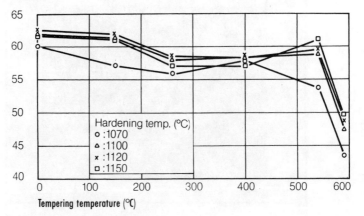

Fig.

Tempering temperature (°C)

Holding time: 2 x 120 minutes
Specimen size: ø 90 mm
Austenizing in vacuum/N₂, 5 bars

The resistance to corrosion depends mainly from the chromium and molybdenum contents in the matrix of the steel. This is why X190CrVMo 21 4 PM offers better corrosion resistance than the plastic mould steels commonly used so far.

As compared to the conventionally produced chromium steels for plastic moulds already available on the market, e.g. X36CrMo 17 or X42Cr 13, the steel X190CrVMo21 4 PM yields better results, particularly in cases where abrasive wear and corrosive attack occur simultaneously (fig.13).

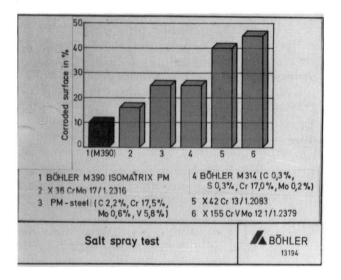

Fig.13

The best heat treatment cycles of this steel is summarized in table 2.

T A B L E 2

Best heat treatment of M390

For excellent corrosion resistance:

Hardening: 1100 - 1150°C (salt bath)
              1150 - 1200°C (vacuum furnace)
              $\geq$20 min holding time, oil or warm bath, $\geq$ 5 bar N_2

Tempering: 1 x 200°C/2^h/air

Hardness: 58 - 62 HRC

For supreme wear resistance:

Hardening: 1150°C (salt bath), 1200°C (vacuum furnace),
              oil or warm bath, $\geq$ 5 bar N_2

Tempering: 2 x 540°C/2^h/air

Hardness: 60 - 62 HRC

For best size stability:

Hardening: 1150°C (salt bath), 1200°C (vacuum furnace),
              oil or warm bath,$\geq$ 5 bar N_2,subzero refrigeration
              (-80 to -120°C)

Tempering: 2 x 540°C/2^h/air

Hardness: ~ 60 HRC

R E F E R E N C E S

1)   Hribernik B., Lenger H., Schindler A.,
     Berg-u.hüttenmänn.Mh.1989, 134, 8

2)   Hribernik B., Hackl G., Lenger H.
     Berg-u.hüttenmänn.Mh.1989, 134, 334

3)   Kulmburg A. et al.
     Härterei-Techn.Mitt.1990, 45, 200

4)   Kulmburg A
     Berg-u.hüttenmänn.Mh.1989, 134, 341

Materials Science Forum Vols. 102 - 104 (1992) pp. 43-52

SURFACE CHEMICAL PROPERTIES OF HEAT-TREATED INDUSTRIAL STEEL

S.K.E. Forghany

Hoogovens Groep B.V., Research and Development Laboratories, Department of Surface Technology,
P.O. Box 10000, NL-1970 CA IJmuiden, The Netherlands

ABSTRACT

The surface finish of cold-reduced industrial steel depends largely on conditions of heat-treatment processes consisting either of continuous annealing (CA) or batch annealing (BA). A satisfactory surface finish implies minimal contaminations and optimal surface chemical activity. During (BA), interrelated extrinsic and intrinsic surface reactions take place concurrently. Intrinsic reactions constitute interfacial enrichment by micro-alloying elements and impurity elements. Such enrichment is due to combined effects of, on the one hand, synergistic forces between metals and non-metal elements in the bulk and, on the other hand, the thermodynamic potential set up between the solid-phase and the gas-phase. The resulting surface/sub-surface chemical heterogeneities have a strong influence on the technical performance of steel. In the work presented, isothermal heat-treatment (at 500 °C, 600 °C, and 700 °C) of steel was simulated in differing mixed H_2/H_2O atmospheres [$21.98 \le -\log pO_2$ (atm) ≤ 27.72] for 15 min to 64 hr. Surface chemical properties of the resulting products were characterized by Glow Discharge Optical emission Spectrometry (GDOS) and by scanning electron microscopy combined with energy-dispersive X-ray micro-analysis. Data obtained for surface reactions, concerning manganese and sulphur, are presented. Kinetics of the segregation of manganese at 700 °C and $\log pO_2$ (atm) $= -25.5$ indicate that the reaction is diffusion-controlled with $\sqrt{(time)}$ rate dependency. An inverse relation was demonstrated between the surface enrichment of manganese and that of sulphur, i.e. the surface activity of S^{2-} declines because of excessive replacement of S^{2-} by O^{2-}. The surface coverage of sulphur was shown to be inversely related to the temperature, i.e. at high annealing temperature, the coverage of sulphur was minimal.

I. INTRODUCTION

Cold-reduced commercial steel, which is widely applied for the manufacturing of varying objects, requires carefully controlled surface preparation. The surface should be free from contaminations, it should possess optimal roughness and minimal porosity and, generally, be chemically active.
Inadequate surface preparation can lead to inferior corrosion resistance, to breakdown in spot welding and to poor adhesion of coating films, all resulting in considerable shortening of the service life of final products. Further, organic coating and electrolytic metal deposition may insufficiently adhere.
The surface finish of cold-reduced steel depends largely on thermal conditions of heat-treatment processes comprising either continuous annealing (CA) or batch annealing (BA). CA steel strips are largely homogeneous in surface composition and contain negligible degrees of surface carbonacious residues and iron fines [1,2]. In addition, they show minimal surface enrichment by minor additives and impurity elements, and no sub-surface oxidation.
In the conventially applied (BA) method, two series of concurrently occurring, yet interrelated, chemical reactions take place, namely extrinsic reactions (A, below) and intrinsic reactions (B, below).

A) Extrinsic reactions are interactions between the reactive gas-phase and surface contaminants which are remnants of preceding processes. Such contaminants are: Cold-rolling oil lubricant, carbon deposits, iron fines, surface oxides, and other impurity species such as Ca^{2+}, Na^+, K^+, Cl^+, etc.

B) Intrinsic reactions comprise the diffusion of micro-alloying elements and impurity elements from the grain interior of the matrix to interfacial sites. Interfacial regions consist of considerable lattice disturbances and such enrichment of foreign atoms is energetically favourable. In practice, when cold-reduced low-carbon steel undergoes prolonged heat-treatment, an influx/outflux of anions, cations and electrons occurs under the combined influence of an electrical potential gradient and a chemical potential gradient set up between the solid phase and the gas phase. The coupling effect between non-metal atoms and micro-alloying elements in the free surfaces, leads to the formation of discrete heterogeneous phases [3,4]. They are epitaxed at regions of high chemical potential such as grain boundaries, and at other sites of crystallographic defects. Such chemical heterogeneities, in surfaces and also along sub-surface layers of steel, give rise to macroscopic stresses [5]. A high stress field in surfaces can lead to the formation of blisters, cracks, and cavities which are detrimental to the technical performance of steel.

Because of the importance of intrinsic reactions on surface properties of steel, These reactions were studied, in the work presented, by isothermal heat-treatment of cold-rolled steel in laboratory experiments. Isotherms applied were 500 °C, 600 °C, 700 °C, and annealing durations ranged from 15 minutes to

64 hours. At 700 °C isotherm, oxygen fugacities of the gas-phase varied within the range 21.98≤-log pO$_2$ (atm)≤27.72 using H$_2$/H$_2$O gas mixtures. Surface chemical properties of the annealed materials were characterized by means of Glow Discharge Optical (emission) Spectrometry (GDOS) and by electron optical methods. Results of the segregation behaviour of manganese and sulphur are presented below.

II. EXPERIMENTAL

Cold-reduced steel samples utilized had a bulk composition shown in Table 1.

Table 1. Bulk Chemical Composition (wt% x10^{-3})

C	Mn	P	S	Si	Al (sol)	Cu	Cr	Ni
40	252	8	11	12	52	6	17	25

The thickness of the sample was 0.3 mm (hot-rolled steel is 2.3 mm thick), its width 820 mm. Degreasing and other cleaning processes had taken place on the production line. Laboratory tests showed that the samples had a residual surface carbon content of about 2 mg.m^{-2} and contained about 2 mg.m^{-2} free iron particles.
Isothermal heat-treatment experiments were carried out at 500 °C, 600 °C, and 700 °C. The duration of the experiments ranged from 15 min. to 64 hr. At the conclusion of heat-treatment, specimens were withdrawn rapidly to the cold zone of the furnace and were cooled down to room temperature in an environment of flowing protective gas.

The protective gas-phase consisted of, both, commercial (HNX) and high purity hydrogen gas (AGA, B.V. Amsterdam). At room temperature, HNX contains 6% H$_2$, and 94% N$_2$, and has a dew point of ca. -40 °C (p$_{H2O}$ ≈ 0.0966 mmHg). HNX generates, at the annealing temperatures applied (500-700 °C), low oxygen fugacities in the range 25.47<-log pO$_2$ (atm)<31.9, which interact with the material. Using mixtures of H$_2$/H$_2$O, oxygen fugacities of the furnace atmosphere at 700 °C varied [6] in the range 21.98 ≤-log pO$_2$ (atm)≤27.72. This permitted the study of effects of differing gas composition on the surface properties of steel.

Surface elemental analysis of heat-treated specimens was carried out using ARL-34000 Glow Discharge Optical emission Spectrometry (GDOS). The voltage across the discharge cell, i.e. sample and anode, was V=500 volts, the electric current rated 73 ± 4 mA, while the pressure of the sputtering gas (argon) was about 10 mbar. Integrated numerical values were based on an arbitrary standard (unity). On each specimen, a total of five (GDOS) experiments were performed and mean average values of measurements were calculated for individual elements. (GDOS) data obtained for the first five sec. of sputtering corresponded to a depth of roughly 40 nm uppermost, while 70 sec. continuous sputtering yielded information from a deeper sub-surface layers (ca. 0.6 μm) and concerning the concentration of differing elements. For the quantitative analysis of bulk concentrations in each sample, the GDOS instrument was calibrated with identical standard steel specimens. Surface topography, texture, and localized micro-elemental composition of specimens were also investigated using a Cambridge scanning (series 4-DV) electron microscope.

III. RESULTS AND DISCUSSION

Sergregation of Manganese

Manganese, in steel, controls deleterious effects of sulphur and oxygen, and also improves the mechanical properties of steel. During the heating/cooling cycle, manganese diffuses to free surfaces and interacts with the protective gas phase [7,8]. Depending on oxygen fugacities of the annealing environment, stable oxides are formed, both, in free surfaces as well as in sub-surface layers. In the system Mn-O, five solid phases exist [9], namely, <Mn>, <MnO>, <Mn_3Q>, <Mn_2Q_3 >, and <MnO_2 >. These oxides are stable at progressively increasing oxygen fugacities and, in the case of <MnO_2>, the pressure at 700 °C increases indefinitely.

Within the range of oxygen fugacities applied, 21.98 ≤ -log pO_2(atm)≤27.38, thermodynamic considerations indicate that <MnO> forms as the only stable intermediate phase in the Mn-O system. Iron is not oxidized at this gas composition range. GDOS measurements for surface enrichment of manganese (see Fig. 1) suggest a parabolic relationship of manganese oxide formation with, both, an increase in temperature and in duration of heat-treatment.

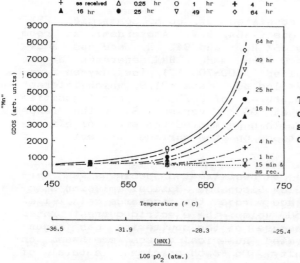

Fig. 1.

Temperature dependency of manganese surface segregation and oxidation.

$$<Mn>_{0.252 \ wtx} + \tfrac{1}{2}(O_2) = <MnO>_{surface} \quad \cdots\cdots\cdots\cdots (1)$$

Figure 1 shows that, with an increase in temperature at time t, the reaction rate is enhanced due to combined effects of, both, an increase in diffusion coefficient of particles in the solid phase, and an increase in oxygen potential of the gas

phase.   It is unlikely that, within   the temperature range and
experimental time applied (max. 64 hr),   an equilibrium distribution   of manganese is fully   attained between   the bulk   and
the   free surfaces   of steel.   At moderate   temperatures,   the
mobility of   atoms through   crystalline lattices   of grains   in
the bulk   is slow. Thus,   an influx/outflux   of particles takes
place   (and for   longer durations   of heat-treatment   a local
equilibrium possibly exists)   only between the surface and   the
near-surface regions   of steel. As   regards the free   surfaces,
because   of abrupt   termination of   grain lattice arrays,   the
Madelung energy   of ions is low   and this   lowers the potential
barrier   for inward/outward migration   of ions.   Moreover, the
gradient of   the chemical   potential, set-up   between the gas-
phase and the   solid-phase, enhances the above described   general effects,   leading to an   asymmetric distribution of   segregants in the free surfaces and sub-surface layers.
The effective sub-surface reaction zone was   found, by analytical electron   microscopy, to   extend maximally 30 µm   deep from
the   surface. The locus   of surface   reactions, thus,   is confined. Four   differing diffusion processes   (1-4, below)   contribute to   the overall reaction   which involves gas   environment,
surface, sub-surface,   and bulk,   as is depicted   schematically
in figure 2.

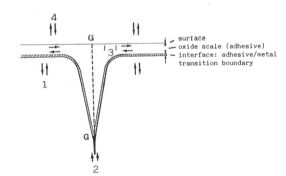

Fig.   2. Schematical
representation of the
"vapour-surface-solid"
diffusion profile
consisting of four
types of mass trans
ports (1-4, below):

1) grain ⇄ surface diffusion
2) grain boundary ⇄ surface diffusion (G-G)
3) surface ⇄ surface diffusion and,
4) vapour ⇄ surface (adsorption/desorption)

Diffusions (1) and (4)   are high energy processes and, in practice, both do not truely achieve an equilibrium. Diffusions   of
type (2) and (3) are   dominant at the isotherms applied in this
study. As   regards reaction (2), the   rate of grain   boundary
mass transport is higher than that of   lattice diffusion. After
boundary planes   become saturated, the   diffusing particles are
adsorbed/desorbed in   adjacent layers, and subsequently   spread
in the mass   of the grains. Such a   process would explain   the
extended   slow uptake   of oxygen   (or   rejection of   manganese)
measured and depicted in   the kinetic curves shown in figure 3.
In addition,   sites of crystallographic   defects (in which   the
chemical   potential of   the matter   is also locally   enhanced)
become   preferential   places of   surface reactions   and   this,
also,   contributes to   the above   explained overall   reactions.
Crystallographic defects in   cold-reduced steel include   a high

density of edge dislocations, point defects, domains of misfit, etc.

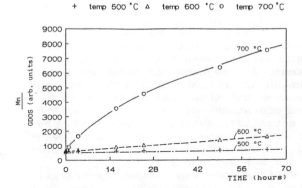

+   temp 500 °C △   temp 600 °C ○   temp 700 °C

Fig. 3.
Time dependency of the
overall surface
reaction of
$<Mn>_{0.252wt\%}/<MnO>$.

In a diffusion-controlled reaction, an exponential rate law is
obeyed:

$$X^\#_{((Mn)0.252\ wt\% \ --)\ (MnO))} = k \cdot t^n + b \quad \cdots\cdots\cdots\cdots (2)$$

Here, $X^\#_{((Mn)0.252\ wt\% \ --)\ (MnO))}$ is the level of manganese coverage at
time (t), k is the temperature dependent rate constant, t is
the isothermal exposure time, and b is the X-intercept of (X
versus t^n) plot. k follows the Arrhenius relation:

$$k = k_o \cdot e^{-E/RT} \quad \cdots\cdots\cdots\cdots (3)$$

Where, k_o is the pre-exponential factor, E is the apparent
activation energy, R is the gas constant, and T the isotherm
applied. In figure 4, a (log-log) coordinate construction is
depicted of the data shown in figure 3. Figure 4 indicates
that, at 700 °C (curve I), there is a parabolic relation, with
exponent (n=½). This would seem to signify that <MnO> forms,
along grain boundary planes and along other surface defects,
by diffusion of particles through a growing film which is
adherent to the base metal.
In general, the total rate of manganese coverage may be expressed by a summation of a logarithmic term and a parabolic
term:

$$X^\#_{((Mn)0.252\ wt\% \ --)\ (MnO))} = K_1 \log(k_2 \cdot t + k_3) + k_4 \cdot t^{1/2} \quad \cdots (4)$$

where, k_1, k_2, k_3, and k_4 are constants. The logarithmic term,
which is characteristic for low temperature surface reactions,
also counts at high temperatures, but only at the commencement
of the reaction i.e. when "t" (time) is small. At the relatively high isotherms and longer durations employed in the present study, the contribution by the logarithmic regime is
negligible and, instead, the parabolic law takes over. The
limiting parameter of surface reactions, here, is diffusion of
particles through the product layers, thus, the rate constant
of the diffusion process is determined by the measured values
of $(X^\#_{((Mn)0.252\ wt\% \ --)\ (MnO))})$ for longer durations.

Figure 4 further shows that at low isotherms, both, the type

of reaction and the reactive area change with time i.e. the
growing film partially covers the surface. At 600 °C, the
reaction follows an average $X^{\ast}_{Mn}=k \cdot t^{0.33}$ relation (curve II),
tentatively interpreted as consisting of two successive steps
(see curve III): At the onset of the reaction, the growth of
Mn-oxide follows a characteristic low exponent ($n=0.15$) rate
law, when saturation is reached ($t>25$ hr), the growth of
the oxide layer again conforms to the square root of time.
Progress of the reaction during the first stage having $n=0.15$
exponent, probably, depends on crystallographic orientation of
grains. In the later stage, under $n=\frac{1}{2}$, the reaction rate is
independent of anisotropy of the system.

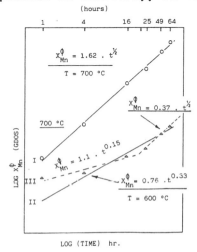

Fig. 4. (log-log) plot of
data shown in figure 3.

Electron microscopy showed that, in the course of heat-treatment, grain growth takes place predominantly along the cold-
rolling direction of steel, with boundary planes being preferential sites of surface chemical reactions. At places where
the chemical potential in the base metal is high, nucleati-
on/growth of segregants seemed to have taken place (figure 5).
The thickness of these nucleations varies randomly from spot
to spot suggesting differences in local reaction rates. Differing local reaction rates are due to: a) anisotropy in poly-
crystalline steel, b) localized differences in surface finish,
and c) the degree of surface cleaness of the material. The
resulting chemical heterogeneity of the surface has a strong
influence on the technological performance of steel, in particular during a thermal cycle. Due to differences between
thermal expansion coefficients of segregants and substrate,
surface stresses rise locally and become significant. Such
high stress fields, as discussed above, account for local fis-
sures, enhanced corrosion rate, and wear. In addition, as the
surface oxides continue to grow, cavities along sub-surface
layers nucleate and grow. Such sub-surface pores arise a) when
kinetics of surface reactions are governed by diffusion,
predominantly, of cations and b) when such a process (a) is
expediated further [10] by a high concentration of sulphides
along sub-surface layers (below). Cavities are detrimental to
the technical properties of steel. For example, they can
decrease the adhesion force of protective layers.

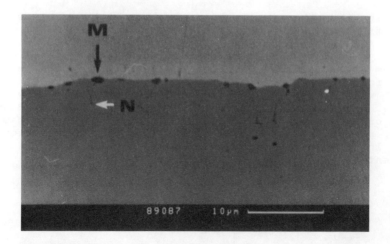

Fig. 5. Cross section through steel specimen annealed at 700
°C for 64 hr in HNX. The micrograph shows localized surface
segregation/oxidation reaction of manganese (↑M surface, ↑N
sub-surface).

As regards the internal oxidation of manganese, the following
results were obtained. At 700 °C and log pO$_2$(atm)=-25.5, there
is a change with the duration of the reaction, in the concentration profile of manganese in sub-surface layers. This is
illustrated in diagrams (A,B) in figure 6.

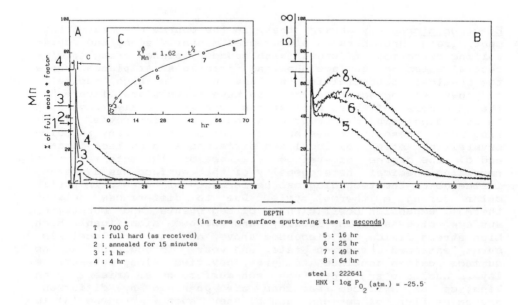

Fig. 6. Change in concentration profile of surface/sub-surface
as gas-metal reaction proceeds in time.

In figure 6 (A,B), a population of peaks develops in which the
intensity of the spectra increases when the duration of heat-
treatment progresses from 15 min to 64 hr. Diagram C (inset)
shows a corresponding time dependency of manganese coverage in
free surfaces, within a depth of ca. 40 nm. Diagram (C) indicates a diffusion controlled reaction with (time)$^{1/2}$ dependence. Through the formation of heterogeneous phases in the
surface/sub-surface layers, the matrix lattice expands internally, thereby generating stresses.

Segregation of Sulphur

GDOS data for time dependency and temperature dependency of
sulphur coverage (Fig. 7) suggest that there is a relationship between the coverages of sulphur and that of manganese.
Figure 7, depicting diagramatically the time dependency of
sulphur segregation, indicates the following: At the onset of
heat-treatment (t<5 hr), the sulphur level in free surfaces
rises sharply, but, with increasing reaction time, the sulphur
concentration decreases to a lower level. Further, there is an
inverse relation between the level of surface coverage of
sulphur and the temperature of the reaction - e.g. at 700 °C
the coverage is lower than at 600 °C, and coverage is again
lower at 500 °C.
The process of segregation of sulphur to free surfaces, possibly, is controlled by quantities of, both, sulphur and micro-
alloying elements in the bulk. Manganese is a strong sulphide
former (<MnS>; $\Delta G°_{r298 K}$=-49.8 Kal.mole^{-1}) which, in all possibility, influences the distribution of sulphur in surfaces and
in sub-surface layers: At the initial stage of heat-treatment,
both, manganese and sulphur atoms enrich the free surfaces. An
increased formation of <MnO> ($\Delta G°_{r298 K}$=-86.8 Kcal.mole^{-1}) at
the surface, removes sulphur to the inner layers and the
surface activity of sulphur is further lowered due to excessive replacement of S^{2-} by O^{2-}.

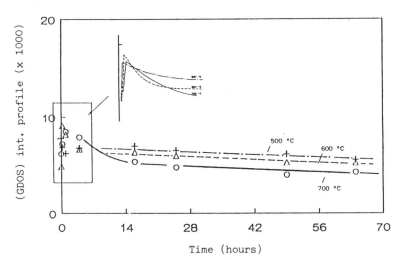

Fig. 7. Time dependency of sulphur coverage.

The above conclusion of sulphur/manganese surface relationship, presently, was further substantiated by GDOS observations of steel samples isothermally annealed for 20 hours at 700 °C in high purity hydrogen. Hydrogen gas consisted of low dew points corresponding to log pO_2 (atm)<-27.34 - i.e. a range of oxygen fugacities which falls below the stability field of <MnO> phase. It was observed that, in the absence of substantial $[O^{2-}]$ in the surfaces, marked sulphur coverage occurred. The degree of sulphur coverage steadily declined when the oxygen partial pressure of the gas phase was increased towards the stability field of <MnO> phase. Accompanying the dissolution of sulphur to sub-surface layers, the manganese level in the surfaces increased, by formation of <MnO>. The increase in manganese coverage continued as far as oxygen fugacities being as high as log pO_2 (atm) = about -24.5; Above these oxygen fugacities, segregation level of manganese fell again. The decline of manganese coverage at log pO_2 (atm)> ca. -24.5 attributed to a high rate of oxidation of manganese at earlier stages of heat-treatment. As a result of this, diffusion paths of particles are blocked.

Acknowledgment

The work presented is part of Hoogovens Groep B.V. project DS-BSK-OVK No. 22140 (Gas-Metal Reactions). The author thanks Dr. W.J. van der Meer and ing. W. van Koesveld for their contribution to laboratory experiments.

References

1) Schön, L., Leijon, W., and Augustsson, P-E.:Scandinavian Journal of Metallurgy, 1984, 13, 220-225.

2) Junius, Hans-Toni:Iron and Steel Engineer, 1989, 45-51.

3) Guttmann, M.:Surface Sci. 1975, 53, 23.

4) Forghany, S.K.E., Cheetham, A.K., Olsen, O.:J Solid Stat Chem. 1987, 71, 305-323.

5) Wood, C.G.:Oxidation of Metals, 1970, Vol.2, No.1.

6) JANAF Thermochemical Tables, 2nd ed., NSRDS-NBS 37. U.S. GPO, Washington D.C., 1971 (by:D.R. Stull and Prophet).

7) Hudson, H.E., Biber, E.J., Oles, Jr., and Warning, C.J.: Metallurgical Transactions A, 1976, Vol. 7A.

8) Hudson, H.E., Biber, E.J., Oles, Jr., and Warning, C.J.: Metallurgical Transactions A, 1977, Vol. 8A.

9) Reed. T.B.:In "the Chemistry of Extended Defects in Non-Metallic Solids", 1969, eds. LeRoy Eyring and Michael O'Keeffe (North-Holland Pub. Comp., Amsterdam).

10) Oudar, J.:Mater. Sci. Eng., 1980, 42, 101.

Materials Science Forum Vols. 102 - 104 (1992) pp. 53-62
Copyright Trans Tech Publications, Switzerland

HEAT RESISTANT ALLOY PERFORMANCE IN THE HEAT TREATMENT INDUSTRY

J.C. Kelly

Rolled Alloys, P.O. Box 310, Temperance, MI 48182-0310, USA

ABSTRACT

Heat resistant alloys used in fixturing for the heat treatment industry need to have good mechanical properties at red heat as well as resistance to chemical attack by the environment. The requirements vary depending upon the equipment in question, being somewhat different for a quenching fixture than for a muffle operating 1150°C. In addition to metallurgical considerations, design and fabrication practice strongly influence life of furnace equipment.

Alloy property requirements include strength, resistance to thermal fatigue, oxidation and carburization resistance. These properties are interrelated in practice. Strength is normally considered to be defined by the creep or rupture behavior, which is improved with coarser grains. Thermal fatigue resistance is also strongly related to grain size but the effect of grain size here is quite opposite the case in creep-rupture. That is, the finer grain alloys, typically ASTM 4 or finer (90 μm or smaller) give superior life in thermal fatigue but have the lower creep strength. In practice, quenching fixtures are not subject to dead weight loads as a creep-rupture bar, but rather a combination of mechanical loads and cyclic thermal strains. Internal thermal fatigue damage in a coarse grained bar may have a more significant effect on deformation in use than does the lower creep strength of fine grained metal.

Oxidation resistance is conferred primarily by Cr, Ni, Si and Al. Laboratory oxidation data emphasizes uniform attack, usually parabolic in nature. Service failures from oxidation are uncommon. Occasionally they exhibit complete perforation from local scale breakdown, roughly analogous to pitting failures in aqueous corrosion.

Carburization resistance depends most strongly on Ni, Cr and Si. The degree of embrittlement from carburization is related to grain size and to whether discrete carbides versus continuous networks form.

Both oxidation and carburization behavior is influenced by mechanical or chemical damage to the protective oxide scale. Thermal cycling and creep deformation mechanically damage the scale and even small amounts of common heat treat salts reduce the protective nature of the scale.

Finally, the performance of very expensive heat resistant alloy equipment is dependent upon knowledgeable design and fabrication practice. Designers should be aware that thermal strains are often more significant than mechanical loads. Fabricators need to pay close attention to weld joint preparation, as incomplete joint penetration is the most common cause of weld failure in service.

INTRODUCTION

The alloys to be addressed in this paper are those wrought solid solution strengthened NiCrFe alloys used as structural elements or containers in the thermal processing industries. Service temperatures range from below 750°C to 1250°C, atmospheres from vacuum to carburizing, and thermal cycling from furnace cool to brine quench. Corrosive environments can be as benign as air or severe as hot chloride fumes and molten metals.

ALLOYS

Some of the grades most commonly used in North America are listed in Table I.

Table I

Grade	UNS	Approximate werkstoff	Equivalent DIN	Cr	Ni	Si	Al	C	Other
RA446	S44600	1.4749 *	X 18 CrN 28	25	--	.5	--	.05	.7Mn .1N
RA309	S30908	1.4833	X 7 CrNi 23 14	23	13	.8	--	.05	--
RA253MA	S30815	1.4893	X 8 CrNiSiN 21 11	21	11	1.7	--	.08	.04Ce 17N
RA310	S31008	1.4845	X 12 CrNi 25 21	25	20	.5	--	.05	--
314	S31400	1.4841	X 15 CrNiSi 25 20	25	20	2.2	--	.10	--
RA85H	S30615	--	--	18.5	14.5	3.5	1.0	.2	--
RA330	N08330	1.4864	X 12 NiCrSi 36 16	19	35	1.3	--	.05	--
600	N06600	2.4816	NiCr 15 Fe	15.5	76	.3	.2	.08	.2Ti
601	N06601	2.4851	NiCr 23 Fe	22.5	61.5	.2	1.4	.05	--
RA333	N06333	2.4608	NiCr 26 MoW	25	45	1.3	--	.05	3Co 3W 3Mo

* except carbon 0.12 maximum

The high chromium ferritic RA446 has very low hot strength, about an order of magnitude less than the austenitic grades. RA446 is used primarily for resistance to oxidation, sulfidation or molten copper, where strength is not a requirement. Large bars (e.g. 5 x 10 cm rectangular) are used for electrodes in neutral salt pots up to 1290°C.

Austenitic grades RA309 and RA310 are used primarily for oxidation resistance to 1040-1100°C. They offer moderate creep strength and, for nickel bearing alloys, good hot corrosion resistance. RA253MA® has exceptionally high strength and oxidation resistance through 1100°C and finds use in annealing covers, radiant tubes, kilns, hot dampers, cyclones and coal fired burners. This cerium bearing grade performs best under oxidizing conditions.

The silicon content of 314 provides good oxidation and carburization resistance, for use in retorts and muffles. In the USA type 314 has been supplanted by RA330®, since the 1960's.

RA85H®, RA330, 600, 601 and RA333® all have good strength and carburization resistance, with oxidation resistance ranging from very good to excellent.

RA85H relies on 3.5%Si and 1%Al for carburization resistance similar to RA330, with oxidation resistance superior to 309. In addition, this high silicon level provides good resistance to attack by chloride salts, Table II.

Table II

Intergranular Attack in Molten Salt Bath	
Grade	Depth of Attack, mm
RA85H	0.11
253MA	0.18
600	0.19
309	0.32
RA330	0.35

Specimens exposed for 21 days, 10-12 cycles per day. A cycle consists of preheat salt at 704°C, transfer to 816°C preheat salt, then to high heat salt 1204°C, followed by quench in 593°C salt, air cool.

RA85H is used for a variety of salt bath fixturing. The one shown in Figure 1 operates just above the salt, in the corrosive fumes.

Figure 1

RA85H drill fixture, used just above salt level in the
fumes, temperature about 820°C.

RA333 has perhaps the best combination of strength and oxidation resistance
of the commercially available materials. Tests at 1150°C show only one third to
one sixth the metal loss, [1] calculated from weight measurement, of other commonly
used heat resistant alloys (Table III). Cyclic oxidation tests [2] run over 7,500
hours at 1038°C demonstrate long-time oxidation resistance superior to the
aluminum bearing alloy 601. The strength and oxidation resistance of RA333
permit this grade to be used as thin as 3mm sheet in copper brazing muffles and
gas fired radiant heating tubes.

Table III

1150°C oxidation, weekly cycles to 3,000 hours	
Grade	Surface Recession, mm
RA333	0.078
RA330	0.24
RA85H	0.49
600	0.53
310	0.78

STRENGTH

Above 550°C a common definition of strength is the stress to cause a minimum creep rate of 0.0001% per hour. The practical effects of creep strength are illustrated in Figures 2a and 2b. At 930°C operating temperature the nitrogen-cerium strengthened RA253MA shows no deformation, as compared with marked creep buckling in 309 stainless. Considering that there is a factor of three difference in measured creep rate between these two grades, this relative service performance is to be expected.

Figure 2a

253MA inner cover,
after 12 months
service up to 930°C

Figure 2b

309 inner cover, same
application as 2a.

Stress to rupture is another commonly used measure of high temperature strength. However it does not necessarily predict the relative resistance to deformation among various alloys. It is quite possible for several grades to have similar rupture strengths but vary widely in resistance to creep deformation. For example in Table II at 980°C, grades 309 and 310 have 90% and 108%, respectively, of the rupture strength of RA330. But the creep strengths of 309 and 310 are both less than 60% that of RA330.

Table IV

Grade	At 980°C, Stress, N/mm^2, for:	
	Rupture in 10,000 hours	Secondary Creep Rate of 0.0001% per hour
RA330	4.34	3.45
310	4.69	2.0
309	3.93	2.0
253MA	7.93	6.14
RA85H	6.20	4.83
RA333	7.24	6.07

High temperature shear strength is important for the 12.7-25.4mm diameter pins used in cast link furnace belts in North America. For the link design of Figure 3 the pins bear the shearing loads imposed by belt and workload weight, and by the force necessary to pull the belt through the furnace.

Figure 3   1 X

Pin bearing link, cast of 35Ni 17Cr .4C alloy.

If the pin lacks shear strength the failure mode is "crank-shafting," as shown in Figure 4. The most common pin alloy in the U.S.A. is RA330HC, a 0.4% carbon version of RA330.

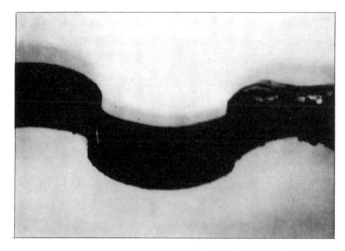

Figure 4

Belt pin deformed in shear, a consequence of improper anneal practice.

With an interlocking design, the cast link itself bears the shearing stresses and a high strength pin is not required. In this case pins of RA330 14.3mm hexagonal bar provide adequate strength. Hexagonal pins are used to continually scrape away carbon deposits which would otherwise tend to prevent the belt from flexing over the drive drum.

THERMAL FATIGUE

Thermal fatigue problems are far more common than creep-rupture, in heat treating furnace equipment. The nature of the business includes thermal shock ranging from air cool to water quench from red heat, with alloys that typically expand 18mm per meter from 0 to 1000°C.

Both thermal fatigue and creep-rupture are strongly influenced by the metal's grain size, albeit in opposing manners. It is well known that a high temperature anneal to produce a coarse grain size improves creep-rupture properties. Such processing is required to develop the full strength of grades such as N08810 and N08811. However in liquid quench applications coarse grains almost invariably result in short life due to thermal fatigue cracking. Specifically, in bar frame heat treating baskets a reasonably fine grain size is required to minimize thermal fatigue damage.[3]

Grain size also affects the life of steel mill annealing covers, when rapid cooling is part of the cycle. Covers of RA330 4.6mm plate, 1.7m diameter by 4.5m high, have typically lasted in excess of 7 years at one North American mill. The maximum cover temperature is reported 815°C, product anneal temperature 745°C. This cover is water spray cooled from about 480°C. RA330 is annealed to develop a grain size typically in the range ASTM 4-7. Other producers of 35%Ni-19Cr alloy prefer to maximize creep strength by a grain coarsening anneal.

When a cover of such material, grain size ASTM 2, was placed into this same service it developed severe through-wall cracks within one year.

The microstructure, Figure 5, shows classic thermal fatigue cracks, intergranular and heavily oxidized. It is preferable to avoid the use of deliberately grain coarsened alloy when the intended service is known to include severe thermal shock.

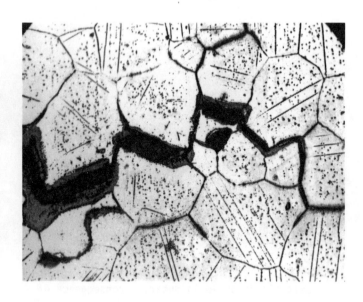

Figure 5   100 X

Thermal fatigue in 35Ni-19Cr alloy. Direction of crack propagation from left to right.

Incompletely penetrated weld joints tend to fail early in thermal fatigue. The unwelded area functions as a crack, and grows outward through the weld during each thermal cycle. Indeed, lack of full penetration is the most common cause of weld joint failure in high temperature service. Figure 6 illustrates an unfortunately common weld joint design used in bar frame basket construction.

While the RA85H bars retained 43% tensile elongation at room temperature, the welds broke. Originally a cross, the completely fractured weld is to the left. The initial unwelded condition may be still observed at the right side. This problem could be minimized by cutting a dull chisel point on the bar ends, to permit the weldor to make a completely penetrated joint.

Figure 6   3 X

Incomplete weld penetration, 12.7mm bar

REFERENCES

1. Rundell, G.R., and McConnell, J. B., "Oxidation Resistance of Eight Heat Resistant Alloys at 870, 980, 1095 and 1150°C," to be published in <u>Oxidation of Metals</u>.

2. Roy, A., Gagen, F. A., and Corwin, J. M., "Performance of Heat-Resistant Alloys in Emission-Control Systems," SAE 740093.

3. Rundell, G. R., "Evaluation of Heat Resistant Alloys in Composite Fixtures," NACE paper No. 377, 1986

Materials Science Forum Vols. 102 - 104 (1992) pp. 63-74
Copyright Trans Tech Publications, Switzerland

COMMERCIAL HEAT TREATMENT TO 2000 AD

T.L. Elliott

Managing Director Senior Heat Treatment Limited

ABSTRACT

Despite the decline in manufacturing in the United Kingdom, there has been tremendous growth both in size and quantity of those companies offering Commercial Heat Treatment services. Three major factors have influenced this growth.

(i)   The decision by "In House" operators to close down their own furnaces in favour of using outside specialists.

(ii)   Customers' demands for improved quality.

(iii)   New Processes and Technology.

In order to meet the above requirements, the modern Commercial Heat Treatment Company has had to make substantial investments in new plant, employing the latest technology and capable of producing competitively priced products of consistent high quality. The company will also have invested in qualified and experienced staff to manage this equipment, which will be operated in accordance with well documented quality assurance procedures.

Future development of the Commercial Heat Treatment Industry will focus on improved automation, to reduce the cost of labour, which is currently the highest element in heat treatment costs, and the introduction of the second generation of surface heat treatment processes, once the design engineer is convinced of the benefits of surface engineering.

"IN HOUSE" OR SUB CONTRACT

The Commercial Heat Treatment industry has made tremendous progress within the last twenty years, moving from a back street industry, tarnished with a reputation for operating second hand plant and providing poor quality products, to a sophisticated, professional organization. Today's Commercial Heat Treater will not only be operating plant using the latest technology, but will be employing qualified and experienced managers to ensure satisfactory returns are obtained from these expensive investments.

With a wide and varied customer base Commercial Heat Treatment companies (in the main) are able to keep plant fully utilised, which is essential if heat treatment costs are to be kept to a minimum. It is no coincidence that in each of the last two recessions, we experienced a substantial increase in enquiries from "in house" operators requesting meetings to discuss the possibilities of them closing their furnaces and sub-contracting out.

Figure 1 shows how heat treatment costs increase with reduced furnace utilization. These costs are not fictitious, they relate to actual operating costs for the operation of Shaker Hearth and Sealed Quench furnaces used for the heat treatment of automotive component parts.

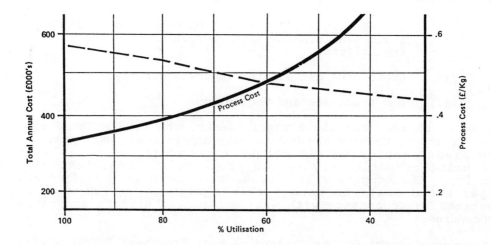

Figure 1:   The adverse effect of reduced furnace utilization on process costs.

The benefits gained by manufacturers' closing their "in house" plant are:-

Immediate cost savings.
They only pay for the work processed - not fixed overheads.
Stable costs.
Expensive floor space and capital is released for their mainstream activities.
A greater choice of heat treatment processes.
Their components are processed in advanced technology equipment by specialists.

Senior Heat Treatment have almost one hundred customers who took the decision to close their "in house" plant and use the facility provided by Senior.

QUALITY

The Oxford English dictionary defines quality as "the degree of excellence of a thing".

Lord Sieff, who was the Chairman of Marks & Spencer, Britain's most famous High Street store, stated four years ago "The largest growing market in the world today is for goods of high quality and good value, not just cheapness". This is probably even more the case today. We all want quality products.

For high quality products to be manufactured, consistent with low manufacturing costs, inevitably heat treatment specifications have been tightened. It is, therefore, more important than ever that process parameters are both clearly specified and closely monitored during processing.

Most Commercial Heat Treatment companies now operate with computers to generate not only invoices but Process Specification cards, which detail the required process parameters.

All procedures for specifying these parameters, monitoring and recording them, calibration of instruments and test equipment, sampling plans and frequency of testing etc., will be defined in the company's Quality Manual. Today in the U.K. most plants operate to BS 5750 (ISO 9002) and those companies working in the Aerospace industry will have the approval of such contractors as Boeing, Dowty, General Electric, McDonell Douglas, Rolls Royce, etc.

Clearly outdated plant, which is often having to operate without adequate maintenance or investment in replacement parts, will not be capable of achieving these higher specifications. In such situations the "In House" operator may elect to sub-contract rather than make expensive investments in new plant or refurbishment costs.

Much has been written about Statistical Process Control, S.P.C., but in effect this is still in its infancy in the Commercial Heat Treatment industry. Many companies practice Statistical Quality Control, S.Q.C., (after the event testing), charting results, e.g. surface hardness, case depth and making adjustment to process parameters once these results show signs of the process going out of control.

Modern furnaces can now be totally computer controlled, including loading and unloading. There are numerous Sealed Quench furnaces operating under computer control, where it is sufficient to dial in the appropriate "recipe" code and then all necessary process parameters are kept within prescribed control limits to ensure treated parts finish with the correct case depth and microstructure. The parameters for each load are kept on record by the computer.

Some Commercial Heat Treatment companies operate as an extension of their customers' manufacturing plants, being linked into their J.I.T. schedules, with heat treatment components going directly onto assembly lines, without any testing by the client.

Inspection both during and post treatment will be carried out by operators as well as the plant's qualified metallurgists. Test equipment is usually digital, linked to a computer for analysis of results. The modern commercial Heat Treatment company will possess a well equipped laboratory, which is often available for consultancy work.

NEW TECHNOLOGY

Whilst much has been written of the apparent virtues of certain of the "new" processes it will do no harm to consider developments in many of the well established heat treatment procedures and equipment, that will continue to be with us well after the year 2000 AD.

SEALED QUENCH FURNACES

These are probably the most popular and versatile piece of plant
operated by Commercial Heat Treaters. Recent developments centre
around the use of computers for total automation and greater
efficiency resulting from the use of improved refractory
materials and reduced heating costs due to enhanced recuperation.

Figures 2 and 3 show a one tonne pay load Sealed Quench furnace
and its computer control system, installed in one of Senior's
West Midland plants, where is it used to heat treat parts for the
automotive and light engineering industries.

Figure 2:   One tonne pay load Sealed Quench furnace at one of
              Senior Heat Treatment's West Midland plants.

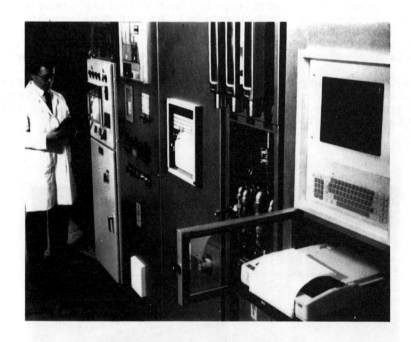

Figure   3:   Computer control system for the furnace shown in
               Figure 2.

VACUUM FURNACES

The growth of vacuum heat treatment has been quite spectacular
but not surprizing considering its advantages in terms of its
improved surface finish, ease of automation and reliability.
Vacuum also has the added bonus of being more acceptable to
current Environmental and Health & Safety Regulations.

Senior Heat Treatment were the first company in the U.K. to
introduce vacuum heat treatment to the tool making industry, when
they installed the horizontal Vacuum furnace, shown in Figure 4,
at one of their West Midlands plants, twenty five years ago.

Today in the U.K. there are almost 100 Vacuum furnaces employed
by Commercial Heat Treaters to heat treat tooling, aerospace
components and stainless steel parts.

Developments again have involved the use of microprocessors for
the total automation of heat treatment cycles as well as the use
of graphite boards for insulation and the introduction of
pressure quenching has enabled lower grade steels to be vacuum
hardened.

Figure 4:   Horizontal Vacuum furnace installed in 1966 by
            Senior Heat Treatment for the heat treatment of
            tools and moulds.

Figure 5 shows a lawn mower die, machined from 5% Chromium hot die steel, weighing 1.7 tonnes, after Vacuum heat treatment at one of Senior's West Midlands plants and Figure 6 shows the Bottom Loading Vacuum furnace, recently installed by Seniors for the heat treatment of Aerospace and Stainless Steel components.

Figure 5:   Lawn Mower casing die, weighing 1.7 tonnes,
            machined from 5% Chromium Hot Work steel, after
            vacuum heat treatment by Senior Heat Treatment.

Figure 6:   3 tonne capacity, Bottom Loading Vacuum furnace,
              recently installed by Senior Heat Treatment.

SALT BATH FURNACES

Although salt baths have largely been superseded for
environmental and economical reasons by vacuum and protective
atmosphere furnaces, they are still used for certain
applications, where they out perform their modern counterparts.
Such applications include the heat treatment of small batches,
marquenching to minimize distortion and the heat treatment of
high speed steel cutting tools, although Pressure Quench Vacuum
furnaces are gradually taking this market over.

At long last cyanide is being replaced as a carburising medium by
low toxic, regenerative salts.

INDUCTION

Induction heat treatment is used to improve wear and fatigue
properties in selected areas of medium carbon steel components.
Recent developments include the use of robots for long runs,
solid state generators replacing thermionic valve sets and
computer monitoring of components whilst being heated, which is
very useful for S.P.C.

PROTECTIVE ATMOSPHERES

Protective atmospheres are generally employed to avoid any change in surface chemistry during heat treatment cycles. In recent years there has been a move to use so called synthetic atmospheres such as nitrogen plus methanol, rather than manufacturing furnace atmospheres by partial or complete combustion of hydrocarbons.

Recent developments include Pressure Swing Absorption P.S.A. generators for the manufacture of nitrogen and the use of pressurised methanol injectors to eliminate sooting on Shaker Hearth and Mesh Belt furnaces.

Oxygen probes are now more reliable and are being used with computerized control systems. With the increased use of computers to control heat treatment plant, one cannot help wondering whether future heat treatment furnaces will need to be managed by a computer engineer rather than a metallurgist.

"NEW PROCESSES" IN COMMERCIAL HEAT TREATMENT

Plasma nitriding has been well established for many years but the high cost of plant is delaying the introduction of plasma carburising. Plasma carburising is very quick and has the advantage of avoiding internal oxidation and thus improving fatigue life, which is considered important for certain critical aerospace applications. It is probably in this industry where plasma carburising will take off.

Pressure Vapour Deposition, P.V.D., is now offered by commercial heat treaters to improve the life of tools and moulds and its specification by end users will gain popularity in the future.

Laser Heat Treatment has been on the scene for some time in the U.K. but its use by the heat treater has to date been slow.

Hopefully it has been demonstrated that the U.K.'s commercial heat treatment industry is aware of the needs of its customers and is making the necessary investments in plant and personnel to ensure that they are provided with a reliable and efficient service.

Just as it led the way in installing Vacuum furnaces in the 70s, and introducing computer controlled furnaces in the 80s, it will no doubt be in the forefront in introducing new processes in the 90s as they become technically proven and commercially viable.

With 1992 approaching the national borders are being broken down and already some of the large groups of heat treatment companies operate internationally.

II. HEAT TREATMENT OF NON-FERROUS ALLOYS

Materials Science Forum Vols. 102 - 104 (1992) pp. 75-84
Copyright Trans Tech Publications, Switzerland

TOWARD THE OPTIMIZATION OF HEAT TREATMENT IN ALUMINUM ALLOYS

C.W. Meyers, K.H. Hinton and Jyh-Sham Chou

The George W. Woodruff School of Mechanical Engineering, Georgia Institute of Technology, Atlanta, GA 30332-0405, USA

ABSTRACT

The major barrier to the increased usage of cast aluminum alloys in critical applications is the unreliability of tensile and fracture properties. Among the many factors contributing to the variability in property levels are the nature and the parameters of the heat treatment protocol and the subsequent microstructural modifications. This paper presents an overview of current research on the heat treatment induced microstructures and their effects on toughness levels. Special emphasis is placed on tailoring the microstructure through heat treatment for acceptable fracture toughness levels.

INTRODUCTION

Excellent castability, resistance to hot tearing and high tensile and fracture property levels are among the attractive characteristics of hypoeutectic aluminum-silicon-magnesium alloys. According to the Aluminum Association [1], the factors regulating these characteristics are alloy and compositional variations, melt quality, solidification rate, metallurgical characteristics and heat treatment. The effects of the first two factors, which may be termed molten metal processing effects, are comprehensively reviewed by Sigworth, Shivkumar and Apelian [2]. Gardner's and Spear's classic work [3] on dendrite effects initiated the quantification of the effects of solidification rate on mechanical properties. And metallurgical characteristics, their evolution, identification, interactions and subsequent effects on property levels have been extensively explored in texts, handbooks and numerous publications. The last of the factors regulating mechanical properties, the heat treatment of the alloy castings, represents one of the major challenges in the total manufacturing cycle [4].

Optimum combinations of strength and ductibility in Al-Si-Mg alloys are reportedly attained through heat treating to the T6 and occasionally the T7 temper [5]. Both tempers involve solutionizing to dissolve soluble constituents, proper quenching to retain these constituents in solution at lower temperatures and artificially aging to facilitate the precipitation of non-equilibrium Mg_2Si within the primary aluminum rich dendrites. The temperatures and the durations of the stages of these tempers are critical governors of the levels of attainable cast properties.

The microstructural changes induced by the tempering process in cast Al-Si-Mg alloys have been the subject of several recent investigations [5,6]. It is now generally accepted that during the solution heat treatment of cast Al-Si-Mg alloys, morphological, numerical density and distribution variations in the silicon-rich eutectic constituent occur [7,8]. Evidence of spheroidization and coarsening have been reported, as have their influences on tensile and fracture properties [9-13]. The continued examination of the effects of the heat treatment induced microstructures on the properties of Al-Si-Mg alloys is the subject of this paper.

PROCEDURE

Three sets of test bars of aluminum alloy A357 were poured in preheated ($100^{\circ}C$) steel permanent molds in the foundry of the George W. Woodruff School of Mechanical Engineering at the Georgia Institute of Technology. No metal treatments, such as grain refinement or eutectic modification, were added.

The bars were then solution heat treated at $540\pm5^{\circ}C$ for 1, 10, 30, 60, 120, 150 hours. After quenching in room temperature water, one set of bars for each solution heat treatment time was artificially aged at $155^{\circ}+5^{\circ}C$ for 4 hours with another set being aged at the same temperature for 16 hours. The third set of test bars was subjected to a dual heat treatment at the same temperature for 12 and 9 hours, separated by a room temperature water quench. Following a 48 hour natural age, the third set of test bars was artificially aged at $155^{\circ}C$ for 8 hours.

Microstructural changes in the silicon-rich eutectic constituent were monitored for each solutionizing and aging combination using conventional point counting techniques. Up to thirty fields of view per specimen were examined to ensure statistical reliability. Using an automated image processor attached to the light optical metallograph, the following data were derived from mathematically sound stereological relationships [14,15]:

1. $(Vv)_{si}$ - volume fraction silicon-rich structures in the interdendritic region
2. N_L - bulk lineal density of silicon-rich structures
3. $(N_L)_{ID}$ - interdendritic lineal density of silicon-rich structures
4. $(N_A)_{ID}$ - local interdendritic areal density of silicon-rich structures
5. A - average area of silicon-rich structures
6. MFP - mean free path between silicon-rich structures
7. $(L_2)_{ID}$ - mean intercept length of silicon-rich structures
8. H - average surface curvature of the silicon-rich structures
9. AR - aspect ratio of silicon-rich structures

Chevron notch short rod fracture toughness tests were performed on samples from each heat treatment protocol in accordance with ASTM E-1304. Extensive statistical analyses were performed to establish the nature of the microstructural feature most significantly affecting fracture toughness levels in alloys subjected to extended solution heat treatments.

RESULTS AND DISCUSSION

Typical microstructural data for the 4 and 16 hour age treatments are shown in Tables 1 and 2, respectively.

Table 1. Typical Microstructural Data - 4 Hour Age

Solution Time (hours)	N_A^+	N_A^-	P_{si}	$(N_{Si})_{ID}$
As Cast	85.9	0.28	60.6	6.3
10	19.7	0.07	32.2	3.6
30	18.1	0.43	34.3	2.7
60	9.5	0.17	30.8	1.9
120	9.1	0.20	33.8	2.1
150	8.8	0.52	32.1	1.9
Special	16.1	0.42	31.2	3.4

Table 2. Typical Microstructural Data - 16 Hour Age

Solution Time (hours)	N_A^+	N_A^-	P_{si}	$(N_{Si})_{ID}$
As Cast	85.9	0.28	60.6	6.3
10	16.9	0.47	32.7	4.4
30	16.7	0.62	29.2	2.7
60	13.5	0.42	31.8	2.3
120	9.0	0.83	33.9	2.2
150	6.8	0.43	30.0	1.7
Special	16.1	0.42	31.2	3.4

The first two columns are used in the determination of average curvature H according to the relationship [14]:

$$H = M_V/S_V \qquad (1)$$

in which M_v is the mean integral curvature given by:

$$M_V = 2\pi\ (\bar{N}_A^+ - \bar{N}_A^-) \qquad (2)$$

and S_v is the surface area per unit volume. In equation (2), N_A^+ represents the average number of loops of an α-B boundary enclosing the α phase per unit area; N_A^- is the average number of loops of the α-B boundary not enclosing the phase. As is typically the case for the divorced eutectic structures in Al-Si alloys, N_A^- is essentially negligible, showing no influence of the duration of solution heat treatment. N_A^+, on the other hand, is sensitive to this phase of the tempering process, showing a 77% reduction in the as-cast volume after 10 hours of solutionizing for the case of the 4 hour age. This reduction increases to 80% for the 16 hour age following the same duration of solutionizing. An additional 50 hours of solutionizing for the 4 hour age case is required to cause an additional 10% decrease in N_A^+. In excess of 110 additional hours of solutionizing is required for 10% further reduction in N_A^+ from as cast levels for the 16 hour age.

Values of P_{si} in Table 1 and Table 2 reflect the stability of the eutectic constituent after an intial 10 hour solution treatment. Both P_{si} and $(N_{si})_{ID}$ are used in the calculations of mean free path, MFP, as follows:

$$\text{MFP} = \frac{1 - (V_v)_{ID}^{si}}{(N_L)_{ID}^{si}} \qquad (3)$$

in which

$$(V)_{ID}^{si} = \frac{P_{si}}{P_T}$$

where P_{si} is the number of grid points coinciding with the silicon-rich phase and P_T is the total number of grid points in the test area. Similarly,

$$(N_L)_{ID}^{si} = \frac{(N_{si})_{ID}}{L_T} \qquad (4)$$

where $(N_{si})_{ID}$ is the number of intercepts of the silicon-rich phase in the interdendritic region and L_T is the test line length.

The effects of the extended solution heat treatment become pronounced when the data calculated from the measured parameters in Tables 1 and 2 are considered. The volume fractions $(Vv)_{si}$ (Table 3) are constant when standard deviations are considered. A non-detectable change in this parameter indicates that appreciable amounts of silicon are not soluble in the aluminum-rich phase at solutionizing temperatures. This is consistent with the 1.65% Si solubility limit of silicon in aluminum at the eutectic temperature. These percentages are unaffected by the interrupted solution treatments and aging delays as is the case for the designated "special" treatment.

Table 3. Calculated Density Parameters - Average Values

Solution Time (hours)	4 Hour Age			16 Hour Age		
	$(Vv)_{si}$ (%)	$(N_L)_{ID}$ $X10^{-2}$/mm	$(N_A)_{ID}$ $X10^{-4}$/mm	$(Vv)_{si}$ (%)	$(N_L)_{ID}$ $X10^{-2}$/mm	$(N_A)_{ID}$ $X10^{-4}$/mm
As Cast		21.5	86.1		21.5	86.1
10	26.6	5.0	19.8	27.0	4.1	16.5
30	28.3	4.4	17.7	24.1	4.0	16.1
60	25.4	2.3	9.5	26.3	3.3	13.2
120	27.9	2.2	8.1	28.0	2.0	8.2
150	26.6	2.1	8.4	24.5	1.6	6.4
Special	25.8	3.9	15.6	25.8	3.9	15.6

Density parameters shown in Table 3 follow exactly reported trends (7-10); that is, interdendritic lineal and areal densities decrease as the duration of solutionizing increases. The first ten hours of solution heat treatment, which falls on the low side of commercial practice, result in an approximate 75-80% reduction in density $(N_2)_{ID}$ of the silicon-rich structures. Extended aging treatment yields an additional 50% reduction from the ten hour values. The "special" treatment (incorporating the interrupted solutionizing and the natural age prior to the artificial one) results in densities which are between those of the 30 and 60 hour solutionizing times for both aging times.

Table 4A. Calculated Size and Shape Parameters - 4 Hour Age

Solution Time (hours)	A $(x10^{-6}mm^2)$	L_2 (1/mm)	AR	MFP (mm)	H
As Cast	0.58	2.33	10.7	2.32	2076
10	1.35	5.32	4.75	14.7	1319
30	1.60	6.43	3.86	16.3	1057
60	2.71	11.0	2.22	32.4	899
120	3.11	12.7	1.93	32.8	670
150	3.19	12.7	1.99	35.0	679
Special	1.65	6.62	3.77	17.3	910

Table 4B. Calculated Size and Shape Parameters - 16 Hour Age

Solution	A	L_2	AR	MFP	H
Time (hours)	$(x10^{-6}mm^2)$	(1/mm)		(mm)	
As Cast	0.58	2.33	10.7	2.32	2076
10	1.64	6.59	5.98	17.8	977
30	1.50	6.03	4.11	19.0	924
60	1.99	7.97	3.14	22.3	847
120	3.42	14.0	1.74	36.0	563
150	3.83	15.3	1.64	47.2	557
Special	1.65	6.62	3.77	19.0	910

The changes in the size parameters, average area A and mean intercept length L_2, are consistent with the decrease in densities. Both increase as the solutionizing time increases for each of the aging times (Tables 4A and 4B). Area data associated with the 16 hour age are slightly larger than those associated with the 4 hour age, indicating that the prolonged exposure to elevated temperatures, even those in the range of the relatively low temperatures associated with aging, is accompanied by increased coarsening. The mean intercept length L_2, on the other hand, when the aging time is short (i.e., 4 hours) starts leveling off at less than 100 hours. The 16 hour aging time data indicates continued coarsening but at a slower rate beyond 120 hours solution time. Again, the "special" treatment results in data (size, A and intercept length L_2) between those of the 30 and 60 hour treatments. Concurrent with the increase in the size parameters is a steady increase in mean intercept length as shown in Figure 1.

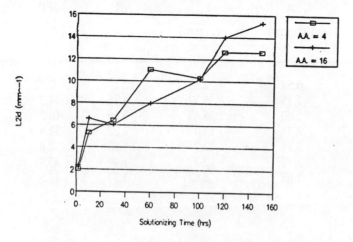

Figure 1. Mean Intercept Length vs. Solution Time

These size changes are consistent with the shape perturbation theory of spheroidization (6) which dictates faceting only after 13 hours solutionizing at 540°C.

Shape changes were quantified by two parameters - aspect ratio (AR) and average curvature (H); data for both also appear in Tables 4A and 4B. The aspect ratios decrease and approach the spherical value of 1.0 as the solution time increases. Curvature values, H, which are inversely related to the radius of an arc describing the curved surface decrease with increasing solution time, with the rate of increase being greater for each solution time for the samples that were aged longer, i.e., 16 hours. Again, the shape parameters describing the silicon-rich structures of the "special" treatment exhibit values between those of the 30 and 60 our heat treatment.

From preliminary regression analyses and analyses of variance, the best microstructural descriptors of the heat treatment induced changes are the ratio of mean free path to mean tangent diameter, MFP/d and the average surface curvature H. These relationships are:

4 hour age:

$$\left(\frac{MFP}{d}\right)^{1/2} = 9.4t + 31.2 \qquad r = 0.82 \tag{5}$$

$$\bar{H} = 57.9t - 0.02 \qquad r = -0.86 \tag{6}$$

16 hour age:

$$\left(\frac{MFP}{d}\right)^{1/2} = 7.8t - 27.1 \qquad r = 0.96 \tag{7}$$

$$\bar{H} = 50.1t - 0.2 \qquad r = -0.93 \tag{8}$$

In equations 5-8, t is the solution heat treatment time in units of hours. Both of the parameters demonstrate coarsening with the longer aging time contributing strongly to the behavior of the data. These results are illustrated in Figures 2 and 3 showing mean free path and curvature effects, respectively.

This relationships between the curvature of the silicon-rich eutectic constituent and solution heat treatment time (equations 6 and 8) represents a significant advance over previous work (7-8) in characterizing the coarsening process. Meyers used an average radius of an equivalent circle, R, to characterize coarsening in these alloys. The parameter, H, possesses no a priori shape assumption and therefore offers a more realistic description of the irregularly shaped 2-D projections of the 3-D silicon-rich eutectic structures.

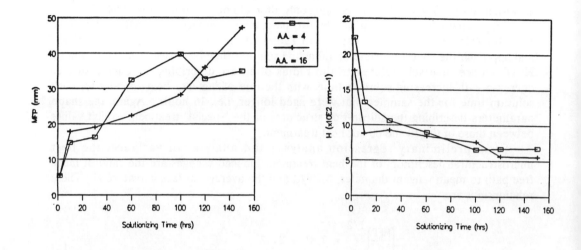

Figure 2.   Mean Free Path
vs. Solution Time

Figure 3.   Average Curvature
vs. Solution Time

Fracture toughness data for each protocol are shown in Tables 5 and 6.

Table 5. Short Rod Fracture Toughness

	4 Hour Age			16 Hour Age	
Solution Time (hours)	No. of Samples	K_{IV} Range MPa m		No. of Samples	K_{IV} Range MPa m
10	3	2.96-39.6		3	28.4-39.6
30	4	35.6-39.6		5	22.5-39.3
60	2	32.0-40.6		4	27.7-36.2
100	2	40.5-42.5		2	36.2-38.2
120	1	-		2	35.5-40.5
150	3	34.8-43.6		4	36.4-41.5

Table 6. Short Rod Fracture Toughness - "Special" Treatment

Sample	K_{IV} (MPa m)
1	24.1
2	25.4
3	25.1
4	25.6
5	22.0
6	21.6
7	22.0

Average Value: 23.7 MPa m
Calculated Value: 24.3 MPa m
(K_{IV} = 1.75 MFP + 21.0)

The ranges of the facture toughness values for each aging time are close, indicating the relatively "passive" effect of this low temperature holding cycle on resistance to crack propagation. This response is expected in light of energetics and kinetics of high temperature post-solidification treatments; the microstructural changes are more sensitive to the higher temperature solutionizing phase of the temper. Aging time only slightly changes the size (MFP) of remaining ductile ligaments between the brittle silicon-rich eutectic constituent.

Average values of the K_{IV} from the chevron notch short rod tests are plotted against solution time for each artificial aging time in Figure 4.

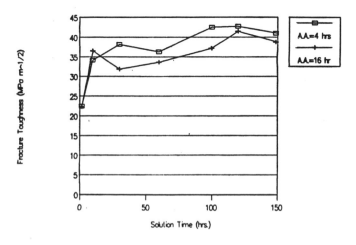

Figure 4. Fracture Toughness (MPA m) vs. Solution Time (hrs.)

As expected, the effect of the extended aging is that the fracture toughness levels appear more predictable and therefore, reproducible, although generally lower. The "special" treatment yielded an average fracture toughness level lower than those associated with the conventional protocols, average $K_{IV} = 23.7$ MPa m. The relationship developed by Meyers [16] below was used to determine the calculated K_{IV}:

$$K_{IV} = 1.75 \text{ MFP} + 21.0 \qquad (9)$$
$$\text{MPa m} \qquad (\text{um})$$

The agreement between the measured average value and the calculated one for the "special" treatment (i.e., $K_{IV} = 24.3$ MPA m) is good. This is again verification of the special treatment producing microstructural changes consistent with those associated with solutionizing between 30 and 60 hours. Equation (5) was developed using solution treatments up to 100 hours. The resulting relationships should not hold for times beyond those limits.

CONCLUSIONS

1) Extended aging has the effects of a) slightly increasing the extent of the coarsening of the silicon-rich eutectic constituent and b) inducing more predictable and regular fracture toughness behavior.

2) The shape parameter H with no geometric restrictions can be employed to characterize the coarsening of the silicon-rich eutectic structures during the temper process.

3) Preliminary examination of the practice of interrupted solution heat treatment and pre-aging suggests that an uninterrupted solutionizing stage of greater than 30 hours and on pre-aging results in the same microstructural refinement and fracture toughness level.

Further study of the ratio of fracture toughness to yield strength, K_{IV}/σ_{ys}, especially at small intervals of solutionizing times between 30 and 60 hours is indicated by this and previous work in the area.

REFERENCES

1. "Special Report on the Mechanical Properties of Sand Cast Aluminum Alloy Test Castings," Jabbing Foundry Division of the Aluminum Association, Washington, D.C., 1985.

2. Sigworth, G.K., Shivkumar, S. and Apelian, D.: AFS Transactions, 1989, 97, 811.

3. Spear, R. E. and Gardner, G.R.: AFS Transactions, 1963, 71, 209.

4. George Bodeen: Heat Treating News, April 1991, 4.

5. Apelian, D., Shiukumar, S., Sigworth, G.: AFS Transactions, 1989, 97, 727.

6. Shiukumar, S. et al: J. Heat Treating, 1990, 8, 63.

7. Meyers, Carolyn W.: Ph.D. Dissertation, Georgia Institute of Technology, 1984.

8. Meyers, Carolyn W.: AFS Transactions, 1985, 93, 741.

9. Meyers, Carolyn W.: AFS Transactions, 1986, 94, 511.

10. Meyers, Carolyn W.: Proceedings of AFS Special Conference on "Mechanical Properties of Aluminum Castings," Chicago, IL, 1987, 167.

11. Zhu, P. Y., Liu, Q. Y. and Hou, T. X.: AFS Transactions, 1985, 93, 609.

12. Saigal, A.: AFS Transactions, 1986, 94, 219.

13. Rhines, R. N. and Aballe, M.: Met Trans A, 1986, 17A, 2139.

14. Underwood, E. E.: Quantitative Stereology.

15. Chou, J. S.: Trans. AFS, 1991, in press.

16. Meyers, Carolyn W., Presentation at the International Symposium on Fracture, Canadian Institute of Metallurgists, Winnepeg, Canada, August 1987.

ACKNOWLEDGEMENTS

The authors express gratitude to the National Science Foundation (Grant Number RII-8506633) for support of this research, to the George W. Woodruff School of Mechanical Engineering for the use of facilities, and to Mr. Jack Glen, President of Glencast for consultations. Appreciation is also extended to Mr. Rick Brown, School of Materials Engineering for assistance in fracture toughness testing and to Ms. Susanne Keiller for the preparation of this manuscript.

Materials Science Forum Vols. 102 - 104 (1992) pp. 85-98
Copyright Trans Tech Publications, Switzerland

HARDNESS, MELTING REACTIONS AND HEAT TREATMENT OF Al-Si-Cu-Mg ALLOYS REINFORCED WITH ALUMINIUMOXIDE PARTICLES

M.J. Starink, V. Jooris and P. van Mourik

Laboratory of Metallurgy, Delft University of Technology, Rotterdamseweg 137,
NL-2628 AL Delft, The Netherlands

ABSTRACT

Results of a study mainly by Differential Scanning Calorimetry (DSC) on an Al-20at%Si-1.5at%Cu-1at%Mg (ASCM) alloy with 0, 2.5, 5, and 10 vol% aluminiumoxide (Al_2O_3) particles are reported. The alloys were produced by gas atomisation, mixing with particles and subsequent extrusion. DSC scans of the as-extruded alloys showed no precipitation reactions. Between 780 and 800 K the melting of the Q ($Al_5Cu_2Mg_8Si_6$) and the Al_7Cu_2Fe (ASCM contains a small amount of iron) phases was observed. On heating the matrix melts between about 810 K to 840 K, subsequently silicon dissolves in the melt. The microstructure of the AE alloys with Al_2O_3 particles (size about 1 to 6 μm) showed no clustering of Al_2O_3 particles. During DSC after solid quenching the formation of GPB zones was observed. After dissolution of the zones, Q phase and θ phase precipitate. Hardness measurements show, that both GPB zones and Q phase precipitates can improve the room-temperature strength of the ASCM alloy by about 75%, compared to the as-extruded alloy. Heat treatments producing GP or GPB zones (ageing at temperatures below about 440 K) do not contribute to the strength of the alloy at elevated temperature. Q-phase precipitates improve the hardness after an additional overageing treatment at 473 K by about 60% (as compared to the as-extruded alloy).

1.   INTRODUCTION

The wear resistance of aluminum alloys can be improved by the introduction of finely dispersed hard particles to the alloy [1,2]. For example, a fine dispersion of silicon particles can be obtained by rapid solidification of a molten aluminum alloy with a high silicon content [3]. A dispersion of ceramic particles in aluminium alloys can be obtained via various production routes, e.g. by compocasting or by mixing particles with aluminium alloy powders followed by extrusion. Metals reinforced by dispersed (ceramic) particles are generally referred to as Metal Matrix Composites (MMCs). Their increased wear resistance, low thermal expansion and improved high-temperature strength, make MMCs attractive for applications like parts of combustion engines.

In this contribution, results of a study on the heat treatment of Al-20at%Si-1.5at%Cu-1.1at%Mg (ASCM) alloys reinforced with 0, 2.5, 5, and 10 vol% aluminiumoxide (Al_2O_3) particles are presented. These alloys combine the presence of two reinforcing components (silicon and Al_2O_3 particles) with the possibility of age hardening of the Al-rich phase. Hence, an appropriate heat treatment is necessary to optimize the mechanical properties at room and elevated temperatures. The presence of dispersed particles in solid-quenched age-hardenable aluminium alloys generally influences the kinetics and sequences of precipitation processes, as compared to corresponding unreinforced alloys [4]. Appropriate heat-treatment procedures for these alloys are, as yet, unknown. This applies for both the solution annealing treatment (temperature) and the age hardening (temperature and time). Therefore, it is necessary to establish appropriate solid-solution temperatures and age-hardening procedures for MMCs. Hence, melting and precipitation phenomena should be investigated. As melting and precipitation involve large enthalpy changes, differential scanning calorimetry (DSC) was used as the main investigation method. Besides, hardness measurements, optical microscopy and X-ray diffraction were applied as well.

2.   EXPERIMENTAL PROCEDURES

2.1. PRODUCTION ROUTE

The alloys under investigation were made available the Japanese firm Showa Denko K.K. The base alloy was rapidly solidified by gas atomisation, yielding fine powder (sizes range from 1 to 100 μm, with a median size of about 24 μm [5]). The cooling rate during gas atomisation is generally about 10^4 to 10^6 K/s [3]. Subsequently, the powder was mixed with ceramic Al_2O_3 particles, in order to obtain mixtures with 0, 2.5, 5 and 10 volume percent of ceramic particles. Finally, the mixtures were extruded at about 670 K into round bars with a diameter of about 20 mm.

The chemical compositions of the extruded alloys are presented in Table 1. The four alloys are indicated by ASCM0, ASCM2.5, ASCM5 and ASCM10, in which the last number refers to the volume percentage Al_2O_3 particles. The main impurities in the base alloy (as measured by X-ray fluorescence) are: Ni (~0.02 at%), Zn (~0.01 at%), Ti(~0.006 at%) and Cr (~0.005 at%).

Table 1 :   Composition of the alloys and the base alloys.

alloy name	Al_2O_3	Si		Cu		Mg		Fe	
		alloy	base	alloy	base	alloy	base	alloy	base
	vol % *)	wt%	at%	wt%	at%	wt%	at%	wt%	at%
ASCM0	-	20.2	19.9	3.47	1.52	0.96	1.10	0.24	0.12
ASCM2.5	2.5	19.7	20.2	3.44	1.56	0.89	1.05	0.24	0.12
ASCM5	5.2	18.9	20.1	3.32	1.57	0.88	1.09	0.23	0.12
ASCM10	10.4	16.9	19.6	2.97	1.52	0.77	1.03	0.20	0.11

*)   Calculated from measured weight percentages using the densities of the ASCM base alloy, 2.67 g/cm^3 and of α-Al$_2$O$_3$, 3.98 g/cm^3 [6].

2.2. DIFFERENTIAL SCANNING CALORIMETRY

A DuPont 910 differential scanning calorimeter was used. Inside the DSC cell a protective gas atmosphere was maintained by flushing with 99.999% pure argon (flushing started 10 min before the experiment). Both specimen and reference were enclosed in an aluminium pan with an aluminium cover. Heating rates between 0.5 and 40 K/min were used. Per specimen three scans were performed; the second after cooling at about 10 K/min and the third after cooling at 2 K/min. The calibration procedures and baseline corrections were described earlier [7].

2.3. HEAT TREATMENT

Experiments were performed on specimens after various production stages and heat treatments. These are denoted as follows :
- LQ: liquid-quenched, gas-atomised powder (stored at room temperature).
- AE: as-extruded alloy (stored at room temperature).

- SQ: solution-annealed (10 min at 768 K) and quenched (in water at room temperature)
                    AE specimen. The SQ specimens were stored at room temperature for 1 hour after
                    quenching.
- SQ NA(x h):   an SQ specimen aged at room temperature for x hours.
- SQ DSC(y K):   an SQ specimen heated to y K inside the DSC cell at a heating rate 2 K/min, and
                    subsequently cooled down to room temperature at about 10 K/min. This cooling rate
                    is sufficiently fast (as compared to the heating rate) to avoid reactions during
                    cooling.
- SQ AA(z K):   an SQ specimen aged at room temperature for 4 days, and subsequently artificially
                    aged at z K for 24 hours.

To investigate mechanical properties after a long-time application at elevated temperatures, hardness measurements were performed on selected specimens which were aged at 473 K for four days (esentially an additional overageing treatment, see section 3.3). All artificial ageing treatments were performed in an oil bath with temperature control within 1.5 K.

2.4. OPTICAL MICROSCOPY AND HARDNESS MEASUREMENTS

Vickers hardness (Hv) was measured on polished surfaces parallel to the extrusion direction. The apparatus used was a Leitz-Durimet hardness tester with an indentation force of 981 N. The values reported are the average of ten determinations; the given errors equal the standard deviation.

Jena Neophot 30 and Jena Neophot 2 optical microscopes were used to characterize the microstructures and the distribution of Al_2O_3 particles. Polished specimens were etched with a 1:1 mixture of Keller and Wilcox reagent and 5% Nital.

2.5. X-RAY DIFFRACTION

For identification of phases, X-ray diffraction was performed using a Guinier camera. Diffraction patterns of the AE ASCM0 alloy at room temperature and of SQ ASCM0 during heating at a constant rate (0.5 K/min) from 293 to 773 K were recorded.

3.   RESULTS

3.1. GAS-ATOMISED POWDERS

DSC scans of the gas-atomised ASCM powder are presented in Fig. 1. During the first scan the heat flow is exothermic over almost the entire temperature range, except for a small endothermic effect in the temperature range 360 to 420 K. After this endothermic effect a clearly defined exothermic effect is observed, with peak temperature 480 K. Since this effect seems to overlap with an exothermic effect located around 570 K, the end temperature of the effect around 480 K is not clearly identifiable. The heat content of the effect around 480 K, ΔQ_{MP}, is estimated by integration of the heat flow from 420 to 535 K. The value thus found is 607 J/mole ASCM powder. Beyond 590 K the heat flow continues to be exothermic. The heat flow in this temperature range does not show a clearly defined peak. In the second and the third runs no exothermic effects are observed. Instead, both runs show an endothermic effect beyond 550 K.

After completion of the three runs the mass of the powder specimen was higher than at the start of the experiment. The increase amounted to 39±5 µg, which corresponds to 0.17% of the original mass of the specimen.

3.2. AS-EXTRUDED ALLOYS

Isothermal X-ray diffraction (at 298 K) reveals the presence of the following phases in the AE ASCM0 alloy: the Al-rich phase, the Si phase (diamond structure), the θ phase (Al_2Cu), the Q phase ($Al_5Cu_2Mg_8Si_6$) and the Al_7Cu_2Fe phase.

In Figs. 2a to c optical micrographs of AE ASCM, ASCM5 and ASCM10 are presented. The following observations are made:

i.   Although bands along the extrusion direction seem to be enriched with Al_2O_3 particles (black colour), no clustering of Al_2O_3 particles is observed.
ii. The sizes of the irregular shaped Al_2O_3 particles range from about 1 to 6 μm, with an average of about 2 μm.
iii. Elongated Al_2O_3 particles are aligned along the extrusion direction.
iv. The Si-phase particles (dark grey), which are homogeneously dispersed in the Al-rich phase (light grey), have a grain size between about 1 and 10 μm, with an average of about 4 μm.
v.   Also a third type of precipitate (grey) can locally be discerned. Considering the phase identification by X-ray diffraction, these are probably Q or Al_7Cu_2Fe-phase precipitates.

By increasing the etching time to about 1 min, the grain boudaries of the Al-rich phase become visible (pictures not presented here). The grain size of the Al-rich phase is about 5 μm.

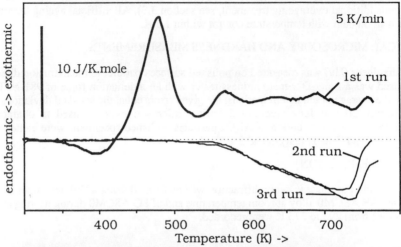

Fig. 1:   DSC scans between 293 and 773 K on gas-atomized ASCM powder.

2a

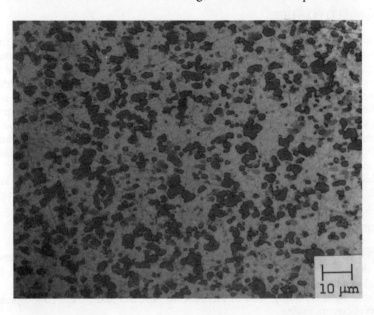

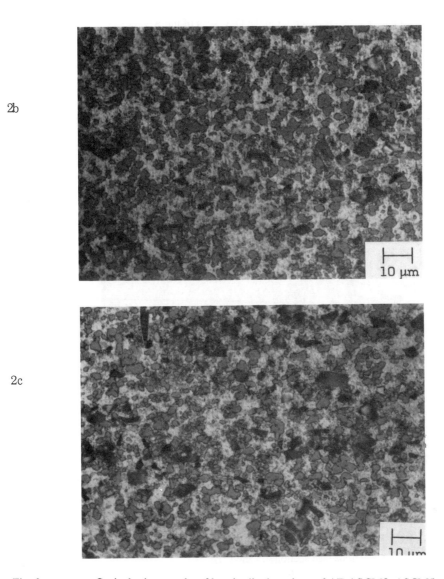

2b

2c

Fig. 2a to c :   Optical micrographs of longitudinal sections of AE ASCM0, ASCM5 and ASCM10.
Extrusion direction is horizontal. Etching time: 10 sec.

In Fig. 3, taken from an AE ASCM0 specimen heated twice to 890 K, severe coarsening of silicon particles and clustering of the Al_2O_3 particles is observed.

DSC experiments were performed on AE ASCM0 and AE ASCM10. These experiments were aimed at determining a suitable temperature for solution annealing before solid quenching. During solution annealing, melting should be avoided and hence the temperature at which melting starts has to be known. Fig. 4 shows the first and the second DSC runs on AE ASCM0. For a temperature range of 750 to 860 K, Fig. 5 shows the second and the third DSC runs on AE ASCM10, which were performed after a first DSC run until 773 K. Note that in Fig. 5 only endothermic effects are observed, and that the effects are much larger than the ones shown in Fig. 4. The effects shown in Fig. 5 are caused by melting reactions, while those in Fig. 4 are caused by dissolution reactions (see section 4.2). The following observations are made:

i.   No substantial exothermic effect is observed in the DSC scan of the AE alloys.
ii. Both the first and the second run of AE ASCM0 (Fig. 4) are dominated by an endothermic effect, which starts at 610 K. This endothermic effect seems to be divided into at least two subeffects: one effect continues upto about 750 K, while the other effect continues beyond 785 K.
iii. The melting of the SQ ASCM10 alloy is subdivided into several seperate stages. Four melting reactions are indicated by A, B, C, and D (see Fig. 5). Effect C is by far the largest. Effect D seems to continue beyond the temperature range studied in this experiment.

Comparison of the DSC runs presented in Figs. 4 and 5 with DSC runs of AE ASCM10 performed at heating rate 20 K/min (not presented here) additionally shows:
iv. The end temperature of the first endothermic subeffect in Fig. 4 is largely uninfluenced by heating rate or Al_2O_3-particle content.
v.   The peak temperatures of the melting reactions are independent of reactions heating rate or Al_2O_3-particle content.

For the temperature of solution annealing before solid quenching, a temperature below the start of melting should be selected. In order to have a safety margin of about 10 K to the start of melting we have chosen 768 K as solution annealing temperature (see section 4.2).

Fig. 3 :   Optical micrograph of AE ASCM10 heated twice to 890K. Etching time 10 sec.

3.3. SOLID-QUENCHED AND HEAT-TREATED ALLOYS

DSC scans of SQ ASCM0 and ASCM10 alloys are presented in Fig. 6. In line with literature data and on the basis of X-ray diffraction, the effects occurring in the DSC of scans these alloys are attributed to the following reactions (see section 4.3 and [7]):
i.   the formation of GPB (Al-Cu-Mg) zones,
ii.   the dissolution of GPB zones,
iii. precipitation of the Q phase ($Al_5Cu_2Mg_8Si_6$),
iv.   precipitation of the θ' and θ phases (both Al_2Cu),
v.   the dissolution of the Q and θ phases.

Only effect i can be considered as a separated effect, the other effects show overlap. The influence of the presence of Al_2O_3 particles on the DSC curves of the SQ alloys is rather small. For a discussion of the influence of Al_2O_3 particles on the precipitation kinetics in SQ ASCM alloys, see Ref. 7.

The room-temperature hardness of SQ ASCM0 was measured after various heat treatments (see section 2.3). The results are gathered in Figs. 7 and 8. In Fig. 7 the room-temperature hardness of SQ DSC specimens are plotted as a function of DSC end temperature, together with the room-temperature hardness of the same specimens after an additional overageing treatment (4 days at 473 K). In Fig. 8, the room-temperature hardness of SQ ASCM0 after various heat treatments is compared with the room-temperature hardness of the same specimens after the additional overageing treatment. The following observations are made.

i.   The maximum hardness, both after initial heat treatment and after the additional overageing treatment, is obtained for the SQ DSC(487 K) specimen. This DSC end temperature corresponds to the end temperature of effect iii observed during the DSC scan.

ii.   Natural ageing of SQ ASCM0 results in a remarkable increase of hardness.

iii. Artificial ageing of SQ ASCM0 for 24 hours at 393 K and at 433 K also yields a remarkable hardness increase.

iv.   The additional overageing treatment considerably decreases the hardness the specimens (with the exception of the SQ DSC(575 K) specimen), but even then, the hardness of the heat treated specimens remains higher than that of the AE ASCM0 specimen.

v.   After the additional overageing treatment the hardness of nearly all solid-quenched and subse-quently aged specimens has dropped to about 120 HV. The only exceptions are SQ DSC(487 K) and SQ DSC (507 K) which retain a hardness of about 140 HV after additional overageing.

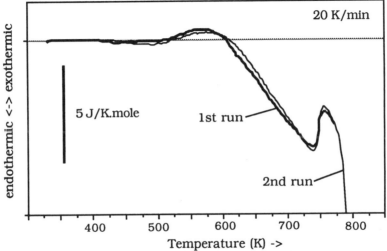

Fig. 4 :   DSC scans of AE ASCM0 in the temperature range below 790 K.

4.   DISCUSSION

4.1. GAS-ATOMISED POWDERS

The liquidus of the Al-Si-Cu-Mg system [8] shows that during slow cooling of the molten Al-20at%Si-1.5at%Cu-1.1at%Mg (the small amount of Fe will be neglected in this part of the discussion) the solidification starts with the following reactions:

liq -> Si   (1)
liq -> Si + Mg$_2$Si   (2)

During the completion of the solidification, the initially formed Mg$_2$Si will disappear again (Mg$_2$Si is not stable below about 790 K [8]), and the Q phase will form. This proceeds via the reaction [8]:

liq + Mg$_2$Si + Si -> Q + Al   (3)

This peritectic reaction occurs at 802 K [8]. Since above this temperature no copper containing phases are formed, it is assumed that above 802 K the solidification process can be described on the basis of the Al-Mg-Si phase diagram. From this diagram [8] follows that during cooling the Al-rich phase will start to form at 828 K with the following reaction:

liq -> Al+Si+Mg$_2$Si   (4)

During gas atomisation, the solidification rate is very fast : about 10^5 K/sec [3]. For the peritectic reaction 3, relatively slow solid state diffusion is necessary. It is thus expected that during rapid solidification the peritectic reaction 3 is suppressed, leading to the presence of metastable Mg_2Si in the rapidly solidified powder, as was observed by Estrada and Duszczyk [3].

Rapid quenching from the melt directly to room temperature, e.g. gas atomisation, can generally result in a supersaturated Al-rich phase. The amount of Si dissolved in the Al-rich phase of rapidly solidified hyper eutectic Al-Si alloys can be as high as 3.3at%, which is much larger than the maximum equilibrium solid solubility [9]. Hence, a high atomic fraction of Si dissolved in the Al-rich phase of the in the gas-atomised ASCM alloys is expected.

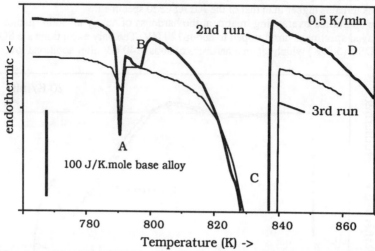

Fig. 5 :   DSC scans of the AE ASCM10 alloy in the temperature range beyond 760 K. The second and the third run are presented; the end temperature of the first run was 773 K. All heat effects are endothermic. Note the large difference in heat flow scale as compared to Figs. 4 and 6.

For hypoeutectic, binary Al-rich alloys, including Al-Cu alloys, it is generally observed that it is not possible to dissolve all the alloying elements in the Al-rich phase by liquid quenching [9,10]. It can thus be expected that in the gas-atomised powder only a limited fraction of the copper is dissolved in the Al-rich phase. This is supported by the presence of θ phase in gas-atomised ASCM powder [3], whereas this phase is absent in SQ ASCM [7].

The above suggests that the Al-rich phase of gas-atomised powder contains a higher fraction of silicon and a lower fraction of copper and magnesium as compared to the Al-rich phase of the SQ ASCM alloys. The increased silicon fraction is expected to result in precipitation of the Si phase during DSC heating. The main exothermic effect in the gas-atomised ASCM powder is observed between 430 and 510 K. This temperature range corresponds closely to the temperature ranges observed for silicon precipitation in hypereutectic rapidly solidified Al-Si alloys [9]. From this it is assumed that this exothermic effect is mainly caused by silicon precipitation. Using the enthalpy of silicon precipitation from the Al-rich phase, ΔH_{Si} (54 kJ/mole Si, see [9]), it is now possible to estimate the atomic silicon fraction in the Al-rich phase after gas atomisation from:

$x_{Si} = \Delta Q_{MP} / (x_\alpha . \Delta H_{Si})$,

in which x_α is the fraction of Al-rich phase (α) in the gas-atomised powder (approximately equal to the fraction of Al atoms in the alloy, 0.78). Using this expression, a value of 0.014 is found for x_{Si}. This is indeed much larger than the maximal equilibrium solid solubility of Si in the Al-rich phase in ASCM alloys which equals about 0.004 [8].

The DSC scan of the gas-atomised ASCM powder shows a small endothermic effect just in advance of the main exothermic effect. Endothermic effects in this temperature range are indicative for

GP-zone dissolution (see section 3.3 and [7]). The GP-zone dissolution effect in the gas-atomised powder is much smaller than the GP-zone dissolution effect in SQ ASCM alloys (in the SQ ASCM alloys GP-zone formation mainly involves the formation of Al-Cu-Mg type GPB zones). This corroborates the above assumption that the atomic fractions of Mg and Cu atoms in the Al-rich phase of the ASCM powders are lower than in the Al-rich phase of the SQ ASCM0 alloy.

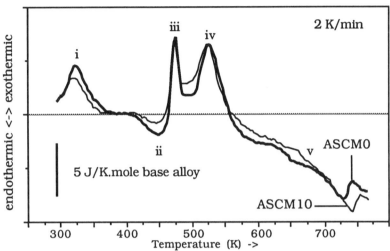

Fig. 6 : DSC scans of SQ ASCM0 and SQ ASCM10.

From about 590 K to the end temperature of the DSC scan, 773 K, the first DSC scan of the gas-atomised powder is dominated by a broad exothermic effect. As the Mg_2Si phase is unstable in the solid ASCM alloys, this phase must eventually transform into Q phase. As this is a thermally activated reaction from a stable to a metastable state, this transformation involves an exothermic reaction with a clearly defined peak temperature. Since no clear peak is observed in the first run on LQ ASCM0 beyond 590 K, the exothermic heat flow beyond 590 K is thought to be largely unaffected by this transformation. The temperature range of the broad exothermic effect in the DSC scan of gas-atomised powder is much broader, and located at much higher temperatures, than the ones usually encountered in the case of precipitation reactions. Indeed, no such effect is observed in DSC scans of SQ ASCM (see section 3.3). The second and the third scans of the gas-atomised powder do not show this exothermic effect. Instead they show only endothermic effects, which, over a large temperature range, resemble the corresponding parts of the first and second run of the AE ASCM0 alloy. These endothermic effects are caused by the dissolution of the Q, Si and θ phases (see section 4.2). Close to the end temperature of the DSC scan, also the second and the third DSC scans deviate from the scans of the AE ASCM0 specimen: beyond 750 K the endothermic heat flow seems to decrease in the powders. These effects, and the broad exothermic effect in the first run, are thought to be caused by oxide formation on the surface of the powders. This will be substantiated below.

During heating of gas-atomised Al-based alloys a series of reactions occur on the surface of the powders. These reactions involve the evolution of gases, and include [11]:

$$Al(OH)_3 \rightarrow AlOOH + H_2O \tag{5}$$
$$2\,AlOOH \rightarrow Al_2O_3 + H_2O \tag{6}$$
$$2/3\,Al + H_2O \rightarrow 1/3\,Al_2O_3 + H_2 \tag{7}$$

The water consumed during reaction 7 can either be provided by reactions 5 and/or 6, or H_2O molecules adsorbed by the oxide layer can be used. If sufficient oxygen is present also the direct oxidation reaction can occur:

$$4/3\,Al + O_2 \rightarrow 2/3\,Al_2O_3 \tag{8}$$

From these four reactions reaction 8 results in the largest enthalpy change (1.68 MJ per Al_2O_3 mole at 700 K, see Appendix). Because of their high specific surface area (0.17 m^2/g), surface reactions can cause a significant heat evolution during DSC experiments on the gas-atomised ASCM powder (see Appendix). Since oxide layers on Al are always in the order of 10 to 100 nm, and since the gas-atomised powder already possesses a 40 nm thick oxide layer [3] it is assumed that the thickness of the oxide layer formed during the DSC run is of the order of 10 nm. The formation of a 10 nm thick oxide layer by reaction 8 produces about 3.1 kJ per mole powder. The formation an identical oxide layer by reaction 7 produces about 1.7 kJ per mole powder. The heats produced by reactions 5 and 6 are about one order of magnitude smaller. The actual heat production during heating of the powder inside the DSC is estimated from the area between the first and the third run in the temperature range where the effect occurs (590 to 773 K). This gives the value 1.6 kJ per mole powder, which agrees fairly well with the above presented estimates for the formation of a 10 nm oxide layer. It is however not possible to distinguish between the two oxide-formation modes (reactions 7 and 8). The mass of the gas-atomised powder increases by 0.17% during the three DSC runs. Since only the direct oxidation (reaction 8) results in a mass increase, this indicates that, at least, part of the oxide formation during the DSC scan is caused by this reaction. Assuming the mass increase to be caused by direct oxidation, a 0.17% mass increase corresponds to a released heat of 1.7 kJ per mole ASCM powder. Thus, the released heat measured can entirely be ascribed to the direct oxidation.

Apparently a significant amount of O_2 is either adsorbed by the oxide layer or trapped between the powder particles forming the DSC specimen. This O_2 is apparently not completely removed by flushing with argon before the start of the DSC experiment.

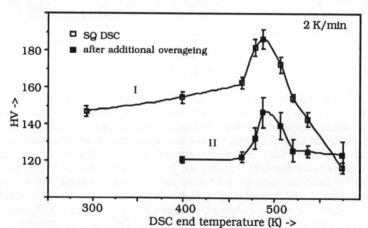

Fig. 7 : Room-temperature hardness of SQ DSC ASCM0 specimens as a function of the DSC end temperature (curve I). Also the hardness of these specimens after an additional overageing treatment (4 days at 473 K) is given as a function of the DSC end temperature (curve II).

4.2   MELTING AND DISSOLUTION REACTIONS IN AE ASCM

The peak temperatures of the exothermic and the endothermic effects appeared to be insensitive to variations of the heating rate. Therefore these endothermic effects are thought to be related to processes that are, to a large extent, equilibrium processes. As Al_2O_3 particles are supposed to be inert, this implies that these effects are insensitive to addition of Al_2O_3 particles, as has been observed.

From phase diagrams of the Al-Cu-Mg-Si system [8,12], it can be deduced that in the ASCM alloy no melting reaction occurs below 780 K. It is well established in literature that endothermic effects occurring between about 350 K and the melting of Al-Cu-Mg, Al-Cu-Si and Al-Cu-Mg-Si type alloys, are caused by the dissolution of alloying elements in the Al-rich phase (see for instance [7]). Accordingly, the endothermic effects in Fig. 4 are ascribed to dissolution of Cu, Mg and Si atoms in the Al-rich phase (since the maximum solid solubility of Fe in the Al phase is lower than 0.01 at% [8], the dissolution of Fe is neglected). From phase diagrams of the Al-Cu-Mg-Si system [8,12], it can be deduced that for the ASCM alloy the dissolution of the θ phase will be completed between 733 and

775 K. This is confirmed by high-temperature X-ray diffraction (see [7]), during which θ-phase lines disappear at 745±10 K. Thus it is concluded that the endothermic effect between 610 and 740 K mainly concerns θ-phase dissolution. As the Si phase and the Q phase are present on the whole temperature range of the high-temperature X-ray diffraction experiment, the remaining part of the dissolution effect is caused by continuing dissolution of the Q and Si phases.

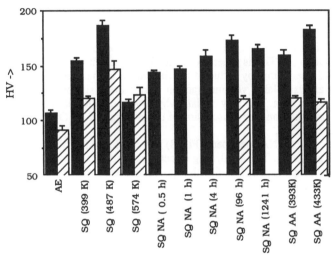

Fig. 8 : Room temperature hardness of various heat treated samples (black bars). For some of the specimens also the hardness after an additional overageing treatment (4 days at 473 K) is given (grey bars).

An AE ASCM10 specimen heated twice to 870 K shows severe clustering of Al_2O_3 particles (see Fig. 3), which indicates that a large part of the alloy (including the Al-rich phase) had been melted. The endothermic effects in Fig. 5 are much larger than those in Fig. 4. Since melting reactions generally involve much larger enthalpy changes as solid state reactions do, and since in Al-rich Al-Cu-Mg-Si alloys the first stage of melting can occur starting from about 780 K [8], the endothermic effects occurring in Fig. 5 are ascribed to melting reactions. From Fig. 5 it is concluded that the melting of the base alloy of ASCM10 is devided into four stages A, B, C and D (the melting point of Al_2O_3 particles, about 2300 K, lies far beyond the DSC temperature range). Considering the solidification reactions denoted by the liquidus surface of the Al-Cu-Mg-Si system (see section 4.1), reaction D is clearly related to the melting of Si phase and Mg_2Si phase. Since effect C represents by far the largest endothermic heat content, this effect must be related to the melting of the (majority) Al-rich phase by the inverse of reaction 4. From the above it is is clear that the remaining reactions (reactions A and B) must be linked to the melting of the Q and Al_7Cu_2Fe phases. From literature [8] it is known that in Al-Cu-Mg-Si alloys various reactions involving the melting of the Q phase can occur in the temperature range 780 to 802 K. It is also known (from [8]) that in Al-Cu-Fe-Mg and in Al-Cu-Fe-Si alloys eutectic melting of the Al_7Cu_2Fe phase occurs at 778 and 793 K, respectively (data on the Al-Cu-Fe-Mg-Si system are unknown until now). So, the known temperature ranges of the Q and Al_7Cu_2Fe-phase melting reactions, while confirming that these two reactions are linked with the observed reactions A and B, give no clear indication which of the two phases melts first. However, in Fig. 5, reaction B is absent during the third run. This is interpreted as follows. During the second run, at the end of which the specimen is almost entirely melted, the melt reacts with the aluminum pan. This decreases the concentrations of Cu, Mg, Fe, and Si atoms in the alloy. From the Al-Cu-Mg-Si phase diagram at 775 K [8] follows that a decrease of the percentage of Mg in these alloys can result in dissolution of the Q phase in the Al-rich phase below 775 K. Then, melting of the Q phase will not occur. So, from the absence of the effect B in the third run of Fig. 5, it is concluded that effect B is most likely caused by the melting of the Q phase. Then, effect A has to be caused by melting of the Al_7Cu_2Fe phase.

4.3. PRECIPITATION IN SQ ASCM

The hardness of an alloy is closely related to its microstructure. From Fig. 6 follows, that the presence of Al_2O_3 particles has a limited influence on the precipitation processes in SQ ASCM (see also [7]). Hence, the optimal heat treatment found for an SQ ASCM0 alloy can be applied for SQ ASCM alloys with Al_2O_3 particles.

For a number of MMCs it has been observed that introduction of ceramic particles alters the kinetics of precipitation [4]. This is generally ascribed to the misfit between Al-rich phase and ceramic particles after solid quenching, by virtue of the difference in thermal expansion [4]. In the Al-rich phases of ASCM alloys, a large volume fraction of misfitting silicon particles (about 22 vol%) is present after solid quenching. This may explain why Al_2O_3 particles, in this case, have little influence on the kinetics of precipitation. For a discussion of the influence of Al_2O_3 particles on the precipitation in ASCM alloys, see Ref. 7.

From hardness of the SQNA specimens (Fig. 8) it is concluded that solid-quenched ASCM0 can be hardened by natural ageing. This hardening effect is caused by the formation of GPB (Al-Cu-Mg) zones [13]. From a comparison of Fig. 6 with Fig. 7, it becomes apparent that the maximum hardness for SQ ASCM0 is reached just after completion of the Q-phase precipitation (see section 3.3). The additional overageing treatment generally lowers the hardness of the SQDSC specimens (see curve I and II in Fig. 7). The initial difference between curves I and II below 473 K can be explained as follows. SQ ASCM0 is kept at room temperature at least one hour before start of the DSC run. During room-temperature storage and during heating upto low DSC end tempertures (lower than 400 K) GPB zones are formed, which contribute to the hardness. During DSC runs on the SQ ASCM alloys the GPB zones start to dissolve from about 400 K, but dissolution is not completed before 473 K (see Fig. 6). The additional overageing treatment is sufficient to allow the dissolution of the GPB zones, leading to a marked drop in hardness. Besides the dissolution of GPB zones, also the annealing out of dislocations (formed during quenching, see [7]) can cause softening of the quenched specimens.

From the foregoing discussion it becomes apparent that two reactions contribute to the hardening of SQDSC ASCM0: i. GPB-zone formation and ii. Q-phase precipitation. Also the hardness changes of the specimens presented in Fig. 8 are interpreted in terms of these two hardening mechanisms.

Natural ageing results in GPB-zone formation. Thus increasing the time of natural ageing results in an increasing hardness (see Fig. 8). The slight decrease in hardness after long-time natural ageing may be related to relaxation of elastic tensions originally present after solid quenching [14]. As Q-phase formation occurs at temperatures around 480 K, the hardness increase observed on artificial ageing at 393 and 433 K is probably also due to GPB-zone formation. In the SQAA(433 K) specimen also θ' and Al-Cu type GP-zones might contribute to the hardness [13,15]. Since both GP zones and θ' are unstable, this explains why after additional overageing the hardness of the SQ AA specimens is reduced to about the same value as is observed for the SQNA specimen and the SQDSC specimens with DSC end temperatures lower than 473 K after the additional overageing treatment.

5.   CONCLUSIONS

- The Q and Al_7Cu_2Fe phases, which are not (completely) dissolvable in the Al-rich phase, start melting from 780 K. Hence, the optimal solution treatment temperature is just below 780 K. The Al-rich phase of the ASCM alloys melts between 810 and 840 K.
- The hardness of the extruded ASCM alloy can be increased by about 75% by heat treatments which produce either GPB zones or Q-phase precipitates (ageing temperatures 433 to 500 K).
- For high-temperature applications, only ageing treatments at temperatures around 480 K, giving rise to the formation of the Q phase, are usefull. These ageing treatments can increase the room-temperature hardness of the ASCM0 alloy by about 60% (as compared to the as extruded alloy).
- The presence of Al_2O_3 particles in the ASCM alloy has little influence on the precipitation after solid quenching.

ACKNOWLEDGEMENTS

The authors are indebted to Mr. J. F. van Lent for performing the Guinier camera experiment and to Dr. J. Duszczyk for providing the alloys. Stimulating discussions with Professor B. M. Korevaar, Dr. J. Duszczyk and Dr. W. H. Kool are gratefully acknowledged.

The financial support of the Foundation for Fundamental Research of Matter and for Technological Sciences (FOM/STW) is gratefully acknowledged. Also the financial support from the European Community program ERASMUS for one of the authors (V. J.) is gratefully acknowledged.

APPENDIX

The exothermic heat effect produced by the growth of aluminium oxide layers of average thickness, d, on powders, is given by:

$$\Delta Q_{ox} = A_S \cdot d \cdot \Delta H \cdot \rho_{ox} \cdot \frac{M_{ASCM}}{M_{Al_2O_3}}$$

in which A_S is the specific surface area of the powders, ΔH is the change in enthalpy of the reacting substances due to the oxide forming reaction, per mole oxide, ρ_{ox} is the average density of the oxide layer, and M_{ASCM} and $M_{Al_2O_3}$ are the atomic masses of the ASCM powder (27.7 g/mole) and Al_2O_3 (102 g/mole). The specific surface area of the gas-atomised ASCM powders equals 0.17 m^2/g [3]. For the enthalpy change of the oxide forming reactions, ΔH, the value of ΔH at 700 K is used (the small variations of ΔH with temperature, about 1% per 100K at 700K [16], are neglected). Then ΔH is given by 0.95 and 1.68 kJ per mole Al_2O_3 for reactions 7 and 8, respectively [16]. The ΔH of reaction 6 is about one order of magnitude smaller [11]. The density of fully dense α-Al_2O_3 (corrundum) equals 3.98 g/cm^3. This value will be used as an estimate for ρ_{ox}.

REFERENCES

1   D. Bialo, J. Duszczyk, A.W.J. de Gee, G.J.J. van Heijningen and B.M. Korevaar, Wear, vol. 141, 1991, pp. 291-309.
2   D. Bialo, T.L.J. de Haan and J. Duszczyk, Internal Technological Report, Oct. 1989, Delft University of Technology.
3   J.L. Estrada and J. Duszczyk, J. of Mater. Sci., vol. 25, 1990, pp. 886-904.
4   J. M. Papazian, Metall. Trans. A, vol. 19A, 1988, pp. 2945-2953.
5   J.H. ter Haar and J. Duszczyk, Mater. Sci. and Eng., vol. A135, 1991, pp. 65-72.
6   D.S.Phillips, T.E. Mitchell and A.H. Heuer, Phil. Mag. A, vol.42, 1980, pp. 417-432.
7   M.J. Starink, V. Jooris and P. van Mourik, Proc. Conf. on Mater. Sci.: MMCs-Processing, Microstructure and Properties, Roskilde, Denmark, 2-6 september 1991, submitted for publication.
8   L.F. Mondolfo: Aluminium Alloys, Butterworth & Co Ltd, London, 1976.
9   M. van Rooijen and E.J. Mittemeijer, Metall. Trans. A, vol. 20, 1989, pp. 1207-1214.
10   Q. Li, E. Johnson, A. Johansen and L. Sarholt-Kristensen, Proc. Int. Conf. on Advanced Al and Mg Alloys, Amsterdam, The Netherlands, 20-22 june 1990, T. Khan and G. Effenberg, eds., pp. 349-356.
11   J.L. Estrada, J. Duszczyk and B.M. Korevaar, J. of Mater. Sci., vol. 26, 1991, pp. 1631-1634.
12   H.J. Axon, J. of the Institute of Metals, vol. 83, 1955, pp. 490-492.
13   W. Bonfield and P.K. Datta, J. of Mater. Sci. Lett., vol. 11, 1977, pp. 1050-1052.
14   J.G.M. van Berkum, R. Delhez, Th. H. de Keijser, E.J. Mittemeijer and P. van Mourik, submitted to Sripta Metall.
15   A.K. Gupta, M.C. Chaturvedi and A.K. Jena, Mater. Sci. and Techn., vol. 5, 1989, pp. 52-55.
16   I. Bahrin: Thermochemical Data of Pure Substances, VCH Verlagsgesellschaft, Weinheim, BRD, 1989.

Materials Science Forum Vols. 102 - 104 (1992) pp. 99-108
Copyright Trans Tech Publications, Switzerland

ADVANCES IN THE HEAT TREATMENT OF 7000 TYPE ALUMINIUM ALLOYS BY RETROGRESSION AND REAGING

B. Cina (a) and F. Zeidess (b)

(a) Metallurgy Group, Tashan Engineering Centre, Israel Aircraft Industries Ltd., Lod, Israel
(b) Formerly of I.A.I. Ltd., presently with the School of Applied Science and Technology,
The Hebrew University, Jerusalem

ABSTRACT

The retrogression and reaging (RR) heat treatment process can be applied to prevent or ameliorate not only the susceptibility to stress corrosion cracking but also that to exfoliation corrosion. The parameters of time and temperature to obviate these two susceptibilities are not necessarily identical. Retrogression can be carried out as effectively by athermal as by isothermal or as by combined athermal and isothermal treatment. The fact that careful allowance has to be made for the rate of heating to the isothermal temperature of retrogression may explain why some of the mathematical formulae offered in the literature to allow for all the above mentioned parameters have not been found to be universally applicable.

RR will be shown to be applicable to a broad range of 7000 type aluminium alloys.

INTRODUCTION

Since the publication [1, 2] of the non-conventional process of heat treatment developed specifically for aluminium alloys of the 7000 type, and designated RR, retrogression and reaging, many workers have confirmed its claim significantly to improve the resistance of these alloys to stress corrosion while maintaining levels of mechanical properties akin to those of the T6 temper, [3-12] i.e. the ability to combine the best of the properties of the T73 and T6 tempers. The original work, while demonstrating the principles of the process, was performed on relatively thin sections. Subsequent work by one of the authors [13] and others [4, 6, 7, 9, 10] showed that sections of considerable thickness could also be successfully treated by variations of the parameters of time and temperature in the stage of retrogression.

It should be recalled that initially the work was based on the hypothesis that the stage of retrogression had two effects, one desirable and one unavoidable. The desirable effect was to dissipate the clusters of dislocations around undissolved particles of intermetallics at the grain boundaries and believed responsible for the sensitivity to stress corrosion of material such as the 7075 aluminium alloy in the T6 temper [14]. It was believed that this effect could be achieved by heating the material to a temperature higher then that at which it had been artificially aged. This, however, now introduces the unavoidable effect, that of retrogression of the precipitation. The immediate effect of heating a structure containing a precipitate to a temperature higher that that at which the precipitate was formed is to bring about some measure of dissolution of the precipitate prior to its reprecipitation on a smaller number of nuclei. Clearly then there would be an upper limit to the retrogression temperature to be employed since the intention was both to avoid any overaging and to restore, eventually, the microstructure of the T6 condition. Recent work has shown that RR does in fact dissipate dislocations at grain boundaries and largely restores the original T6 microstructure [15] in accordance with the original hypothesis.

Consideration ot the matter suggest d that while it might be theoretically simpler to talk of specific isothermal temperatures of retrogression, the latter phenomenon and that of diffusion of dislocations would occur to varying degree also as a function of the rate of heating above the initial T6 temperature of artificial ageing. This could possibly have significant practical importance both from the point of view of the means of heating employed and for the retrogression of thick sections. Accordingly it was decided to investigate this aspect of the subject per se and in conjunction with predominantly isothermal retrogression, the work to be carried out on sections of material considerably thicker than those employed in the original work.

EXPERIMENTAL WORK

The work was carried out on three commercially available alloys of the 7000 series as follows:-

(1)   Plate material, $1^3/_4$" (45 mm) thick, of the 7075 aluminium alloy in the T651 temper, purchased as per U.S. Federal Specification QQ-A-250/12.

(2)   Plate material 5" (127 mm) thick of the 7050 aluminium alloy in a non-standard T651 temper. The plate was obtained by special order from the manufacturer (the Aluminium Company of America, Pittsburgh, U.S.A.) after solution treatment and cold stretching for stress relief. It was subsequently artificially aged at Israel Aircraft Industries Ltd. for 24h/120°C. Its chemical composition conformed with the requirements of U.S. Aerospace Material Specification 4050.

(3)   Extrusions of T shape, approximately 2.3 and 4.8 mm thick, made from a 7278 aluminium alloy and supplied in the T651 temper. Its nominal chemical composition was 1.9%Cu, 2.9%Mg, 7.0%Zn, 0.21%Cr, 0.03%Ti and 0.05%V.

The 7075 aluminium alloy

For plate material of the 7075 aluminium alloy, testing was carried out for susceptibility to both stress corrosion and exfoliation corrosion. For the former all testing was carried out on test specimens in the short transverse (ST) direction of the plate, this being the direction of maximum sensitivity in the T6 or T651 tempers. Specimens for stress corrosion testing were of tensile type of dimensions as per Fig. 1 and were loaded in tension employing a Unisteel stress corrosion testing machine manufactured by Distington Engineering Co. Ltd. of Workington, Cumberland, U.K.

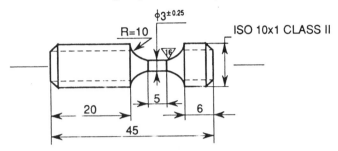

Fig. 1: Specimen for stress corrosion testing.

The specimen was loaded through a double lever system to a stress corresponding to 75% of its actual yield stress as measured in the ST direction. Testing was carried out in a boiling solution of 6% NaCl in water until rupture occurred, or for at least 96 hours without rupture which is the time specimens should endure without rupture for the 7075 aluminium alloy in the T73 temper [16]. To avoid possible effects of galvanic corrosion all mechanical attachments to the specimen likewise subject to the corrosive environment were manufactured from the same 7075 aluminium alloy as that being tested.

For susceptibility to exfoliation corrosion, the test procedure followed was the EXCO test as per the U.S. Standard A.S.T.M. G34-79. For plate material test specimens were taken from locations at a tenth, quarter and half thickness of the plate, hence forward $^T/_{10}$, $^T/_4$ and $^1/_2$ respectively.

The severity of the exfoliation corrosion was assessed visually aided where necessary by a metallographic section taken perpendicular to the exposed $^T/_X$ surface where X = 2, 4 or 10.

Static mechanical properties were measured in the ST direction.

Heat treatment for retrogression was carried out either in a molten salt bath containing 50% $NaNO_3$ and 50% $NaNO_2$ or in a fluidised bed furnace employing fine particles of Al_2O_3 as the medium for heat transfer and a stream of compressed air to provide the fluidising action.

Since one of the intentions of the work was to determine the effect of the heating rate to the selected maximum temperature of retrogression, the experiments were designed to obtain different heating rates by varying both the dimensions of the block to be heated and the temperature of the furnace. In the latter case blocks were sometimes withdrawn from the furnace on reaching a preselected temperature which was less than that of the furnace.

The temperature of each block was determined by means of a thermocouple inserted into a small hole which bottomed at the centre of the block. The space round the thermocouple was sealed with a flexible thermal insulating compound.

After every experiment with retrogression, the block so treated was held for seven days at room temperature and then artificially aged for 24 or 48 hours at 120°C, as indicated.

The 7050 aluminium alloy

Since it was desired to simulate the heating of large sections of material without the necessity of actually using such large sections, the concept of equivalent thickness was employed. The latter is defined as twice the ratio of the volume to the sum of the areas of the lateral surfaces of a block of material [3]. Thus, for example, a block of dimensions 30x30x45 mm has an equivalent thickness of 11.3 mm.

Block specimens were prepared of 3, 15, and 125 mm thick in the S.T. direction of the plate and of 125 and 130 mm in dimension in the T and L directions respectively. The 15 and 125 mm thick specimens had equivalent thicknesses of 12.1 and 42.2 mm respectively. All the retrogression heat treatments were carried out in a molten salt bath. All mechanical testing was carried out in the original L direction of the plate as per A.S.T.M. Specification E-8. Testing for resistance to exfoliation corrosion was carried out by the EXCO test as previously and at $^T/_2$ and $^T/_{10}$ locations.

The 7278 aluminium alloy

These extrusions were retrogression heat treated in a molten salt bath in lengths of 30 cm and were subsequently tested for resistance to exfoliation corrosion by the EXCO test, for hardness by Rockwell B testing and for electrical conductivity, %I.A.C.S., using standard commercial equipment.

RESULTS

Retrogression by athermal heating only

Specimens of dimensions 30x30x45 mm were taken from the 7075-T651 aluminium alloy plate of 45 mm thickness such that the 45 mm dimension of each specimen represented the full thickness and therefore ST direction of the plate. From the point of view of rate of heating, these dimensions are equivalent to an infinite plate of 11.3 mm thickness. [3]. These specimens were heated in a fluidised bed furnace as described previously. The specimens were withdrawn from the furnace on reaching the required temperature, aged at room temperature and then for 48h at 120°C as described previously and tested for mechanical properties and resistance to accelerated stress corrosion, all testing being in the S.T. direction. The conditions of the athermal heating and the several results obtained are given in Table 1. From this the following conclusions can be reached.

(1)   Allowing for the direction of testing, all four athermal retrogression trials retained the static mechanical properties required of the original T651 condition by U.S. Federal Specification QQ/A/250-12.

(2)   All four trials showed an improvement in resistance to stress corrosion of at least two orders of magnitude.

(3)   The above improvement in resistance to stress corrosion was accompanied, in the case of trials nos. 1 and 2, by no improvement in the resistance to exfoliation corrosion and, in the case of trials nos. 3 and 4, by complete immunity to exfoliation corrosion.

Retrogression by solely athermal and by combined athermal and isothermal treatment.

The specimens were of dimensions as in the previous section and were heated in a fluidised bed or in a molten sath bath. Ageing and testing were likewise as previously. The results obtained for the two types of heating medium are detailed in Tables 2 and 3 respectively.

Table 1: Properties obtained after athermal retrogression
in a fluidised bed furnace and reaging

Trial No.	Temperature, °C		Time of heating from 120°C to withdrawal, secs.	Time to failure in accelerated S.C. test, h.	Susceptibility rating in EXCO test	Mechanical Properties, MPa [1, 2]	
	Furnace	Specimen on withdrawal				Y.S.(0.2%)	U.T.S.
1.	210	200	185	79.9, 66	ED	483	538
2.	270	220	90	135UB, 138UB[3] 147UB	ED	486	529
3.	270	240	133	96UB, 96UB, 163UB	N	488	521
4.	270	260	190	58, 74, 115UB	N	447	527
Minimum requirements for the original T651 condition in the longitudinal direction as per U.S. Fed. Spec. QQ-A-250/12.				0.5[4]	ED[4]	462	525

Notes: (1) As measured in the S.T. direction.
      (2) Had these properties been measured in the longitudinal
           direction they would have been about 15-20 MPa higher.
      (3) UB = unbroken.
      (4) The result as obtained in this work. There is no requirement.

Table 2: Properties obtained after athermal and athermal plus
isothermal retrogression in a fludised bed furnace and reaging.

Trial No.	Temperature of furnace, °C	Time of heating from 120°C to temperature of furnace, and holding, mins. secs.	R(T) [3]	Time to failure in accelerated S.C. test, h.	Susceptibility rating in EXCO test	Mech. Prope. MPa. [1, 2]	
						Y.S.(0.2%)	U.T.S
16	200	9'00" (A) [4]	0.95	90.1, 123.4, 166UB [5]	ED	471	537
18	200	13'00" (A+I) [4]	1.18	77, 157, 133UB	ED	485	530
33	200	19'00" (A+I)	2.38	138UB, 138.2UB	N	473	518
34	200	29'00" (A+I)	4.37	116UB, 116UB	N	458	500
40	210	4'30" (A)	0.77	21, 113.3UB	ED	468	508
41	210	8'10" (A+I)	1.99	75, 110.1	N	461	508
42	210	12'14" (A+I)	2.33	91.4UB, 91.4UB	N	451	500
13	220	4'38" (A)	1.58	143UB, 163UB, 163UB	N	453	516
14	220	6'05" (A+I)	2.52	159UB, 159UB, 232UB	N	446	506
15	220	8'23" (A+I)	3.93	-- -- --	N	450	496
Minimum requirements for the original T651 condition in the longitudinal direction as per U.S. Fed. Spec. QQ-A-250/12.				0.5 [6]	ED [6]	462	525

Notes: (1) Mechanical properties were measured in the S.T. direction.
      (2) Had these properties been measured in the longitudinal direction of the plate,
           they would have been about 15 to 20 MPa higher.
      (3) Value of R(T) calculated as per U.S. Patent No. 4,189,334.
      (4) A = Athermal, I = Isothermal.
      (5) UB = Unbroken.
      (6) As measured in this work. These is no minimum requirement.

From Table 2 it will be seen that allowing for the difference in direction of testing between retrogressed specimens and that for which there are minimum requirements of strength, almost all the former conform with these requirements. With regard to resistance to stress corrosion, athermal retrogression alone gave very significant improvement consistently better than two orders of magnitude in life for furnace temperatures of 200° and 220°C and a mixed one to two orders of magnitude improvement for 210°C, all as compared with the life for the T651 condition. This slight inconsistency for 210°C was not investigated and is possibly due to some abnormal experimental factor. The improvement in life for athermal retrogression to 200°C is akin to that observed for the same temperature for slightly different athermal conditions of heating as reported in Table 1. The stress corrosion results for combined athermal and isothermal retrogression are all at least two orders of magnitude better than that for the original T651 condition, many specimens in fact remaining unbroken after 96 hours, or more, of testing. This would represent a resistance to stress corrosion as good as that of material in the T73 temper.

The results of testing for resistance to exfoliation corrosion were most interesting. For furnace temperatures of 200° and 210°C, athermal retrogression alone gave no improvement in resistance to exfoliation corrosion, whereas for 220°C the athermally retrogressed specimen was now immune. Even combined athermal and isothermal retrogression at 200°C, as in trial no. 18, failed to improve the resistance to exfoliation corrosion, all this despite the significant improvement in resistance to stress corrosion displayed by specimens solely athermally retrogressed to 200°C or given combined conditions of retrogression as in trial no. 18. For longer periods of isothermal retrogression at 200°C, and for all the conditions of combined athermal and isothermal retrogression at 210°C, all specimens displayed immunity to exfoliation corrosion. For a furnace temperature of 220°C, athermal retrogression alone, or combined athermal and isothermal retrogression gave immunity to exfoliation corrosion. The values of R(T) in Table 2 will be discussed later.

The results for employing a molten salt bath as the heating medium for retrogression are given in Table 3.

First by comparing the rate of heating to 200°C in Table 3 with that in Table 2, it will be seen that heating in a molten salt bath is more rapid than that in a fluidised bed furnace. Again, allowing for the difference in direction of testing for mechanical properties, virtually all the trials resulted in retention of the minimum requirements for the original T651 condition.

The resistance to stress corrosion was particularly good, with all the specimens remaining unbroken after 96 hours or more of testing, with one exception. The latter, trial no. 46 at 200°C, nevertheless showed an improvement of more than two orders of magnitude.

With regard to resistance to exfoliation corrosion, simple athermal retogression had no beneficial effect as for specimens athermally retrogressed in a fluidised bed furnace and again, as for the latter specimens, despite the very significant improvement in resistance to stress corrosion. Combined athermal and isothermal retrogression gave a progressive improvement in resistance to exfoliation corrosion until immunity was reached. Specimens retrogressed in a fluidised bed furnace had shown a much more abrupt improvement.

Table 3: Properties obtained after athermal and athermal plus
isothermal retrogression in a molten salt bath and reaging.

Trial No.	Temperature of furnace, °C	Time of heating from 120°C to temperature of furnace and holding, mins. secs.	R(T)[3]	Time to failure in accelerated S.C. test, h.	Susceptibility rating in EXCO test	Mechanical properties MPa [1, 2]	
						Y.S.(0.2%)	U.T.S
51	190	9'30" (A)[4]	·	96.2UB[5], 96.3UB	ED	462	516
52	190	19'15" (A+I)[4]	1.08	137UB, 137UB	EC	461	510
53	190	29'30" (A+I)	2.16	96UB, 96UB	EA	454	504
54	190	39'20" (A+I)	3.23	--------	N	-	-
47	200	2'21" (Almost A)	0.23	137.5UB, 137.5UB	ED	464	506
43	200	5'17" (A+I)	0.83	--------	ED	-	-
44	200	7'46" (A+I)	1.33	114UB, 115UB	EC	466	504
45	200	10'34" (A+I)	1.83	--------	N	-	-
46	200	13'18" (A+I)	2.42	58.9, 148.9UB	N	445	501
Minimum requirements for the original T651 condition in the longitudinal direction as per U.S. Fed. Spec. QQ-A-250/12.			-	0.5 [6]	ED[6]	462	525

Notes: (1) - (6) See footnotes to Table 2.

Table 4: Results of RR treatment on a 7050 aluminium alloy

Thickness of specimen, mm	Conditions of isothermal retrogression		Hardness R_B	Electrical conductivity, % IACS	Susceptibility rating in EXCO test	Mechanical Properties, MPa	
	Temp., °C	Time, mins.				Y.S.(0.2%)	U.T.S.
3	180	90	89	39.5	P	495	533
15	180	90	86	40	P	491	528
42[1]	180	60	88	38.5	P	511	540
Properties of original T651 temper.			90	35	B/C	529	555
Minimum requirements for T7651 temper, 37.5- 75 mm thick, per U.S. Aerospace Material Specification 4201A[2]			84[3]	37 min.	B, max.	455	525
Minimum requirements for T7451 temper per U.S. Aerospace Material Specification 4050D			82[3]	38 min.	B, max.	440 (up to 50 mm) 420 (125 mm)	510 (up to 50 mm) 490 (125 mm)

Notes: (1) Equivalent thickness.
(2) There is no A.M. Specification for plate 5" thick. Had there been, the mechanical properties would have been somewhat lower than those quoted.
(3) Value as per AMS 2658 for material <3" thick.

The values of R(T) in Table 3 will be discussed later.

THE VERSATILITY OF THE RR PROCESS

It has been shown in previous work [2, 6, 9, 10] and in the present work that the RR process can be applied to sheet and thin plate made from the 7075 aluminium alloy. It has also been successfully applied to plate of 75 mm thickness [13, 17]. The 7475 aluminium alloy has likewise been shown to respond to RR treatment, [3, 7, 8, 11, 17] as has the 7049 aluminium alloy [3].

Table 4 shows some results obtained for a 7050 aluminium alloy for specimens of 3 and 15 mm actual thickness and 42 mm equivalent thickness taken from a plate 5" (125 mm) thick as described previously. It will be seen that for a retrogression temperature of 180°C, specimens of all three thicknesses in the range 3-42 mm became immune to exfoliation corrosion. From this and from their significantly increased value of electrical conductivity it would be expected that their resistance to stress corrosion would be correspondingly improved. It is to be noted that the mechanical properties, especially the yield stress, of all the retrogressed specimens very closely approach those of the original T651 condition for which there are no requirements this being a non-standard condition. These mechanical properties were significantly higher than the minimum requirements for the T7451 and T7651 tempers. Thus one of the potentially strongest 7000 type alloys can be seen to benefit from the RR heat treatment, even after allowing for the differences in thickness between the specimens examined and those covered by the Aerospace Material Specifications.

RR has also been applied to extrusions made from a 7278 aluminium alloy in the T651 temper of thickness 2.3-4.8 mm. The material had an initial hardness of 93RB, electrical conductivity of 30% I.A.C.S. and rating B/C in susceptibility to exfoliation corrosion. Retrogression for 35-45 mins. at 190°C for an R(T) value of 4 or higher, followed by ageing for 48 h at 120°C resulted in a hardness of 94 RB, i.e. no reduction compared to that of the T651 temper, an electrical conductivity of 37% I.A.C.S. and a rating of N/P in susceptibility to exfoliation corrosion. The susceptibility to stress corrosion should likewise have been very low but was not investigated.

DISCUSSION

The present work shows that the retrogression obtained under certain conditions of athermal heating can be as effective as retrogression carried out by isothermal heating at least from the point of view of improving the resistance to stress corrosion. Furthermore very significant improvement in resistance to stress corrosion is not necessarily accompanied by a corresponding improvement in resistance to exfoliation corrosion, whether the retrogression is carried out by athermal or combined athermal and isothermal retrogression. For an improvement to be obtained in resistance both to stress corrosion and to exfoliation corrosion, specimens had to be retrogressed athermally by heating to a higher final temperature or by combining athermal with isothermal retrogression. The explanation for these results is not immediately obvious. Whereas the susceptibility to stress corrosion in the 7075 aluminium alloy in the T6 temper has been attributed to the presence of dislocations adjacent to grain boundaries [14] and it has recently been shown [15] that the improvement in resistance to stress corrosion obtained by RR treatment is due to the dissipation of those dislocations, there is no generally accepted explanation for the susceptibility to exfoliation corrosion. The latter has been considered as a form of stress corrosion, the stress being residual or produced from the pressure due to the larger volume of corrosion product [18, 19]. The relative density of dislocations in specimens retrogressed solely athermally and by combined athermal and isothermal retrogression may well be profitably examined in an attempt to find an explanation as above.

Attempts have been made mathematically to assess the favourable parameters of time and temperature in retrogression [3, 4].

Thus the optimum conditions of retrogression of 7000 type aluminium alloys for specimens up to 15 mm thick are claimed by Dubost and Bouvaist [3] to be given by the formula $R(T) = \frac{10^{16}}{1.5} \int_0^T e^{\frac{-13,400}{\theta(t)}} dt$, where T is the total retrogression time in seconds after the specimen has reached 190°C, $\theta(t)$ is the temperature in °K above 463°K (190°C) and t is the time in seconds. For material initially in the T651 temper, the value of the function R(T) should be within the range 0.5 to 1.25 and preferably in the range 0.5 to 1.0 [3]. The values of R(T) were calculated for the several trials of retrogression carried out in a fluidised bed furnace as detailed in Table 2 and are in face given in that Table. It will be seen that for combined athermal and isothermal retrogression at 200° and 210°C, when the value of R(T) is less than 1.5, there is the expected improvement in resistance to stress corrosion but there is no improvement whatsoever in the resistance to exfoliation corrosion. For retrogression temperatures of 200°, 210° and 220°C, only when the value of R(T) is greater than 1.25 are improvements observed in the resistance both to stress and exfoliation corrosion. In fact, in contrast to the narrower range of the value of R(T), 0.5 to 1.0, preferred in the French work [3], the present work shows that progressively better results are obtained as the value of R(T) increases beyond 1.

Similar results were obtained for retrogression trials carried out in a molten salt bath as detailed in Table 3. Values of R(T) less than 1.25 for retrogression temperatures of 190° and 200°C failed to give simultaneous improvement in resistance to stress and exfoliation corrosion.

The inconsistency of the present results with those of Dubost and Bouvaist [3] is emphasised when it is noted that these workers give no results of t sting for resistance to stress corrosion but only to exfoliation corrosion. The actual thickness of the sheets they tested was 2, 5 and 8 mm only. Certainly sheet material of 2 and 5 mm thick would have had only very slight, if any, susceptibility to stress corrosion especially in the L or T directions, the ones convenient for testing. Sheet of 8 mm thick may have had some susceptibility. In the present work, although some of the specimens had an equivalent thickness of 11.3 mm, they derived from plate of 45 mm thickness which, in the T651 temper, had very considerable susceptibility to both stress and exfoliation corrosion.

The French work [3] was carried out on the 7475 and 7049 aluminium alloys and the present work with which comparison is made was carried out on the 7075 aluminium alloy. This may be a further cause for the difference in results.

CONCLUSIONS

1. Solely athermal heating for retrogression can be as effective as isothermal heating in improving the resistance to stress corrosion.

2. For certain conditions of retrogression, improvement in the resistance to stress corrosion is not necessarily accompanied by improvement in the resistance to exfoliation corrosion.

3. Conditions of retrogression can be found to improve the resistance to both the above modes of corrosion.

4. Mathematical formulae offered in the literature for determining optimum parameters for retrogression have been found to be inadequate for the present materials and experimental conditions employed.

5. The RR temper has been shown to be applicable to a wide range of 7000 type aluminium alloys including the strongest.

ACKNOWLEDGMENTS

The authors wish to thank Mr. I. Eldror and Dr. F. Pozarnik for their assistance with the experimental work.

REFERENCES

1. Cina, B.: U.S. Patent No. 3,856,584, Dec. 24, 1974.
2. Cina, B. and Ranish, B.: Aluminum Industrial Products, American Society for Metals, Pittsburgh, PA., U.S.A., 1974.
3. Dubost, B. and Bouvaist, J.: U.S. Patent No. 4,189,334, Feb. 19, 1980.
4. Pechiney Aluminium: U.S. Patent No. 4,200,476, April 29, 1980.
5. Kaneko, R.S.: Metal Progress, 1980, 118, 41.
6. Wallace, W., Beddoes, J.C. and de Malherbe, M.C.: Can. Aeronaut. Space J., 1981, 27 (3), 222.
7. Islam, M.U. and Wallace, W.: Metals Technology, 1983, 10, 386.
8. Islam, M.U. and Wallace, W.: Metals Technology, 1984, 11, 320.
9. Tankins, E.S., Agarwala, V.S., Neu, C.E. and Bethke, J.J.: Report No. NADC-84096-60, Dept. of the Navy, Washington, DC, U.S.A., 1984.
10. Thompson, J.J., Tankins, E.S. and Agarwala, V.S.: Corrosion, Paper No. 204, Houston, TEXAS, U.S.A., March 17-21, 1986.
11. Ohnishi, T. and Shiota, H.: J. Jpn. Inst. Light Metals, 1986, 36, 10, 647.
12. H. Nam, H.H., Lee, J.Y. and Park, J.K.: J. Korean Inst. Met., 1988, 26(2), 134.
13. Cina, B.: 2nd Israel - Norwegian Technical and Scientific Symposium Electrochemistry - Corrosion, Trondheim, NORWAY, June, 12-15, 1978.
14. Jacobs, A.J.: Trans. Amer. Soc. Met., 1965, 58, 579.

15. Talianker, M. and Cina, B.: Met. Trans. A., 1989, 20A, 2087.
16. Sprowls, D.O.: Private communication.
17. Cina B.: to be published.
18. Liddiard, E.A.G., Whittaker, J.A. and Farmery, H.K.: J. Inst. Metals, 1960-61, 89, 337.
19. Shreir, L.L.: Corrosion, Vol. 1, 1963, p. 412, George Newnes Ltd., London, U.K.

Materials Science Forum Vols. 102 - 104 (1992) pp. 109-114
Copyright Trans Tech Publications, Switzerland

EFFECT OF HIGH TEMPERATURE TREATMENT ON THE STRUCTURE AND MECHANICAL PROPERTIES OF ALUMINIUM FOUNDRY ALLOYS

V.S. Zolotorevsky, N.A. Belov, A.A. Axenov and Das Goutam

Department of Physical Metallurgy of Non-Ferrous Metals and Alloys

ABSTRACT

The microstructural changes in aluminium foundry alloys of the Al-Si-Cu-Mg, Al-Mg, Al-Mg-Zn and Al-Zn-Mg-Cu systems have been investigated. The heat treatment carried out at a temperatures above the non-equillibrium solidus. It was found that particles of Fe and Si containing phases (Si, Mg_2Si, $Cu_2Mg_8Si_6Al_5$, $LiSiAl$, $CaSi_2Al_2$, $(Na,Al)Si_2$, Fe_2SiAl_8, $FeNiAl_9$), with complex morphology, have a tendency to spheroidization during heat treatment. These changes have positive influence on mechanical properties, especially ductility (δ) and fracture toughness (K_{1c}). The values of δ and K_{1c} can be improved on 20 to 50%. It was shown that tendency to spheroidization may be intensified by the increasing fineness of microstructures which depend on cooling rate during crystallization and alloy composition.

I. INTRODUCTION: To a considerable extent the properties of cast Al-alloys are determined by the morphologies of crystallized surplus (secondary) phases, which may be in large amount in the cast structure. These phases may be formed by the impurities (Fe and Si), as well as by alloying elements and it depends on the composition of the alloys and its applications. As a rule, the presence of needles and branched type particles of these phases in the structure results in decreasing the mechanical properties, particularly plasticity and fructure toughness. On the contrary, the presence of compact, spherical crystallized secondary particles in the structure, even in large quantity, can be granted for high level of mechanical properties.

The data obtained from literature, including that of the author's work, points the possiblity to control the morphology of crystallized secondary phases with the help of supplementary elements, modifying elements or special technological methods[1,2,3]. If, by the above method the change of morphology in the cast alloy is not enough, then there is a possibility to control it by heat treatment. It is well known, that dendritic Si-crystals in Al-Si eutectic are spheroidised after heating above 500°C, in order to increase the plasticity and fracture toughness [4].

The present work was dedicated to investigate the possible application of heat treatment for changing the non-favourable morphologies of different secondary phases to improve the mechanical properties of Al-casting alloys.

Our earlier investigations and data from literature showed that the influence of heat treatment in standard regimes before quenching on alloy system of Al-Mg,Al-Mg-Zn,Al-Zn-Mg-Cu is not sufficient for changing the morphologies [3,4,5].Probably,this is connected with a slow diffusion rate of elements (Fe,Si,Ni) , which formed non-soluble phases during homogenization.So,the investigation have been carried out at a temperature near the equllibrium solidus temperature.It is well known,that the various cast alloys of Al-system(Al-Mg,Al-Zn-Mg, Al-Si-Cu-Mg,) contained different phases(Si,Mg_2Si,$Cu_2Mg_8Si_6Al_5$, LiSiAl, $CaSi_2Al_2$,(NaAl)Si_2, Fe_2SiAl_8, $FeNiAl_9$) with various concentration of Fe and Si[1,2].

II. EXPERIMENTAL: The structure and the mechanical properties of the alloys have been studied after heat treatment at different temperatures . The Structure of the alloys have been investigated with the help of Optical and Scanning Electron Microscope. Phase composition was determined by Electron Probe Microscope (JSM-35CF model). The hardness(HB), ultimate tensile strength (UTS), yield strength (YS) and percentage elongation (δ) have been measured and the fructure toughness (K_{1c}) with the help of the specimens containing notch with pre-cracks. Three points bending method has been been used to measure the fracture toughness.

III.RESULTS AND DISCUSSION: The harmfull Mg_2Si phase appeares in Al-Mg-Zn and Al-Zn Mg -Cu alloys very frequently, due to the impurities of Si. It follows from the phase diagram analysis of Al-Mg -Si , the equllibrium solidus temperature is high (more than 530 C) over a wide range of concentration (Mg up to 8%,Si - 0.2 to 1.5%).It is confirmed by experimental data for alloy systems of Al-Mg, Al-Mg-Si and Al-Cu-Si-Mg (Table.1).

TABLE.I.

Composition of Alloys	Equilibrium Solidus Temp. in ° C.	Annealing Temp. in ° C
Al-6Mg-0.5Cu-0.8Zn-1Fe-1Si	560	540
Al-12Si-0.6Mg-0.2Cu -0.1Fe	555	540
Al-8Si-3Cu-0.3Mg -0.3Fe-0.3Ni	530	515

The structural investigation showed that with increasing time and temperature of annealing,the skeleton type phase of Mg_2Si gets fragmented and becomes globular (Fig.1). It was observed that,if the annealing temperature is above 520°C and the time of exposure is more than 5 hours then strong

spheroidizing process takes place.It has been shown in Fig.1b
for alloy system Al-8%Mg-1%Fe-0.8%Cu-1%Si. Fig.1a shows ,
the branch of Mg_2Si precipitation in cast structure.But
after 3 hours heat treatment at 520°C fragmentation takes
place .

On increasing the time of annealing at 520°C up to 10 hrs.
the Mg Si particles start to coagulate(Fig.1c).Such effect has
positive influence on plasticity indices.Similar effect
described above, has been observed in alloy systems AL-Mg-Zn
and Al-Zn-Mg-Cu. However,in high strength casting alloys of
these systems,the presence of Si impurities resulted is a
decrease in strength,which is connected with the decreasing of
Mg concentration in solid solution,because of the appearence of
Mg Si phases during crystalisation.

After eliminating these effect and also the harmfull
influence of Fe impurities,the authors have devoloped new serise
of Al-casting alloys which contains Fe and Si up to 0.5% each as
impurities.Ca,Li,Na have been used to eliminate these harmfull
effect.

For example, one of the element from Ca,Li,Na can be added
with the alloy Al-6%Zn-1.5%Mg-1%Cu-0.5%Si-0.5%Fe-0.2%Na. These
elements help to restore the deficit Mg by forming $CaSi_2Al_2$,LiSiAl
or $(NaAL)Si_2$. In this case when new phases appeares as needle
shaped, a high temperature heat treatment gives an ability to
eliminate these harmfull effects (Fig.2a,2b,2c,2d).

In general, the refined structures have more influence on
fragmentation and spheriodisation processes.With fine structure
these processes occur quickly.In case of dispersed structure, it
is confirmed that Fe,Ni containing phases are spheroidised
quickly (Fig.3a,3b).

Most of the interest have been concentrated on alpha (Fe_2SiAl_8)
phase, which may connect Fe without supplementary addition of
modifying elements.At a certain concentration, the cast structure
can be obtained with eutectic colonies of Al+alpha (Fe_2SiAl_8)
where alpha have very refined structure(Fig. 4a).

In the process of heating before quenching for Silumins at
temperature(510-540 ° C) , a large number of globular particles
are formed instead of initial skeleton type alpha
phases(Fig.4b).

With the help of these effects a new group of cast
Al-alloys, with high mechanical and technological properties
better than secondary standard silumins have been devoloped.

REFERENCES

1.L.F.Mondolfo:Structure and Properties of Al-alloys.,
M.Metallurgy
2.V.S.Zolotorevsky,A.A.Axenov and N.A.Belov:Proceedings of the
International Workshop held in Balatonfured,Hungary,May 1989.
3.N.A.Belov,Y.B.Evseev and V.S.Zolotorevsky: Izv. V.U.Z. Tsvetn.
Metall.,No.5,1985,pp-71.
4.V.S.Zolotorevsky,N.A.Belov and T.A.Kurdumova:Izv. V.U.Z. Tsvetn.
Metall.,No.1,1989,pp-78.
5.Metals Science of Al and its Alloys , Handbook, Ed.
J.N.Fridljander, M.Metallurgy,1983.

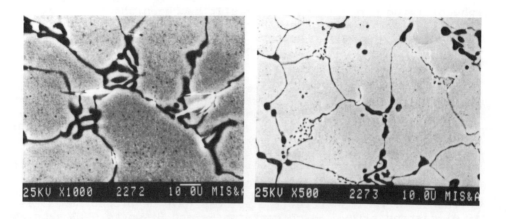

a.   b.

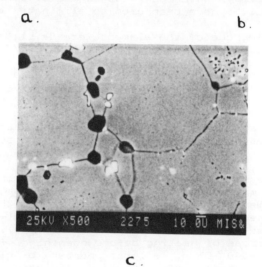

c.

Fig.1. Microstructure of Al-8%Mg-0.8%Cu-1%Fe-1%Si Alloy

    a. After casting.
    b. After annealing at 520°C for 5 hours.
    c. After annealing at 520°C for 10 hours.

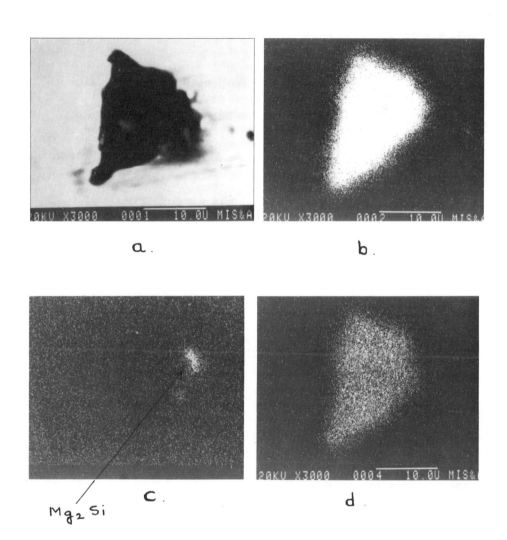

Fig.II. Mg$_2$Si and (NaAl)Si$_2$ phases in Al-1,5%Mg
-1%Cu-6%Zn-0.5%Fe-0.5%Si-0.2%Na alloys.

a. by secondry electrons
b. by K$_\alpha$Si
c. by K$_\alpha$Mg
d. by K$_\alpha$Na.

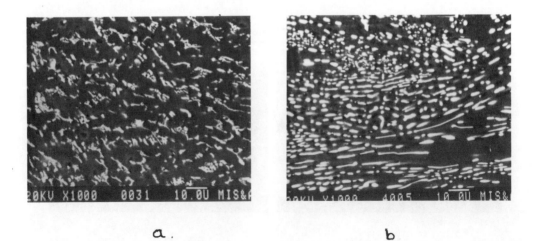

a.   b

Fig.III. Microstructure of alloy Al-6%Zn-
        1.6%Mg-1%Cu-1.3%Fe-1.3%Ni

        a. cast structure and
        b. after annealing at 520 C for 10 hours.

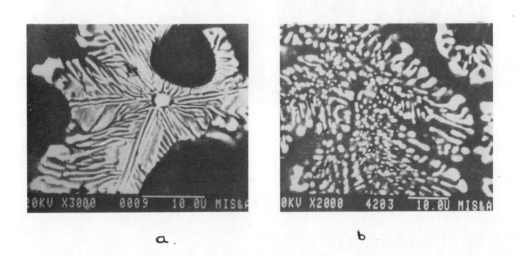

a.   b

Fig.IV. Microstructure of alloy Al-2.5%Fe-2.5%Si

        a. cast structure
        b. after annealing at 520°C for 10 hours.

Materials Science Forum Vols. 102 - 104 (1992) pp. 115-124
Copyright Trans Tech Publications, Switzerland

HEAT TREATMENT AND PHASE INTER-RELATIONSHIPS OF THE SPRAY CAST Cu-15 wt%Ni-8 wt%Sn ALLOY

E.W.J. van Hunnik, J. Colijn and J.A.F.M. Schade van Westrum

Shell Research Arnhem

ABSTRACT

The aging behaviour of the spray cast Cu-15 wt%Ni-8 wt%Sn spinodally hardenable alloy has been investigated. The TTT-diagram noses of the discontinuous and continuous gamma precipitation, and the nose of the spinodally decomposed phase combined with metastable pre-precipitation have been determined. From a theoretical point of view, it is explained why the continuous gamma phase most likely grows in needle form in a <111> direction along the matrix <110> direction. This means that, before continuous gamma precipitation, DO_{22} pre-precipitation is more likely to occur than Ll_2 pre-precipitation.

INTRODUCTION

Presently CuBe alloys are commonly used for highly demanding applications in electronic equipment. Although these alloys suffer from environmental and health related problems, no alternative alloy is widely accepted thus far. Billiton Research Arnhem is, therefore developing a useful alternative: the spinodally hardenable Cu-15 wt%Ni-8 wt%Sn alloy. The Cu15Ni8Sn alloy is one of several alloys that exhibit properties attractive to the designers of electrical springs and high performance electrical connectors. The advantageous mechanical properties of the Cu15Ni8Sn alloy stem mainly from the ability of the supersaturated alpha phase to undergo spinodal decomposition strengthening followed by precipitation at intermediate temperatures i.e. 350-475 °C. Rapid solidification of the alloy by means of the Osprey spray deposition technique, eliminates the incidence of macrosegregation of tin (and nickel), a major problem with conventionally produced copper-nickel-tin alloys, and yields a homogeneous product with a fine scale microstructure.

The CuNiSn alloy system

The CuNiSn alloy system attracted a great deal of interest when, mainly through the work of the Plewes group [1] at Bell laboratories, in the seventies, it was discovered that between 5 and 25 percent nickel, hardenable alloys of commercial interest could be produced. These alloys appeared to have good mechanical properties for various electromechanical applications, like contact springs and connectors. The good mechanical properties are due to spinodal decomposition of the supersaturated alpha phase. Since presently, spinodal decomposition is a well known phenomenon [2], it will not be further discussed here. A significant additional advantage of spinodal decomposition in this system is the extremely small volume change that occurs upon decomposition in comparison with normal precipitation. Therefore, the shape stability upon aging is much better for spinodally hardenable than for conventionally precipitation hardenable alloys.

Although the CuNiSn alloys were shown to be very promising, they could not be commercialised effectively in the past. The main reason is the large Sn segregation during solidification. Figure 1 shows the quasi binary phase diagram (Cu + 15% Ni - Sn). From this diagram it is clear that, due to large solidification range at 8% Sn, segregation is inevitable with conventional casting processes. This Sn macrosegregation leads to inferior mechanical properties and poor workability. Because solution annealing of these segregated structures takes an unrealistic amount of time (resulting in unacceptable grain growth), these CuNiSn alloys never became successful.

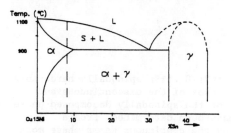

Figure 1. Quasi binary phase diagram of (Cu + 15% Ni - Sn).

New metallurgical routes to Cu15Ni8Sn

Today there are two relatively new processing routes to produce CuNiSn alloys successfully without severe segregation problems. First the powder metallurgical route which produces CuNiSn out of the elemental powders, and secondly the spray deposition route. Since 1986 Billiton has been developing a Cu15Ni8Sn alloy by means of the second technology, known as the Osprey technology, therefore, only this route will be discussed. The spray deposition process itself has been treated at great length in literature [3] and therefore only a short description of the process will be given here.

The spray deposition process is one of the routes for the production of rapidly solidified metal alloys. In this process molten metal is sprayed on a substrate by means of a gas flow. The gas flow supercools the metal droplets. When the supercooled droplets strike the substrate they solidify very rapidly. This way, rapidly solidified slabs can be produced. Figure 2a shows the microstructure of one of these slabs. In this microstructure the alpha grains containing relatively small lamellar discontinuous gamma precipitates are clearly visible. The gamma phase precipitates because, after a rapid solidification the slabs cool relatively slowly, allowing the gamma phase to precipitate. However, this precipitation was successfully suppressed by extra

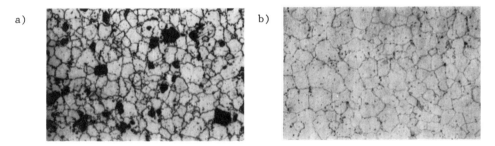

Figure 2. Microstructure of spray cast Cu15Ni8Sn without (a) and with (b)
         extra cooling (average grain size 20 μm)

cooling of the slabs. The result of this extra cooling on the microstructure is
shown in figure 2b. An additional advantage of this production method is the
relatively small grain size achieved after spraying (approximately 20 microns).

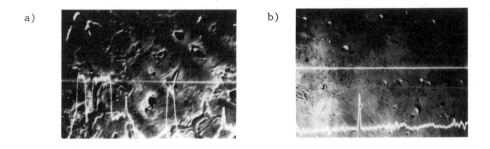

Figure 3. EPMA Sn line scan on (a) conventionally cast and (b) spray cast
         Cu15Ni8Sn

The success of this route is confirmed by electron probe micro analysis (EPMA).
Figure 3a clearly illustrates that the tin is homogeneously distributed in the
supersaturated alpha matrix, which is in contrast to the large tin segregation
observed in the conventionally cast alloy (figure 3b). Obviously we have found
a good production route for the Cu15Ni8Sn alloy, which can be further processed
after spray deposition without macrosegregation problems.

Phase inter-relationships and heat treatment

At present it is not quite clear which metastable phases form in the Cu15Ni8Sn
alloy upon heat treating. However it is known that below approximately 500-
550 °C spinodal decomposition [4] of the supersaturated matrix takes place.
Upon further heating at least one metastable phase (the DO_{22}-phase or the
$L1_2$-phase) is formed [5,6,7] in the modulated structure. Finally the
equilibrium gamma phase (DO_3 structure) precipitates discontinuously. Both the
equilibrium phase and the metastable phases have the same formula: $(Cu_x Ni_{1-x})_3$-
Sn. Above 550 °C no metastable phases have been found; the discontinuous gamma
phase precipitates directly from the supersaturated matrix. Above approximately

700 °C this phase precipitates continuously but no research has been done on possible pre-precipitation phases in this temperature range.

It is impossible to consistently produce a good alloy without understanding the phase inter-relationships and heat treatment performance of the alloy. Therefore, an extensive research programme has been initiated to examine the behaviour of this alloy during heat treatment. The objectives of this study are to determine the discontinuous and continuous gamma nose, the combined spinodal/metastable precipitation nose in the TTT-diagram, and the aging treatment to obtain maximum hardness. Presently the first part of this programme has been finished and the results of this part will be discussed.

EXPERIMENTAL

A Cu15Ni8Sn alloy (0.1 wt% V addition) was spray cast and cold rolled, with intermediate annealing, to a thickness of 0.4 mm. Subsequently all specimens have undergone a heat treatment consisting of the following steps:

- a solution heat treatment (825 °C, 15 min)
- subsequent water quenching
- an aging treatment at 350-775 °C for 5-120 min
- subsequent water quenching.

The first two steps are designed to create a supersaturated undeformed alpha matrix. This matrix structure is the starting structure for the aging heat treatment. A light microscopy study combined with micro Vickers hardness measurements has been carried out to investigate the influence of time and temperature on the behaviour of this alloy.

The continuous and discontinuous gamma phase could be studied directly by means of light microscopy, and indirectly by hardness measurements. The aging due to spinodal decomposition and DO_{22} pre-precipitation was followed indirectly by hardness measurements.

RESULTS AND DISCUSSION

Continuous gamma precipitation

From the microstructures (figure 4) it becomes clear that continuous gamma precipitation mainly occurs above approximately 650 °C. Below this temperature discontinuous gamma precipitation can occur because the nucleation can only start at high energy sites like grain boundaries. Above this 'boundary' temperature nucleation inside the grains is also possible. This homogeneous nucleation is believed to take place only at higher temperatures because the precipitate/matrix interface energy plays a minor role at these temperatures. The hardness measurements (figure 5) in the continuous gamma temperature range indicate that the TTT-diagram nose lies at about 700 °C and that this phase gives a relatively small hardness increase. Therefore, this heat treatment on its own is not commercially attractive. However, a continuous gamma precipitation heat treatment followed by a spinodal decomposition heat treatment (as discussed in [8]) might give good mechanical properties.

Analysis of a lot of structures with continuous gamma precipitates, showed that this phase precipitates in certain discrete directions. Therefore a TEM study

has been planned to determine the habit planes and growth directions of the
continuous gamma phase.

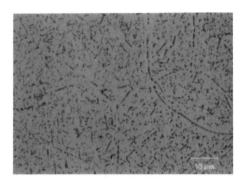

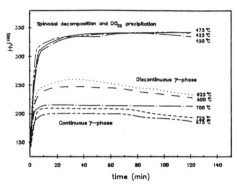

Figure 4. Continuous gamma precipitates   Figure 5. Hardness measurements of
          in a sample aged at 700 °C   non-deformed Cu15Ni8Sn at
          for 30 min.   different aging temperatures

Discontinuous gamma precipitation

The microstructures (figure 6) combined with the hardness measurements indicate
that the nose of the discontinuous gamma precipitation lies at 625 °C. Not only
the nucleation of this gamma phase differs from the previously described
homogeneous gamma phase, the orientation is also different. In the
discontinuous phase no favoured orientation has been discovered.

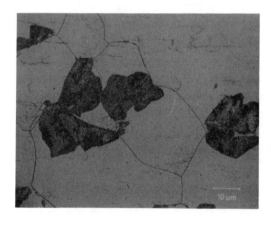

Figure 6. Microstructure of sample
          aged at 550 °C for 30 min
          showing discontinuous gamma
          precipitates at grain
          boundaries.

Spinodal decomposition and metastable precipitation

The experimental techniques used cannot distinguish the difference between the
spinodal decomposition and the metastable precipitates (DO_{22} and/or Ll_2).
Therefore, the metastable precipitation will in most figures be indicated as
DO_{22} precipitation since this is the most often found precipitate in
literature. The maximum hardness found is approximately 350 HV after 60-100 min
heat treating at 425-475 °C. Heating times longer than 120 min have not been
tested yet so it is possible that slightly better results can be obtained after

longer times at 425-475 °C. However, this is not expected for the higher
temperatures since discontinuous gamma precipitation is visible at 475 °C for
annealing times above 40 minutes. The negative influence of the discontinuous
gamma phase on the hardness is clearly seen in the 600 °C-curve.

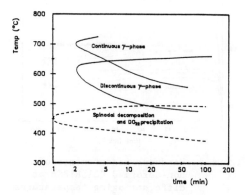   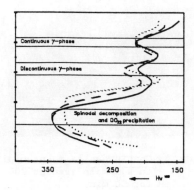

Figure 7. Estimated TTT-diagram (5%   Figure 8. Hardness distribution after
transformation lines) of   10 (   ), 60 (- -) and 120
Cul5Ni8Sn.   (—) min aging at different
temperatures.

All the hardness and optical microscopy results discussed above are summarised
in the TTT-diagram shown in figure 7. In figure 8 the hardness distribution
after 10, 60 and 120 min aging at different temperatures is given. When curves
are fitted through the measured points, the same TTT-like noses are found.

Phase inter-relationships

Besides the matrix phase there are several other phases, mentioned in the
literature, in the Cul5Ni8Sn alloy; the spinodal decomposed phase, the
$L1_2$-phase and/or the DO_{22}-phase and finally the discontinuous and continuous
gamma phase. There is still much confusion about the metastable
pre-precipitation phases $L1_2$ and DO_{22}. In this study no differentiation can be
made between these two phases and the spinodal phase, however an attempt will
be made to discuss the various inter-relationships on the basis of their
lattice parameters and crystal structure.

First consider the crystallographic relationship between the continuous gamma
phase and the matrix. From the microstructure of this gamma phase it is clear
that this phase has a preferred orientation in the matrix. Studying the gamma
(DO_3) crystal structure, reveals that the DO_3 fcc unit cell actually consists
of eight bcc unit cells (figure 9), half of them having a Sn atom in the
central position, the other half a Cu or Ni atom. Looking at one of these eight
cells the crystallographic inter-relationship (figure 10) between this cell and
the matrix is simple. The lattice parameter of the matrix is 0.362 nm (average
literature value) and that of the gamma phase is 0.592 nm (average literature
value). A relatively simple calculation shows that the length of the edge
diagonal of the matrix is almost exactly equal to the length of the body
·diagonal of the 'bcc gamma cell' (= one eighth of the unit cell). Assuming that

the gamma phase is bcc, then the so-called Kurdjumov-Sachs relationship is found:

$$(1\bar{1}0)_{bcc} \;//\; (111)_{fcc}, \quad [1\bar{1}1]_{bcc} \;//\; [0\bar{1}1]_{fcc}$$

Although the misfit in the <110> lattice direction is very small (< 1%), the misfit in the other directions is relatively large (up to 18%). This indicates that the precipitation shape will preferably be needle-like growing in the <111> gamma direction in order to minimize the misfit strain. This theoretical prediction agrees well with the microstructures seen.

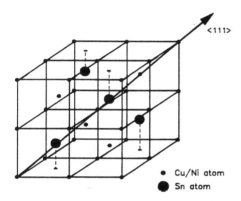

Figure 9. Crystal structure of the DO_3 phase.

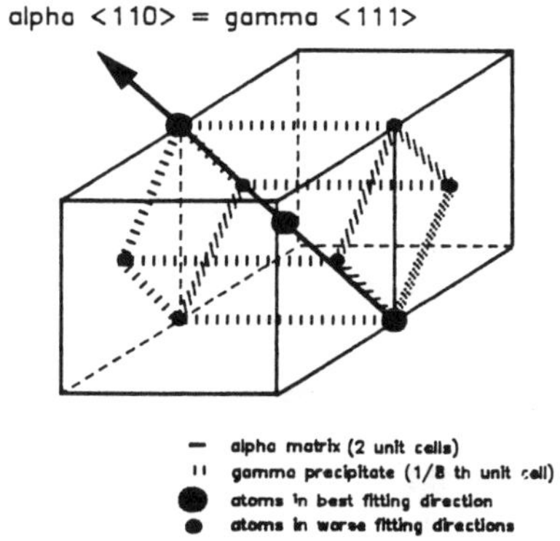

Figure 10. Crystallographic relationship between the alpha matrix and the gamma phase.

Assuming the above prediction is correct, then further conclusions can be drawn about the precipitation sequence of the continuous gamma phase. In literature two possible pre-precipitation phases are mentioned, the DO_{22} and the $L1_2$-phase, before discontinuous gamma precipitation. Therefore, only these two phases will be discussed as possible pre-precipitation phases before continuous gamma precipitation. First, looking at the atomic arrangement of the continuous

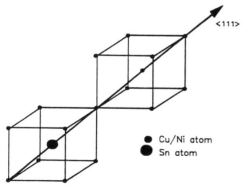

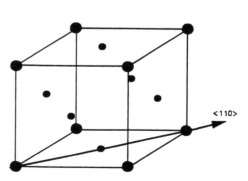

Figure 11. Atomic arrangement in the <111> direction of the gamma phase (2/8 unit cell).

Figure 12. Crystal structure of the $L1_2$ phase showing the atomic arrangement in <110>

gamma phase in the <111> direction (figure 11), three Cu/Ni atoms are followed by one Sn atom etc. Since this <111> direction should fit on the <110> direction of a possible pre-precipitation phase (the DO_{22} or Ll_2 phase), this metastable phase should have the same atomic arrangement in this direction. The fcc Ll_2 phase (figure 12) has the same atomic arrangement in all three <110> directions; one Cu/Ni atom followed by one Sn atom etc. Since this atomic arrangement is essentially different from the arrangement of the <111> gamma direction, it is unlikely that if pre-precipitation of a metastable phase occurs, the Ll_2 phase forms.

The other metastable phase is the tetragonal DO_{22} phase (figure 13), which has the same atomic arrangement as the Ll_2 phase in the [110] and [110] direction. Therefore gamma precipitation in this direction is also very unlikely. Besides it is known that the DO_{22} phase grows into needle like particles with the c axis aligned along the long axis for minimum strain energy [9]. Therefore, it is very unlikely that the needle-like gamma phase nucleates and grows along the [110] or [1$\bar{1}$0] direction of the DO_{22} phase, since this would mean that the new gamma needle would grow perpendicular to the old DO_{22}-needle which is, in principle, possible but not very likely since this would require very long diffusion distances.

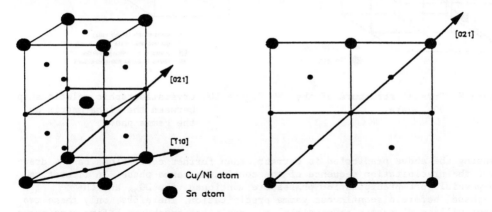

Figure 13. Crystal structure of the DO_{22} phase.   Figure 14. (100) or (010) plane of the DO_{22} phase.

Fortunately there are four other <110>-like directions (figure 14) in the DO_{22} phase which have exactly the same atomic arrangement as the gamma phase in the <111> direction. These four directions are the [021], [02$\bar{1}$], [201] and [20$\bar{1}$] directions of the tetragonal lattice. Therefore, if pre-precipitation of a metastable phase occurs before continuous gamma precipitation, it is more likely that the DO_{22} phase (with the gamma phase growing in the above mentioned directions) forms than the Ll_2 phase.

CONCLUSIONS

- Maximum hardness is found after 60-100 min aging, between 425 and 475 °C

- Discontinuous gamma precipitation is found between 450 and 650 °C with the peak of the nose in the TTT-diagram at 625 °C.

- Continuous gamma precipitation is mainly found above 650 °C with the TTT-diagram nose at 700 °C.

- The growth direction of the continuous gamma phase is the <111> direction. This direction is along the <110> matrix direction.

- If pre-precipitation occurs before continuous gamma precipitation, the most likely precipitate is the DO_{22} phase.

ACKNOWLEDGEMENTS

The authors wish to thank C.J.R. Groenenberg for metallographic preparation and examination of the samples and M.A. Ouwerkerk for his contribution to the experimental work.

REFERENCES

1) Schwartz, L.H., Mahajan, S., Plewes, J.T.: Acta Met., 1974, 22, 601

2) Cahn, J.W.: Acta Met., 1963, 11, 1275

3) Leatham, A.G., Brooks, R.G., Coombs, J.S., Ogilvy, A.J.W.: Proceedings of the International Conference on Spray Forming, 1990, 4.1

4) Plewes, J.T.: Met. Trans., 1975, 6A, 537

5) Baburaj, E.G., Kulkarni, U.D., Menon, E.S.K., Krishnan, R.: J. Appl. Cryst., 1979, 12, 201

6) Ray, R.K., Chandra Narayanan, S., Devraj, S.: Scripta Met., 1982, 14, 1181

7) Miki, M., Ogino, Y.: Trans. Japan Inst. Metals, 1984, 25(9), 593

8) Louzon, T.J.: Trans. ASME, 1982, 104, 234

9) Spooner, S., LeFevre, B.G.: Met. Trans., 1980, 11A, 1085

Materials Science Forum Vols. 102 - 104 (1992) pp. 125-148

INFLUENCE OF HEAT TREATMENT ON THE MICROSTRUCTURAL AND CHEMICAL EVOLUTION OF $Au_{80}Sn_{20}$ ALLOYS FOR WELDING APPLICATIONS

G. Moulin and F. Charpentier

Laboratoire de Métallurgie Structurale, Bât. 413, URA 1107, Université Paris XI, F-91405 Orsay, France

ABSTRACT

Pure $Au_{80}Sn_{20}$ eutectic alloys are used for brazing applications in microelectronic field. No contamination and good mechanical properties are particularly required for to elaborate specific pieces of AuSn material for further braze realization. During annealing in different conditions (temperature, time, oxygen partial pressure) structure evolution occurs from close lamellae of ξ and δ phases to equiaxe structure, usually composed of different gold enriched phases together with δ phase (at least at high temperature of 260°C). Accurate analyses showed that diffusion affected zones are dependant with the growth process of SnO and SnO_2 oxide scales on the surface and with the inner oxidation of bulk phases with high tin content. In low oxygen partial pressure $(pO_2 < 10^{-22} atm)$ inner oxidation can take place especially on δ phase into the strip of $Au_{80}Sn_{20}$. For suffisent oxygen partial pressure, a thicker and more protective stannous and stannic oxide scale is developped, which keeps down further inner oxidation and helps the dissolution of tin enriched phases such as δ phase on account also of an important outward flow of element. However, after a long annealing without any further oxidation process, in agreement with the higher stability of δ with respect to ξ, the two phases are again observed, no more as close lamellae, rather as equiaxe structure.

I) INTRODUCTION

Braze moulds in alloys with low melting point, high purity, good adherence are required for braze realization in microelectronic field (1). Both the purity and the element content are needed in order to obtain a melting point as low as possible and to avoid contamination during the further welding process. But such an eutectic composition usually induces alloy's brittleness (2) so much the more inconvenient than previous moulded pieces must be elaborated before brazing use in some parts of electronic circuit.

The present study deals with AuSn system at the eutectic composition , ie $Au_{80}Sn_{20}$. Usually literature data refer to AuSn systems with rather Sn as solvant and gold as solute atoms. Also elaboration almost corresponds to thin fims obtained from vapor deposition of Sn onto Au (and reverse process) (3). For massive alloys, many solidification ways are used for instance such as directional solidifications carried out in a Bridgman type furnace (4) or such as splat cooling of liquid alloys on substrates (5). Besides accurate composition in alloy's phases is uneasy to reach on account of very fast element diffusion into the AuSn alloys (6) even at room temperature for tin segregation (7). Also Sn and Au diffusivities are very dependant with the tin content or the crystallography of phases (ex c or a axis of the hexagonal close packed structure) (8), (9). The problem is complex because the phase equilibrium diagram is almost under discussion, for example concerning ξ' phase at high temperature or the formation ofmetastable amorphous phase.

Then tin oxidation is a very favourable thermodynamic reaction as reported elsewhere (10).

The knowledge of structure evolution mechanisms, especially under the influence of environment, is then point out. For these specific $Au_{80}Sn_{20}$ eutectic alloy, the morphology shape of close lamellae of δ and ζ phases as well as the very narrow temperature gap between each phase domain need for very accurate examination of the strip samples.

The aim of the study is connected with the structure evolution in various oxygen partial pressures for different annealing times in the two distinct phase domains according to Massalski's work (11) (i e below or above 190 °C; figure 1).

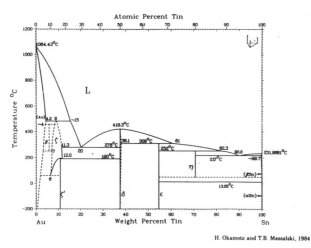

H. Okamoto and T.B. Massalski, 1984

Figure 1 : equilibrium phase diagram of $Au_{80}Sn_{20}$

II) MATERIAL AND EXPERIMENTAL PROCEDURE

Two distinct pure $Au_{80}Sn_{20}$ materials are elaborated:
_ as strip, after fast solidification + hot rolling at 200 °C (thickness = 0.05 mm)
_ as wire, after extrusion, then hot rolling (diameter = 1 mm)

Heat treatment:
Sample annealing for 4 hours or 1 month takes place in a sealed quartz capsule in various oxygen partial pressures, i e $PO_2 = 10^{-5}$ atm (argon) or 10^{-22} atm. Two temperatures below eutectic temperature are explored : 175 °C and 260 °C.
For further structural examination, on account of their small thickness, samples are previously wrapped in Wood alloy. They are mechanically polished up to alumina powder and diamond paste (3 µm)

Experimental procedure:
Main technics are used : optical and electron microscopy, electron microprobe analyses with wave lenght or energy dispersive analyses (ZAF correction), X ray diffraction analyses (Cu anticathode; $\lambda = 1.542$ A), X ray photoelectron spectroscopy (XPS) with argon sputtering (P $_{Ar+}$ = 2 x 10^{-5} Torr, V = 3eV, i = 3 mA; i e a sputtering rate of 3 A per minute); Vickers microhardness (load = 15 g).

III) RESULTS.

III . 1) AS RECEIVED Au$_{80}$ Sn$_{20}$ ALLOY.

III . 1.1) MICROSTRUCTURAL STUDY.

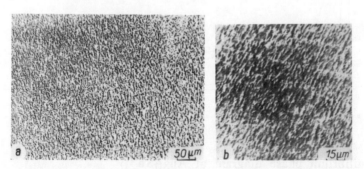

Figure 2 a and b : Broken lamellae observed in the structure of the as received Au$_{80}$Sn$_{20}$.

After hot rolling at 200 °C (strip) or extrusion then hot rolling (wire) the thin foil structure is composed of close black and white brocken lamellae in agreement with the very high cooling rate (4) (Figure 2 a and b). The structure is a little bit thinner and more homogeneous in the case of the wire than in the case of the strip. The structure sometimes contains white bands of large width (Figure 3)

Figure 3 : Extended large white area of ζ' phase observed in some places in the structure of as received Au80Sn20 strip.

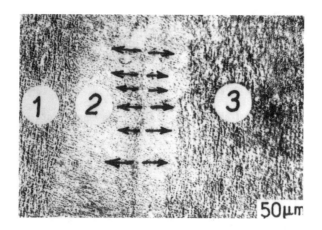

Figure 4 : Observation of distinct orientations 1, 2, 3 in the structure of as received $Au_{80}Sn_{20}$ strip. A frontier shares region 2 in the two parts.

Grains with different lamellae orientations are also observed, as sometimes quasi linear frontiers inside a block of same oriented structure (Figure 4). Then, some large extended cracks and small oriented cracks randomly distributed in some places are noted in the case of the strip (Figure 5)

Figure 5 : random distribution of cracks (arrows) in the structure of as received $Au_{80}Sn_{20}$ alloy

III 1.2 ANALYTICAL STUDY:

Sn and Au average content , using electron microprobe analysis, are reported in the table 1 for every phases observed in the strip and in the wire.

Sn content on different phases as a function of annealing. (wt % ± 10 %).

Phase \ Annealing	ξ or ξ'	δ	others oxidized phases
Received	10.5	36-37	—
$(P_{O_2}=10^{-2}torr)$ 175°C	10.5	—	4-7
260°C	10.2-11.3	—	7-9
$4h$ $(P_{O_2}=10^{-19}torr)$ 175°C	11.8	—	50-56
260°C	8.5-10.6	—	3-10
$1 month$ 175°C	10.7	36-37	—
$(P_{O_2}=10^{-2}torr)$ 260°C	9-12	36-37	—

Table 1 : Main element content in every phases in as received strip or after   annealing.

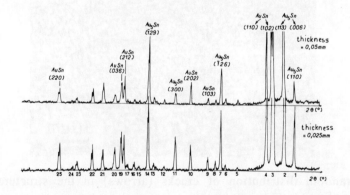

Figure 6 : X ray diffraction patterns (λ = 1.542 A) of AuSn and Au5Sn in as received $Au_{80}Sn_{20}$ (thickness = 0.05 mm) or in the $Au_{80}Sn_{20}$ sample after hot rolling (thickness = 0.025mm).

Then white phase can be identified as ζ phase (close lamellae) or primary ζ phase (wide white band) on account of the main analyzed content around 10.5 wt %. With regard to their content of $=$ 35 wt %, black lamellae agree with δ phase. As a confirmation, Au_5Sn and $AuSn$ are well evidenced in the X ray diffraction pattern and no texture is observed, at least for the as received strip*. (Figure 6).

Sample ratio	$Au_{80}\,Sn_{20}$ (as received)	$Au_{80}\,Sn_{20}$ (hot rolled)
$\dfrac{Au_5Sn\,(100)}{Au\,Sn\,(103)}$	3.6	4.4
$\dfrac{Au_5Sn\,(110)}{Au\,Sn\,(202)}$	1	2.3

Table 2 : Evolution of the X ray diffraction peak of Au_5Sn on $AuSn$ after hot rolling at 200 °C.

However after wire drawing or after hot rolling up to 0.025 mm thickness of the strip, modifications of the element percentage can be noted, especially in the outer surface zone where the tin content decreases up to 8 wt%. In addition to a slight textural effect for Au_5Sn growth, it is observed that the proportion of Au_5Sn is enhanced with respect to $AuSn$ probably as a result of the outer surface modification after hot mechanical treatment (Table 2).

III 1.3 MICROHARDNESS EVOLUTION

For the structure composed of a mixture of broken lamellae of $AuSn$ (δ phase) and Au_5Sn (ζ phase), microhardness data are in the order of 175 + 5 Hv; i e in the case of the as received strip of $=$ 0.05 mm thickness (table 3). When the δ phase content decreases after hot rolling of the strip with the same lamellae shape morphology (table 4), or inside primary ζ phase (i e Au_5Sn) microhardness data spread between 110 - 120 Hv (Figure 7). Then, there is a slight hardening effect of the δ phase itself, rather than a true contribution of the broken lamellae structure.

*Any small peaks of the X ray diffraction patterns could also be interpretated as a contribution of Au or SnO oxide according to (12).

Vickers Hardness

Au Sn (0.05 mm)	Au Sn hot rolled (up to 0.025mm)	Au Sn (0.05mm)+4h30 at	
		175°C	250°C
175 ± 6	110 ± 10	110 ± 6	25 ± 3

Table 3 : Microhardness data versus mechanical or annealing treatment through the white structure of ζ

At the end or from each side of cracks, a continuous evolution of microhardness proves stress evolution, may be as a result of any diffusionnal process insofar as the lamellae morphology is not largely disturbed (Figure 8 and Figure 5).

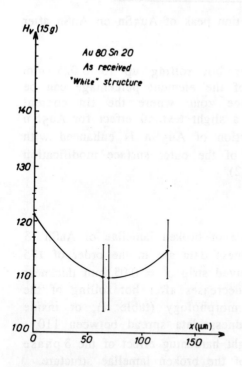

Figure 7 : Microhardness evolution through ζ phase in as received Au80Sn20.

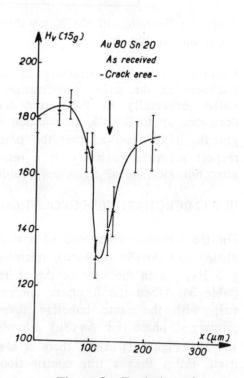

Figure 8 : Evolution of microhardness data from both side of a crack.

III 2) ANNEALED Au$_{80}$Sn$_{20}$ STRIPS

III 2.1 MICROSTRUCTURAL STUDY.

Effect of both annealing and oxidation in various oxygen partial pressures (pO2 = 10 $^{-5}$ atm or P$_{O2}$ = 10 $^{-22}$ atm) are then studied with respect to the transition temperature of 190 °C indicated in the equilibrium phase diagram.

At 175 °C, annealing for 4 h in P$_{O2}$ = 10 $^{-5}$ atm does not influence morphology which remains roughly as successive close lamellae, but with a white microstructure becoming more and more extended on the surface (Figure 9).

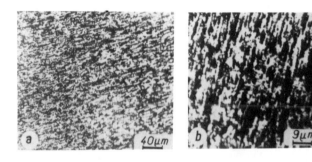

Figure 9 : Lamellae observed in the outer structure of Au80Sn20 strip annealed for 4 h at 175 °C in P$_{O2}$ = 10 $^{-5}$ atm.

However when decreasing oxygen partial pressure (P$_{O2}$ = 10 $^{-22}$ atm) a structural evolution takes place in the outer surface zone of the thin foil where white phases look like "leaves" (Figure 10). Also the lamellae shape no longer remains after a long annealing, for instance 1 month in P$_{O2}$ = 10 $^{-5}$ atm, black particles are surrounded with white intricated filaments (Figure 11).

 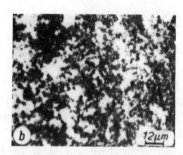

Figure 10 : "Leaves" of white structure observed in the outer zone of Au 80Sn20 strip annealed for 4 h at 175 °C in $P_{O2} = 10^{-22}$ atm.

.At 260 °C, the structure is much more equiaxe as observed for instance on the Figures 12, 13, 14 whatever the atmosphere ($P_{O2} = 10^{-5}$ atm or 10^{-22} atm) or the annealing time (4h or 1 month). Also grain-boundaries enriched with a black structure section primary ζ' bands (Figure 12)

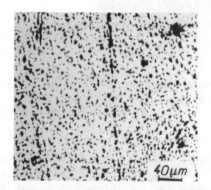 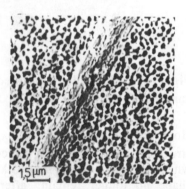

Figure 11 : White intrecated filaments and black particles in the structure of Au80Sn20 annealed for 1 month at 175°C in $P_{O2}=10^{-5}$ atm.

Figure12: Equiaxe structure observed in the structure of Au80Sn20 annealed for 4 h at 260°C in $P_{O2}=10^{-5}$ atm.

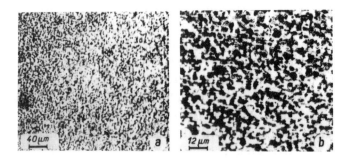

Figure 13 : Equiaxe structure observed in the structure of $Au_{80}Sn_{20}$ strip annealed for 4 h at 260 °C in $P_{O2} = 10^{-22}$ atm.

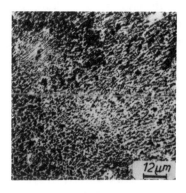

Figure 14 : Equiaxe structure with white filaments in Au80Sn20 annealed for 1 month at 260 °C in $P_{O2} = 10^{-5}$ atm

III 2.2 ANALYTICAL STUDY

Bulk observation

The large structural evolution in the case of annealing for 4 hours whatever the oxygen partial pressure and temperature is confirmed by microprobe analyses : the phase containing $\cong 36$ wt % of tin (i e δ phase) no more exists. But distinctions are noted below or above the transition temperature of 190 °C of the phase diagram. At 260 °C, the phase containing within 10.2 to 11.3 wt % of tin looks like unoxidized (i e ζ'

phase) while there is a sharp tin content decrease on the other black phase (i e δ' phase) probably as a result of any tin oxidation or segregation process (Table 1). At 175 °C, black and white ζ and δ phases are oxidized and tin content spreads over a large range of values, i e from 4 wt % (white structure) to 7 wt % (black phase).

. In the case of low partial pressure of oxygen ($P_{O2} = 10^{-22}$ atm), oxidation is easy. For instance at 260 °C, δ phase is no more observed and besides unoxidized ζ phase (wt % of Sn = 8.5 to 10.6 wt %), unoxidized and oxidized white β phase (wt % of Au = 92 wt %) is analyzed. A large oxidation phenomenon is noted straight in the cracks. At 175 °C, ζ' phase (wt % of Sn = 11.8 wt %) is observed together with some oxidized phases with high Sn content (between 50 - 56 wt % of Sn) (Table 1).

. After annealing 1 month in $P_{O2} = 10^{-5}$ atm, phases in agreement with the equilibrium phase diagram again form, i e δ' phase or AuSn (= 36 - 37 wt % of Sn) and ζ' or Au_5Sn phase (= 10.7 wt %) at 175 °C or ζ solid solution (9 - 12 wt % of Sn) at 260 °C (Table 1). Then, cracks enriched with Sn up to 25 wt % are always visible.

Surface analysis :
On the very outer part of the structure, gold, tin, oxygen and carbon can be analyzed by X ray photoelectron spectroscopy analysis, whatever the temperature, for example after annealing 4 hours in $P_{O2} = 10^{-22}$ atm (Figure 15, 16, 17, 18).

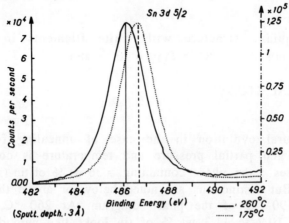

Figure 15 : Sn $3d_{5/2}$ binding energy analyzed by X ray photoelectron spectroscopy in $Au_{80}Sn_{20}$ strip annealed for 4 h in $P_{O2} = 10^{-22}$ atm at 260 or 175 °C.

The ratio of O 1S on Sn 3d$_{5/2}$ is 0.308 at low temperature and only 0.235 (at high temperature, showing then a highest surface oxidation at low temperature. This result agrees with the macroscopic observation of a bright surface at high temperature and a dull aspect at low temperature. Moreover, Au4f$_{7/2}$ peak is uneasy to distinguish at 175 °C (Figure 16).

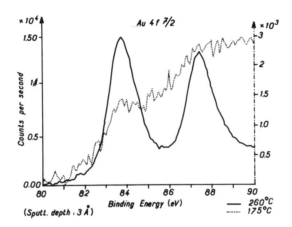

Figure 16 : Au 4f$_{7/2}$ binding energy by X ray photoelectron spectroscopy in Au$_{80}$Sn$_{20}$ strip annealed for 4 h in P$_{O2}$ = 10^{-22} atm at 175 °C or 260 °C.

The binding energy of Sn 3d$_{5/2}$ corresponds to a main contribution of oxidized Sn as SnO or SnO$_2$ (486.6eV according to literature data (13)). The contribution of metallic Sn (484;8 eV) is very low and mainly observed at 260 °C (Figure 15). The shift of the binding energy of Sn 3d5/2 from b.e. = 486.1 eV at 260 °C to b. e. = 486.7 eV at 175 °C (Δ b. e. = 0,6 eV) is in agreement with a more oxidized state of Sn at low temperature (13).

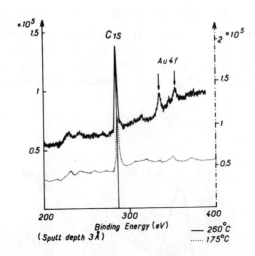

Figure 17 : C $1s$ binding energy analyzed by X ray photoelectron spectroscopy in $Au_{80}Sn_{20}$ annealed for 4 h in $P_{O2} = 10^{-22}$ atm at 260 °C or 175 °C.

Both C1s binding energy (be = 284.8 ev) and O1s binding energy (be = 532.2 eV) prove that C- O associations are formed (14) especially after annealing at high temperature (Figures 17 and 18).

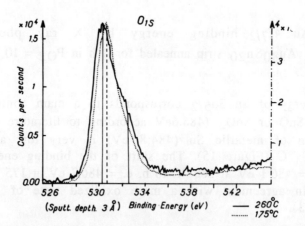

Figure 18 : O 1s binding energy analyzed by X ray photoelectron spectroscopy in Au80Sn20 annealed for 4 h in $P_{O2} = 10^{-22}$ atm at 260 °C or 175 °C.

III 2.3. MICROHARDNESS EVOLUTION

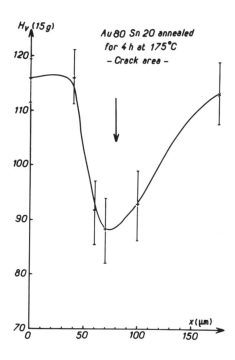

Figure 19 : Evolution of microhardness data from both side of a crack area in Au80Sn20 strip annealed for 4 hours at 175 °C in $P_{O2} = 10^{-5}$ atm

After annealing for 4 hours in $P_{O2} = 10^{-5}$ atm, microhardness strongly decreases with respect to the received strip. At 175 °C, the microhardness value of 110 + 0.5 ev is the same than in the case of for white band (cf fig 7), which is in agreement with the disparition of δ phase. (Table 3). At 260 °C, there is a strong softening and now the lamellae morphology no longer remains : an evolution into equiaxe structure with voids takes place within the strip (cf Figure 12).
The same microhardness gradient is observed from both side of an area containing cracks as in the case of received strip (Figure 19).

IV) DISCUSSION.

In spite of the stability of AuSn in front of Au_5Sn (zeta phase), especially at high temperature (10), AuSn is no longer observed after a short annealing, at least near the surface of the strip. At high temperature (260 °C) or for long annealing times at low temperature (175 °C), the morphology composed of successive lamellae of AuSn and Au_5Sn is destroyed. In some places also (at the end of cracks, inside primary ζ' phase, at both side of large grain-boundary) microhardness evolution was observed usually attributed to element gradient evolution as a result of diffusion phenomena. (15)

.Literature works show a sharp influence of the tin percentage on the pseudo diffusion parameters of gold and tin : interdiffusion of gold and tin are favoured on intermetallic phases such as AuSn (and $AuSn_2$) with respect to gold enriched phases (6). For instance at 175 °C, gold diffusion into intermetallic compounds is found 10^8 or 10^9 times greater than into gold enriched phases (6). Diffusion parameters of Au or Sn are calculated at both temperatures of 175 °C and 260 °C and reported in table 4 : the diffusion depth of 100 µm deduced from microhardness data evolution after 4 h at 175 °C is in the same range of depth than after diffusion calculations.

alloy or compound $T(°C)$	$Au(1.5\ Sn)$ $(Sn)\ (6)$	$Au\ Sn$ (20)	$Au\ Sn_2$ $(Au)\ (6)$	$Sn\ O_2$ $(Sn)(18)$
175	1.3×10^{-19}	9.5×10^{-12}	2.9×10^{-9}	6.7×10^{-52}
260	6.3×10^{-17}	1.1×10^{-10}	3×10^{-8}	2.9×10^{-42}

Table 4 : Pseudo-diffusion parameters of Au and Sn in several (AuSn) compounds at 175 °C and 260 °C.

A crystallographic influence on diffusion is also reported all the more important because of a preferential texture for AuSn intermetallic compound at least in the case of films obtained from successive evaporation of Au and Sn (16). Such a crystallographic result as a matter of fact is not verified for our massive strips but it always remains that Sn diffusion is 3 times greater than Au in a direction perpendicular to the c axis of the hexagonal cell of AuSn (8). Also, the similitude of the crystalline hexagonal structure between AuSn and Au_5Sn makes easy interdiffusion between the two compounds inasmuch as the morphology is composed of close lamellae of AuSn and Au_5Sn phases before annealing (6).

Others driving forces for the phase dissolution can be considered such as preferential surface segregation of one species or oxidation processes. Then, very fast enrichment of tin atoms over the surface are reported for gold surface using electron spectroscopy prospection (7). Moreover Somerjai indicates that surface diffusion coefficients for most materials fall in the range of 10^{-4} - 10^{-7} cm^2/sec for the temperature range of room temperature to the melting temperature of Sn (17). This corresponds to usual atoms displacement of the order of a few tens of microns. Also one of the most significant modification of the strip, as mentionned above, is the oxidation of tin into stannous and/or stannic oxide according to the temperature range and oxygen partial pressure. Then a very fast oxidation process takes place, even under very low oxygen partial pressure (down to $pO_2 = 10^{-22}$ atm), straight on the tin enriched phase (for instance AuSn) due to the highest tin activity.

With the assumption of a continuous stannous oxide scale developed on the surface, an oxide thickness of 14 A, in agreement with literature data of $\cong$ 10 A (18), could be deduced from the inelastic escape depth of gold electron through the stannous oxide layer formed at 260 °C in pO_2 $\cong 10^{-22}$ atm for instance. Effectively Au 4 f 7/2 peak is visible together with Sn 3 d 5/2 in oxidized state after X ray photoelectron spectroscopy analysis. The inelastic mean free path of gold, λ_{Au}, is calculated from the Seah and Quench formula (13) (equation 1):

$$\lambda_{Au} = 0.41 \, a^{1.5}_{Au} \times E^{0.5}_{Au} \qquad (1)$$

where λ_{Au} and a_{Au} are in nanometer and E_{Au} en eV. E_{Au} is the binding energy of gold and a_{Au} is the atom size of Au derived from equation 2 :

$$a = (A/\rho_{Au} nN) \times 10^{24} \qquad (2)$$

with A the atomic weight of Au (= 197), N the Avogadro number (= 6.02 x 10^{23}), ρ_{Au} the density of Au (= 193 00 kg/m^3) and n the number of atoms (= 1).

A thickness of the order of 12 A is confirmed using Ar+ sputtering together with X ray photoelectron analyses, at least at 260 °C. At 175 °C the oxide thickness is a little bit thicker and a slight covered Au $4f_{7/2}$ XPS peak is observed together with Sn $3d_{5/2}$ in an oxidized state. In this case, O1s peak strongly decreases after sputtering = 30 A of the scale.

The effective oxidation time to obtain the present oxide thickness of nearly 14 to 30 A extends to a few seconds using the parabolic kinetic of oxidation, k_p, of Sn into SnO (19) according to equation 3 from (19):

kp ($g^2 cm^{-4} sec^{-1}$) = k_c (cm^2/sec) = 2.73 x 10^{-6} exp(-18700/RT)   (3)

The correlative diffusion affected area into the underlying alloy can be calculated from some microns or less in the case of intermetallic compounds to a few Angströms for gold enriched phases, with a slight effect of temperature (table 5).

Compound Temperature °C	Au (1.5 Sn) (Sn*)	Au Sn (**)	Au Sn_2 (Au*)
175	10^{-5}	< 01	≃ 1
260	(1-2) 10^{-4}	≃ 02	≃ 3

    * Diffusing species
    ** Planar growth

Table 5 : Diffusion affected depths in several (AuSn) compounds after annealing for 4 h at 175 °c and 260 °C. (μm)

Such a calculated diffusion depth does not really agree with the experimental observation of a diffusion disturbed area over several microns , neither with the influence of the strip thickness on the X ray diffraction peaks ratio Au_5Sn on AuSn.*

-> In fact a continuous protective scale of SnO_2 on the surface is probably only observed after annealing in air according to the work of Bevolo et al (16). Due to lower oxygen partial pressure, carbon pollution and temperature modification , more complex oxide scales are probably formed.

Then the mixture of SnO, SnO_2 and the small content of Sn could result from many reduction processes more or less suitable. In fact the reduction of SnO_2 by CO usually takes place near 400 °C (21) and the dissociation of SnO according to 2 SnO -> Sn + SnO_2 is unlikely in vacuum below 370 °C according to (21). But studies reported that at T > 200 °C, SnO_2 really decomposes to SnO by annealing in vacuum (22). Then the most striking effect is related to the evolution of the

*Tompkin's experiments (7) show that for segregation of Sn on the surface, bulk diffusion is probably ruled out due to the rapid growth of the Sn signal during Auger spectroscopy analysis. Others experiments (20) point out the influence of grain boundaries for Sn diffusion, as observed sometimes in this study (cf Figure 12).

oxide scale morphology and nature in relation with the oxygen partial pressure evolution.

,At low oxygen partial pressure (pO$_2$ < 10^{-4} Torr), in addition with metallic Sn, Bevolo et al (18) concluded at a growth of islands of SnO and SnO$_2$ until the surface is covered by the oxide. This conclusion could be confirmed both by their experimental shift towards the lowest oxidized state of tin, at 260 °C (Δ be = 0.5 eV) with respect to 175 °C and by the slight decrease of the ratio of O 1s on Sn 3d5/2 XPS peaks with increasing temperature (0.308 at 175 °C and 0.235 at 260 °C). SnO2 content   increase with the temperature lowering agrees with the suported mechanism of surface adsorption of oxygen in the low temperature range (23). At the opposite oxygen dissolution into the metal becomes evident at T > 232 °C (24). As a confirmation experimental observations at high temperature proves that many SnO particles scatter in the alloy's bulk straight on the Sn enriched phases (mainly δ phase with respect to ξ or ξ' phases). Inner oxidation probably quickens δ phase dissolution into the alloy's bulk, increases the depth of the disturbed zone in the underlying alloy and favours the structural evolution into equiaxe grains morphology. But, up to now, no thermodynamic, crystallographic and metallurgic data are collected in literature about the selective oxidation of phases with respect to tin content.

. For higher oxygen partial pressure (pO$_2$ = 10^{-2} Torr) oxidation works are more numerous. SnO and SnO$_2$ oxides are present in the thickening oxide layer,
with the SnO$_2$ phase enriched at the outer layer of the oxide (25). Only a few atomic layers of stannic oxide on the surface can interrupt the oxidation of stannous oxide on the surface (18). When the scales become protective, a deviation is found in the parabolic plot of oxidation, i e for 5 - 10 minutes at 260 °C and 35 - 40 minutes at 175 °C according to further works (19). The corresponding scale thickness is around 1000 A and more.
In the underlying alloy, taking an account Au diffusion from Au Sn phase (6), the affected area could extend between a few tens   microns (260 °C) to some microns (175 °C).
According to the high oxygen activity, in spite of the good protective behaviour of the scale, a first transient step induces largely the phase dissolution (except for stable ζ phase) due to easy oxygen dissolution and important outward flow of Sn and Au. The lamellae easily evolve to equiaxe grains with voids formation associated to interdiffusion process (25), (26).

-> After a prolonged annealing for one month at 750 °C (in $PO_2 \cong 10^{-5}$ atm) the equilibrium phase composed of both ζ (or ζ') and δ phases is again observed, in sofar as the oxidation stop after formation of the protective scale. This result agrees with the metastability of zeta phase (ζ) with respect to delta phase (δ) according to the thermodynamical study of A.K.Jena et al (10), whatever the annealing temperature. The formation of δ takes place within ζ (or ζ') phase which leads to a mixture of discontinuous black δ phase within equiaxe fibers of white ζ phase (26) and (Figure 11). The amount of δ phase, which increases with temperature, can be interpretated in term of growth kinetic of AuSn phase (δ) according to the growth rate constant of Hugsted and al (20) ; the calculated affected depth in the alloy is nearly 200 μm at 260 °C and 60 μm at 175 °C, i e superior than the initial strip thickness. Such a growth kinetic is estimated in the case of Au/Sn evaporated films, for a planar growth of AuSn taking into account a preferential diffusion of gold atoms along the grain boundaries of AuSn followed by diffusion into the tin grains (20). Probably for massive strip the diffusion length will be smaller. Also void regions around islands of pure gold grains are mentionned in the case of Au/Sn thin film couples as a result of a random nucleation of AuSn intermetallic phase inside the original large tin islands (16) (25). Then microhardness data decrease as observed in table 3. A same mechanism due to interdiffusion is supposed to take place for the massive Au80Sn20 alloy.
- For the mechanical point of view, the most important result is the formation of cracks enriched with tin (around the eutectic content) associated with an element content evolution as a result of any diffusion phenomena; But it is not easy, up to now, to associate this brittleness to metallurgical, chemical or mechanical effects. After annealing formation of phases enriched with tin ($AuSn_2$, then $AuSn_4$) are often mentionned (27).

V) CONCLUSION.

As received strips of Au80Sn20 alloys get a structure of close lamellae of both ζ ($Au_5 Sn$) and δ (AuSn) structures. Morphology and phase content evolve during annealing with a sharp influence of the temperature and time of heat treatment and with oxygen partial pressure.
Phase dissolution in the outer zone of the strip appears very dependant with the formation of stannous and stannic oxides, on account of the

surface enrichment of tin diffusing from the tin enriched phase like AuSn (especially via intergranular paths). Tin oxidation is then favoured.

While under high oxygen partial pressure (10^{-5} atm) a protective oxide scale forms, in reduced oxygen partial pressure a mixture of stannous and stannic oxides is observed. The protection against further inner oxidation is rather bad, especially at 260 °C when the oxygen dissolution is enhanced. According to the partial pressure of oxygen, the diffusion affected depth is modified because the phase dissolution is under the control of inward flow of oxygen ($pO_2 < 10^{-22}$ atm) or of outward flow of tin to form a rather thick and continuous scale of stannic and stannous oxides ($pO_2 = 10^{-5}$ atm).

Especially in the case of low partial pressure of oxygen, the morphology of the phases evolves from close broken lamellae of both ζ and δ phases to an equiaxe structure especially composed of ζ (or ζ') phases, gold enriched phases and, may be, stannic oxide particles in the outer zone of the annealed strip. After a long annealing without inner oxidation ($pO_2 = 10^{-5}$ atm), on account of the higher stability of δ phase with respect to ζ phase, again the two phases ζ and d are observed together, in agreement with the phase diagram.

A strong evolution of the microhardness data is noted from each side of cracks containing nearly the same tin content than the eutectic point.

ACKNOWLEDGMENT:

We would like to thank Mr J. M. Siffre of the laboratoire de Physico-Chimie des Surfaces, E.N.S.C.P., Université P et M Curie, Paris, for his help with X ray spectroscopy analysis.

REFERENCES.

1) Humpston, G. and Jacobson, D.M.:Gold Bull., 1990, 23, 3, 85.
2) Nevit, M.V., Rudman, P.S., Stringer, J., Jaffee, I.R.:Phase Stability in Metals and Alloys, Materials Science and Engeenering series, Mc. Graw Hill Book Comp., 1966, 281.
3) Nakahara, S. and Mc. Coy, R.J.:Thin Solid Films, 1980, 65, 247.
4) Favier, J.J. and Turpin, M.:Acta Met., 1979, 27, 1021.
5) Giessen, B.C.:Z. Metallkd., 1968, 59, 10, 805.
6) Nakahara, S., Mc. Coy, R.J., Buene, L. and Vandenberg, J.M.:Thin Solid Films, 1981, 84, 185.
7) Tompkins, H.G.:J. Vac. Sci. Technol., 1975, 12, 2, 650.
8) Dyson, B.F.:J. of Applied Physics, 1966, 37, 6, 2375.
9) Chang, C., Callcott, T.A. and Arakawa, E.T.:J. of Applied Physics, 1982, 53, 11, 7362.
10) Jena, A.K. and Bever, M.B.: Met. Trans B, 1979, 10, 5.
11) Okamoto, H. and Massalski, T.M.:Binary Alloys-Phase diagrams, American Society for Metals, 1987, 264.
12) Donaldson, J., Moser, H., Simpson, R.J.:J. Chem. Soc., 1961, 830, 141.
13) Seah, M.P.:Practical Surface Analysis by Auger and X ray Photoelectron Spectroscopy, Ed. by Briggs, D. and Seah, M.P., John Wiley and Sons Ltd., 1983, 187, 5.
14) Moulin, G., Rousselet, J.M., Huntz, A.M.:J. Micros. Spectros. Electron., 1987, 12, 229.
15) Nakahara, S. and Mc. Coy, R.J.:J. Electrochem. Soc., 1981, 128, 1781.
16) Buene, L., Falkenberg-Arell, H. and Tafto, J.:Thin Solid Films, 1980, 72, 457.
17) Somerjai, G.A.:Principles of Surface Chemistry, Prentice Hall, Englewood Cliffs, NJ,1972, sec 3-5.
18) Bevolo, A.J., Verhoeven, J.D. and Noack M.:Surface Science, 1983, 134, 499.
19) Nagasaka, M., Fuse, H. and Yamashina, T.:Thin Solid Films, 1975, 29, 2, L29.
20) Hugsted, B., Buene, L., Finstad, T., Lonsjo, O. and Olsen, T.: Thin Solid Films, 1982, 94, 81.
21) Barawy, K.A. and Warczok, A.:Reactivity of Solids, 1990, 8, 9.
22) Hoflund, G.B., Asbury, D.A., Kirszensztezn, P. and Laitinen, H.A.:Surface and Interface Analysis, 1986, 9, 169.
23) Paria, M.K. and Maiti, H.S.:J. of Materials Science, 1983, 18, 8, 2101.
24) Baird, T., Fryer, J.R. and RIddell, E.V.:Surface Science, 1971, 28, 525.
25) Powell, R.A.:Appl. Surface Sci., 1979, 2, 397.
26) Nakahara, S. and Mc. Coy, R.J.: Appl. Phys. Letters, 1980, 37, 1, 42.
27) Vefuji, T., Shimomura, Y. and Kino, T.,Jap. J. of Appl. Phys., 1977, 16, 6, 909.

III. CARBURIZING

Materials Science Forum Vols. 102 - 104 (1992) pp. 149-154

GAS FLOW REGULATION IN A CARBURIZING HEAT TREATMENT OF STEELS

P. Casadesus and M. Gantois

Laboratoire de Science et Génie des Surfaces (URA CNRS 1402), Ecole des Mines, Parc de Saurupt, F-54042 Nancy Cédex, France

ABSTRACT

The carburizing power of an argon-methane plasma obtained through glow discharge is determined. The methane introduced in the furnace dissociates to produce on the whole carbon and hydrogen. In a steady state, the flux of carbon transfered through the treated surface is constant for a given temperature and gas flow and relative to a set furnace configuration. It is independant of the composition of the iron alloys treated and in particular of the superficial carbon content.

These experimental results permit the establishement of the global carbon balance of the reactor. The carbon flux transfered through the sample is given as a function of: The inlet gas flow, the methane dissociation rate, the temperature, the treated surface size and the size of the furnace walls . With the flux condition at the surface thus determined we calculate the carburizing profile by resolving the equations of diffusion.

INTRODUCTION

The argon-methane plasma carburizing process is closely related to ionic nitriding. However, some specificities of carburizing relative to the nature of the plasma itself, to a higher work temperature and to the methane flux through the treated area, which is higher than the nitrogen flux, makes absolutly necessary the complete knowledge of the carbon transfer mechanism in the furnace, so that it would be possible to efficiently master and control the process. Most of the work spent on the setting up of this process, among which we mention those of P.Collignon(1), P.Cousinou(2), J.L.Medina Picazo (3), has mainly allowed the establishement , by the means of metallurgical analysis, subsequently performed, a correlation between the carbon concentration profiles and the carburizing conditions.

The work presented here has the same aim, but with an approach based on experimental elements obtained in-situ during the treatments, with the help of an investigating apparatus conceived to this end.

I-EXPERIMENTAL APPARATUS - RESULTS

I-1. Mass spectrometry and thermogravimetric analysis

The apparatus that allows us to measure the carbon transfer during the treatment is made up of a thermobalance adapted to work in an ionized environement and a mass spectrometer to analyse the gas taken from the shaft furnace (figure(1)). The sample is heated to the carburizing temperature in the homogeneous austenitic phase range (>910°C), in an argon plasma. When the working conditions are established, the methane flow is turned on as is shown in figures (2) and (3). We analyse the gas by sampling from various points (inlet, inside, outlet). The response time during the sampling, of the order of the second, is negligible on the scale of the operating time of the process and the mass tranfers studied. In particular, the time lap between the inlet of the methane flow and its outlet can not be significantly apprehended.

The methane and hydrogen outlet concentrations, given in arbitrary units on figure (2), increase and stabilize after a time interval of a few minutes. They stay constant during the whole period of time that the methane inlet flow is kept constant. Therefore the methane concentration is independant of the treatment's state of advance which is characterized by the sample's carbon surface content. When the electric discharge is turned off while the gas flows are maintained, we observe an important increase of the methane concentration, up to the level corresponding to the inlet gas concentration. The hydrogen signal tends towards the background noise. Other analysis, made while varying the carburizing temperature and the iron base alloys composition, and keeping the argon and methane flows constant, have shown that the concentration of the non dissociated methane is independant of the furnace temperature and the carburized alloy composition.

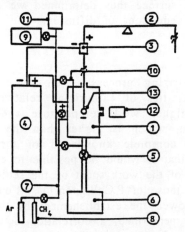

1--Reactor
2--Thermobalance
3--Electric adaptor
4--Generator
5--Valve
6--Blade pump
7--Gas inlet
8--Flowmeters
9--Mass spectrometer
10-Sampling probe
11-Pump
12-Pyrometer
13-Sample

Figure1. Experimental apparatus

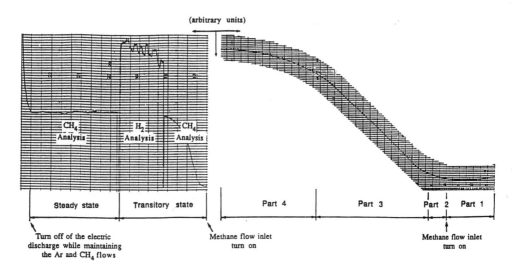

(arbitrary units)

| Steady state | Transitory state | Part 4 | Part 3 | Part 2 | Part 1 |

Turn off of the electric
discharge while maintaining
the Ar and CH₄ flows

Methane flow inlet
turn on

Methane flow inlet
turn on

Figure2. Mass spectrometry analysis of the
methane and hydrogen outlet.

Figure 3. General configuration of a plasma
thermogravimetry curve.

The plasma thermogravimetry curves, figure (3), generally show the following main parts :
-Part 1 corresponds to the temperature increase in the argon plasma. We observe a small decrease of the sample's mass due to the surface erosion by cathodic pulverisation.
-Part 2 translates the transitory state following the beginning of the methane introduction. The duration of this state, of the order of a few minutes, equals the necessary time to obtain a constant outlet gas composition.
-Part 3 shows a lineary mass increase. The slope of this straight line corresponds to the transfered carbon flux. Hence the carbon transfer between the plasma and the sample is carried out at constant flux independently of the surface carbon content which increases in this linear part of the mass intake curve as a function of the imposed carbon flux. The slope of the straight line, and therfore the carbon flux transfered, increases with the inlet methane flow and with the carburizing temperature for a given furnace configuration set.
-Part 4 is characterized by a decrease of the flux, tending to a saturation limit. The carbon solubitilty limit in iron is reached at the beginning of this zone. The continuation of production of carbon by the plasma at the surface is shown through the modification of the electric discharge working conditions. The electric discharge role, as creator and accelerator of reactive species toward the surface, is not correctly carried out.

I-2 Methane dissociation rate and carburizing power of the plasma

As it was shown by the preceding results the methane introduced in the furnace dissociates under the action of the electric discharge to produce mainly hydrogen and carbon, a portion of this carbon being transfered to the sample. We have established that the outlet hydrogen molar flow is double the dissociated methane molar flow (4), according to the reaction:

$$CH_4 \longrightarrow 2H_2 + C$$

The carbon molar flow available for carburizing is:

$$F_C = X_{CH_4} \cdot F_{CH_4}^E \tag{1}$$

where X_{CH_4} and $F_{CH_4}^E$ represents respectively the methane dissociation rate in the furnace and the inlet gas flow.

This relationship defines the carburizing power of the plasma. The electric induced methane dissociated rate X_{CH_4} is experimentally determined for the furnaces used(4). The working conditions of the electric discharge being fixed, the dynamic aspect related to the speed of the gas flow in the furnace is alone susceptible to act on the dissociation rate. This dissociation rate is between 0.7 (for high flows) and 0.95 (for weak flows), within the limits of the speed of the gas flow that is compatible with the process functionning.

I-3. Expression of the carbon flux transfered to the sample.

A fraction of the carbon flow $F_C = X_{CH_4} F_{CH_4}^E$ produced by the plasma corresponds to the flux of carbon transfered to the sample. This fraction increases with temperature and with the treated surface size. It decreases with an increase of the furnace size represented by its walls. In the temperature interval studied, between 910 and 1050°C, and for the furnaces used in this study, we have established the following formula for the carbon flux transfered to the sample:

$$F_C^t = \alpha_C^t X_{CH_4} F_{CH_4}^E \tag{2}$$

where,

$$\alpha_C^t = 0,533 + 4,25 \ 10^{-3}(T-1213) + 20 \ (A_s - 10,36 \ 10^{-4}) - 0,975(A_w - 0.12)$$

represents the carbon efficiency transfer.

T : Carburizing temperature   (K);

A_s : Carburized surface (m^2);

A_w : Surface of the furnace walls (m^2);

The interval of validity of formula (2) experimentally established is as follows:

A carburizing temperature interval from 1183 to 1323K; 1213K corresponds to the mean value of this interval.

The carburized surface is less than 30.10^{-4} m^2. The furnace walls surface is between 0.12 and 0.32 m2.

The constants related to the conditions mentioned above and which appear in formula (2) are only valid for the experimental furnaces used in this work.

We deduce from expression (2) the carbon flux density on the carburized surface:

$$f_C^S = (F_C^t / A_S) = (\alpha_C^t X_{CH_4} / A_S) F_{CH_4}^E \qquad (3)$$

This expression gives the correlation between the carbon flux density at the surface on the one hand and the carburizing parameters on the other.
The flux density f_C^S is the one value which controls the elaboration of the carbon concentration profile. Hence relashionship (3) allows us to act on the profile, particularly with the help of the methane flow, the other parameters being kept fixed.

EXAMPLES OF GAS FLOW REGULATION

We have used argon-methane mixtures under pressures of the order of 10^3 Pa.
Argon allows the electric discharge functionning and also heats up the sample to the carburizing temperature.
The study of the influence of the argon flow having shown a methane dissociation rate variation between 0.70 (high flows, short stay time) and 0.95 (weak flows, long stay time), subsequently we have used a mean argon flow which gives a methane dissociation rate of 0.80. Then for furnaces of various sizes we choose argon flows which gives us stay times of about the same order. We therefore obtain an identical methane dissociation rate.
To regulate the methane flow, most of the time we try to master the carbon flux density at the surface during the treatment, because the surface carbon content is controled by the flux density. The treatment can start with a high carbon flux density and then continues with a weak one (figure. 4(a)). We recreated these same conditions with an other furnace, and with an other carburized surface size(figure. 4(b)).

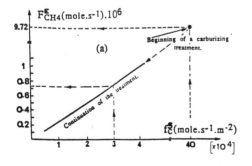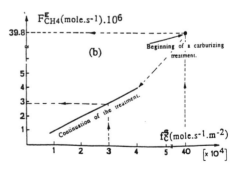

Figure 4. Methane flow adaptation to carbon flux density at the surface.

(a) : $X_{CH_4} = 0.80$;   (b): Same conditions as in (a) but

$T = 1213K$;   using the following area sizes:

$A_S = 10.36 \cdot 10^{-4} \, m^2$; $A_W = 0.12 m^2$. $A_S = 30.10^{-4} \, m^2$; $A_W = 0.32 m^2$.

CONCLUSION

This experimental study has permited us to characterize the carburizing power of the plasma and to formulate the carbon flux density at the surface as a function of the inlet gas flows. The formula obtained, valid for the used furnaces, having been deduced from the global carbon balance, does not for this reason take into account all the other carburizing process specificities.

References

(1) P. Collignon, thèse, Nancy, (1978).
(2) P. Cousinou, thèse, Nancy, (1977).
(3) J.L. Medina Picazo, thèse, Nancy, (1985).
(4) P. Casadesus, M. Gantois, A study of an argon-methane plasma for steel carburizing, 10 th International Symposium on Plasma Chemistry, Bochum,1991.

Materials Science Forum Vols. 102 - 104 (1992) pp. 155-168
Copyright Trans Tech Publications, Switzerland

PLASMA OVERCARBURIZING OF CHROMIUM STEELS FOR HOT WORKING AND WEAR APPLICATIONS

J.P. Souchard (a), P. Jacquot and M Buvron (b)

(a) BMI (HIT Group), Rue du Ruisseau, Z.I. Chesnes Tharabie, St. Quentin-Fallavier,
F-38290 La Verpilliere, France
(b) Innovatique (HIT Group), 25 Rue des Frères Lumière
F-69680 Chassieu, France

ABSTRACT

This paper presents the metallurgical study concerning the overcarburizing of chromium steels by using the plasma carburizing process.

Plasma overcarburizing is a high carbon carburizing process where the case carbon content is significantly higher than austenite saturation in order to produce into the superficial layers of alloyed steels with carbide formers, great quantities of very fine globular carbides in a martensitic matrix.

The produced layer containing a high quantity of complex carbides type $M_7 C_3$ on the alloyed steel, is wear and erosion resistant.

Applications increasing the life of the hot work tools like forging dies, punches, inserts, glass molds are envisaged.

1. INTRODUCTION

Plasma enhanced surface treatments present a substantial field of new developments in the scope of materials. One example is carburizing in a glow discharge of steels. This technique offers significant advantages over conventional carburizing techniques (gas, salt, fluidized bed) and can be considered as an alternative or complementary method to gas

Fig. 1 : View of the plasma carburizing furnace with the control
panel.

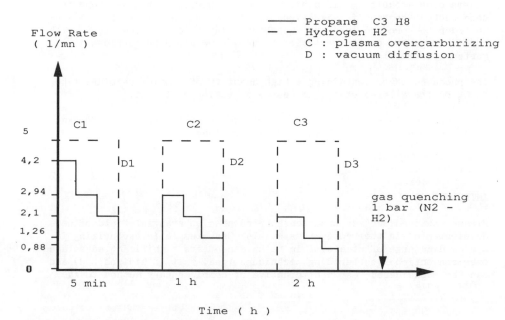

Fig. 2 : Evolution of propane and hydrogen flow rates during a
plasma overcarburizing treatment.

carburizing.

Many works have been published on plasma carburization. Howewer, those
works concerned only carbon enrichment below austenite saturation and
applied on low alloyed steels in order to have surface carbon levels
acceptable to obtain after martensitic quenching with optimal
characteristics (% C   0,8) for different case depths.

In this paper, we will present the results obtained with plasma over
carburization of hot working steel type H11. The goal of this
metallurgical study is using a carbon level significantly higher than
the about 0,8 % of conventional carburizing one. The depth of the
diffused layer containing great quantities of small and spheroidal
carbides is fixed to 0,6 - 1 mm.
The special carburizing, gas quenching and tempering parameters with
or without a final plasma nitriding treatment were defined. Some
industrial application tests on hot forging dies will be presented.

2.   **EXPERIMENTAL PROCEDURES**

The material chosen for this study is a hot work tool steel type
H11 - H13 with 5 % of chromium used in forging and casting dies.
The samples are cylindrical parts of 40 mm diameter and 10 mm
thickness.

The carburizing treatment is performed in a vacuum heat treatment
furnace manufactured by BMI (FRANCE) which is shown in **FIGURE 1.**
The useful sizes of the vacuum chamber are 500 X 600 X 900 mm and the
maximum weight of the load is 400 Kg. The specimens are connected
to the cathodic bias of a pulsed power supply (20 KW - 6-7 KHz)
and the heating elements to the anode. After the loading of
the parts, the chamber is pumped down to a pressure level of 2.10^{-5}
mbar by using a primary, roots and diffusion pumps. The parts are
heated by an additional heating system. It consists of graphite
resistance elements connected to a 120 KW power supply. The hot
zone is thermally insulated with graphite fibers insulator. The
temperature is controlled mainly by the additional heating power
and the electrical glow discharge. During heating up of the load a
soaking temperature (750°C - 30 mn) is used to obtain uniform
temperature throughout the load. The heating rate is 15°C/mn and
10°C/mn respectively before and after the first soaking temperature
(15 mn). The pressure of the working chamber is raised to 3 mbar by adding
controlled amounts of propane and hydrogen mixture.

The pulsed power supply is connected to the parts (cathode), isolated
from the chamber and the heating elements (anode). The power density
applied to the load is 0,75 W/cm². The carburizing phase is carried
out usually at 1000°C during 3 "boost" sequences of 15 min and followed
by 3 diffusion sequences of 30 min (**FIG. 3**). Surface carbon content and
case depth can be adjusted by fixing several boost-diffusion sequences.
The control unit manages and regulates the various process parameters.

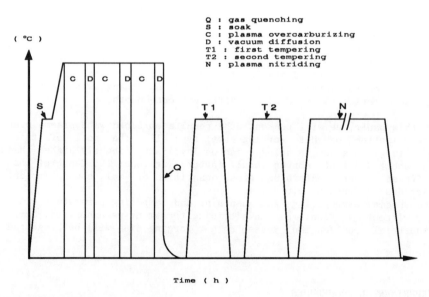

Fig. 3 : Temperature versus time for typical plasma overcarburizing
treatment with or without plasma nitriding.

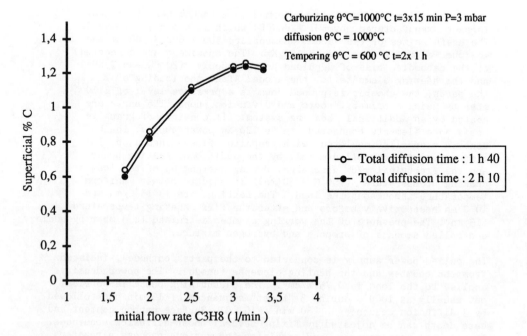

Fig. 4 : Plasma overcarburizing of H11 steel.
Variation of carbon content at the surface in function of
initial flow rate of propane.

In our test conditions, we have studied the influence of different
propane initial flow rates between 1,7 to 4 1/mn. During the carburizing
phase, the flow rate of propane decreases sequence after sequence
and the initial flow rate of propane decreases in the time during each
sequence. The flow rate of hydrogen is maintained constant (5 1/mn)
during all the cycle (**FIG. 2**). At the end of the carburizing cycle,
the load being heated at 1000°C in order to have a homogeneous
austenitisation temperature is gas quenched by a 1 bar positive pressure
(N_2 - H_2). After quenching, two tempering cycles are carried out in
a vacuum furnace between 500 and 600°C.

3.   RESULTS

3.1.- Plasma overcarburizing "Surdiff" Process

High carbon carburizing is characterized by its carbide
precipitation which is due to higher carbon contents than in
conventional gas carburizing.

The process can only be used on low or preferency alloyed steels
in which there are substantial amounts of carbide formers such as
chromium, molybdenum or vanadium.

It is the reason why we have applied this treatment on H11, H13,
AISI 420, D3 steel grades. Some recent applications was made on
different steels : 5 % Cr steel (1), special steels (2), low alloy
steels (3) and stainless steels (4).

During a preliminary survey, we have studied different parameters
to better know the main parameters of the process and the best way
to realize our technical objectives as : case depth about 0,8 - 1 mm
carbon content 0,6 % and globular carbides formation in the case.

The selected constant parameters are :
- the propane flow : 4,2 1/min et hydrogen : 5 1/mn
- the time of carburizing : 3 sequences of 15 mn
- the time of vacuum diffusion : 3 sequences of 30 mn
- the working pressure : 3 mbar
- the electrical parameters : intensity : 12 amps, power density
  0,75 W/cm²
- the surface treated : 1 m2
- the temperature of the first tempering : 500°C - (1h).

In a second study, we have examined the influence of the carburizing
temperature, the second tempering temperature, nitriding temperature
and time on the case depth, the surface and core hardness, the
morphology and repartition of the produced carbides.

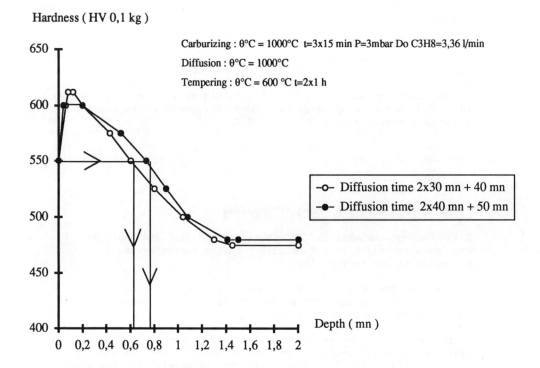

Fig. 5 : Plasma overcarburizing of H11 steel.
Hardness profile.

3.2.   Influence of the propane initial flow rate on the superficial
        carbon content

We have examined the influence of various propane initial flow
rate : 1,7 1/mn, 2,5 1/mn and 3,36 1/mn on the superficial carbon
content of plasma overcarburized H11 steel.

The results are shown **FIGURE 4**. It can be seen that the superficial
carbon content increases with the propane flow rate. The maximum carbon
level obtained is 1,2 %. On the optical micrographs, we can see that
the size of carbides increases and the quantity of carbides decreases
with an increase of propane flow rate. In the same way, the surface
hardness and the case depth decreases with a decreasing propane flow rate.
The average surface hardness is about 600 HV 0,1 and the case depth is
0,6 mm (at 550 HV) (**FIGURE 5**).

3.3.   Influence of the carburizing temperature

We have studies the influence of different carburizing temperatures :
1000, 1020, 1040°C for fixed parameters :
        carburizing time : 3 X 15 mn
        diffusion time : 3 X 30 mn
        $C_3 H_8$ flow rate : 4,2 1/mn
        first tempering : 500°C

When the carburizing temperature increases the surface and core
hardness increases, certainly because the carbon solubility in austenite
increases with temperature and the produced martensite is more rich in
carbon and so more hard.

The case depth obtained (0,82 mn) and the superficial carbon content
(1,3 - 1,4 %C) are not depending on the carburizing temperature.
However, the carburizing temperature dependance on the microstructure
is shown Fig. 6. When the temperature increases the carbides number
decreases and their sizes increases. At high temperature, carbides
growth follows grain boundaries to form laminated and coarse intergranular carbides.

At 1000°C, carbides are very fine, globular and homogeneously scattered.
X-Ray analyses indicate that complex carbides of (Fe, Cr) 7 C_3 type
are created. In order no to decrease toughness or resiliency of the
matrix, we have to avoid the formation of the laminated intergranular
carbides and residual austenite. It is the reason why we have chosen
the optimal carburizing temperature at 1000°C.

(1)

(2)

(3)

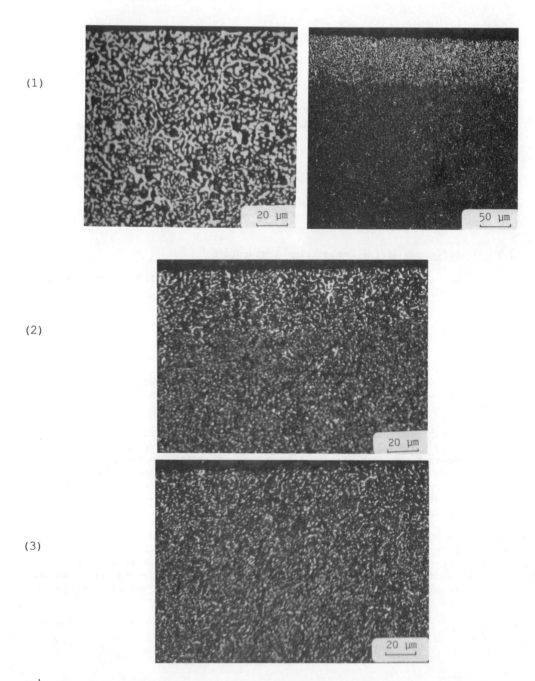

Fig. 6 : Carburizing temperature dependance on the microstructure.
Micrographs of the carbide layer for various carburizing
temperature.
(1) 1040°C   (2) 1020°C   (3) 1000°C

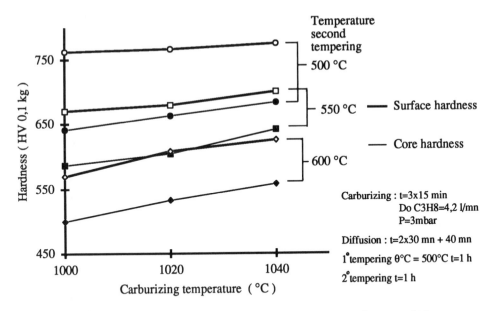

Fig. 7 : Evolution of the core and surface hardness with
      carburizing and second tempering temperatures.

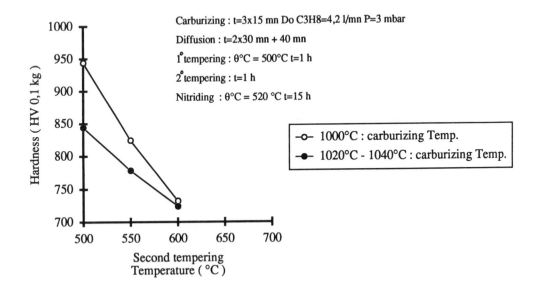

Fig. 8 : Plasma overcarburizing and nitriding of H11 steel.
      Evolution of the superficial hardness with second tempering
      and carburizing temperatures.

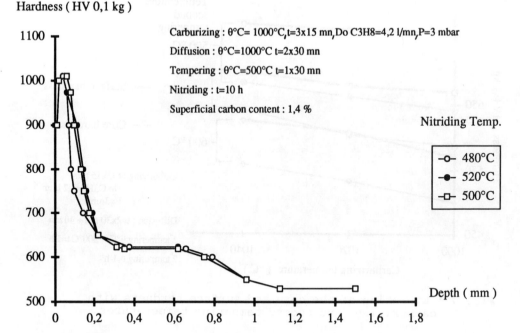

Fig. 9 : Plasma overcarburizing and nitriding of H11 steel.
Evolution of the superficial hardness with nitriding
temperature after overcarburizing at 1000°C.

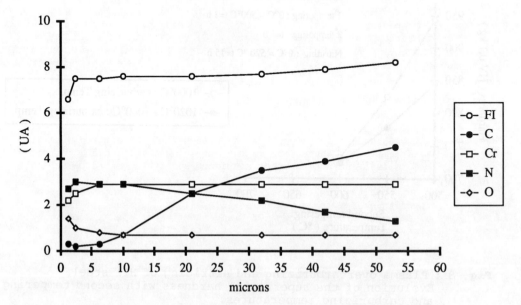

Fig. 10 : Surface analysis of an overcarburized and nitrided H11
steel in depth on the first 55 microns.

3.4.   Influence of the tempering temperature

         After carburizing at 1000, 1020, 1040°C, we have carried out a
         first tempering at 500°C and a second tempering at 500,550 and
         600°C. For a fixed carburizing temperature, an increase in the
         second tempering temperature decreases the core and surface
         hardness. So, an increase in the tempering temperature results
         in a coalescence of the produced carbides. The maximal surface
         hardness decreases from 760 HV 0,1 to 570 HV 0,1 when the second
         tempering temperature increases from 500 to 600° C (**FIG. 7**).

3.5.   Plasma nitriding after overcarburizing of H11 steel

         In order to increase the superficial hardness and the wear
         resistance, we have tried some plasma nitriding treatments after
         plasma overcarburizing cycle. We have carried out a plasma
         nitriding cycle at 520°C during 15 h on overcarburized samples
         at 1000, 1020, 1040°C and tempered at 500,550 and 600°C. The
         results are showed **FIG. 8.** It can be observed that after nitriding
         the superficial hardness decreases when the parts are overcarburized
         for the highest temperature (1020 - 1040°C) and when the second
         tempering temperature increases, the core hardness is about 500 HV 0,1
         and the superficial hardness is about 720 to 950 HV 0,1.

         When the second tempering temperature increases, the nitrided depth
         decreases. That is explained by the fact that carbides formed at
         high temperature are big and very few of them are present. It is the
         reason why these carbides are very difficult to be dissolved during
         plasma nitriding and the chromium available to form nitrides is
         probably limited.

         The microhardness profile obtained after overcarburizing at 1000°C,
         2 tempering at 500°C and nitriding during 10 h for different
         temperature of H11 steel is shown **FIG. 9.** We can see a peak on the
         curve at 1000 HV 0,1 from the nitrogen diffusion layer with chromium
         nitrides precipitation and a flat at 600 HV 0,1 from the carbon diffusion
         layer.

         In order to verify the influence of the temperature and the duration
         of the nitriding cycle, we have selected 3 temperatures : 480, 500,
         520°C and 3 times : 10 h, 15 h, 20 h using a gas mixture with 5 %
         nitrogen and 95 % hydrogen. We have seen that for a fixed nitriding
         treatment time, increasing in the nitriding temperature decreases
         the maximal surface hardness and increases the nitrided depth.

         When the nitriding time increases, the surface hardness decreases and
         the nitrided depth increases a litle. It is necessary to limit the
         plasma nitriding at about 500°C and the time at about 10 h. We obtain
         always compressed layers in the surface without to form brittle
         structure as lamelar iron carbonitrides.

         The nitrided layer is composed of iron nitrides type $Fe_4 N$ and very fine
         precipitates of chromium nitrides. Surface analysis realized by glow
         discharge optical spectroscopy shows that the nitrogen profile is slowly
         decreasing, like it is for carbon (**FIG. 10**).

	TEST 1	TEST 2	TEST 3
Slug Temp.	1050°C	920°C	1250°C
Forged material	AISI 1042	Ti 6 Al V	AISI 5120
Die preheat Temp.	220°C	500°C	200°C
Cycle rate (parts / h)	150	120	450
Lubrification	graphite water	grease	graphite water
Machine	drop forging press 1250 t	hydraulic press 600 t	drop forging press 300 t
Average life time of the die before surdiff	6000	700	3500
After surdiff	20000	1100	5500
Breakdown cause before surdiff	erosion	deformation hot wear	binding
Breakdown cause after surdiff	deformation	deformation	breaking

Table 1 : Examples of hot forging tests with data.

Near the surface, carbon is pushed back to the substrate by the nitrogen. That explains the quasi elimination of the carbides at the surface and the nitrides formation.

4. APPLICATIONS

In order to valid the overcarburizing treatment, we have tried some pre-industrial trials on hot work toolings and wear resistant tools.

The objectives is to increase 1) the life of the hot work tools like dies, punches, inserts, molds, rollers in the field of hot and half hot forging, stamping, steel mills, glass industry and 2) the life of parts submitted to high wear action like sand blasting parts, crusher parts and tiles tooling.

In hot forging, wear is due to :
    - hard particules abrasion (oxydes) drained by the forged metal
    - adhesion wear due to sticking at the interface forged metal / tool
    - friction of the forged metal on the surface of the dies.

Usually hot forging dies are quenched, tempered and nitrided and sometime, plated with hard chromium coating.
Plasma overcarburizing and nitriding treatments applied on forging dies can significantly reduce adhesion wear, abrasion and friction.
We have obtained very good results for the forging of stainless steel, titanium alloys, and midle steel (see **TABLE 1**).

Life time gain can be very important (X2 to X3). Some applications on glass molds are envisaged.

5. CONCLUSION

The results obtained during this metallurgical study, concerning the "surdiff" process, confirm the possibility to obtain with plasma overcarburizing, superficial layers containing a great quantity of fine carbides on alloyed steels type H11, with hot wear resistance.

The other results can be summarized as follows :
    - a special vacuum heat treatment furnace with plasma equipment
      was manufactured to perform the overcarburizing process.
    - the optimal carbon content necessary to obtain carbide layer
      on H11 steel is about 1,4 %. But there is a limitation to the
      carbon enrichment : carbides depletes the matrix of chromium,
      molybdenum and can results in reducing case hardenability.
    - the overcarburizing temperature must be fixed at 1000°C + or -
      10°C.
    - the tempering temperature must be adapted to the final
      application (500 - 600°C).
    - the nitriding temperature must be fixed at about 500°C for a
      duration of 10 h.
    - some applications in hot forging and glass molding have been
      tested.

REFERENCES

(1) Y. POURPRIX, O. CHIGNARD
Overcarburization of chromium steels with glow discharge
International seminar on plasma heat treatment 21-23 Sept 1987
Senlis FRANCE (PYC Edition, Paris)

(2) K. NAMIKI, K. NAKAYAMA, T. IWANO, T. KENMOKU
Application of supercarburizing to hydraulic lash adjuster
Society of automotive Engineers
Publications division (USA) 1988 n°88 0416

(3) R.F. KERN
Super carburizing
Heat treating Oct 1986 P. 36-38

(4) K. AKUSTSA, M. NAKAMURA
Practice and experience with plasma carburizing furnace
Proceedings of ASM'S 2nd International Conf on ion nitriding /
carburizing, CINCINNATI (USA) 18-20 Sept 1989.

Materials Science Forum Vols. 102 - 104 (1992) pp. 169-182
Copyright Trans Tech Publications, Switzerland

INFLUENCE OF THE BASE MATERIAL HARDENABILITY ON EFFECTIVE CASE DEPTH AND CORE HARDNESS

B. Vandewiele

Surface Treatment Company

ABSTRACT

In the large volume production of highly performant mechanical components, carburising is still the most widely used heat treatment.
In order to continuously improve and guarantee the quality there is a need for tighter controls on both dimensional and heat treatment specifications.
Therefore it is imperative that the variables affecting case depth and core hardness are thoroughly understood and controlled.
This paper shows the relative significance of the critical parameters : carburising time and temperature, carbon potential, cooling rate and steel hardenability. The relationship between base material hardenability and case depth and core hardness is very strong. Based on this relationship, it is easily understood that the use of steels with controlled and reduced hardenability is really necessary to obtain the general goal of less variation in the after heat treatment state.
Establishing an optimal combination between carburising practice and the use of steel with controlled hardenability also offers a significant potential to reduce the heat treatment cost. To realize this a close cooperation between all people that are involved is necessary.

INTRODUCTION

In recent years, there has been considerable competitive pressure to reduce cost and increase reliability and durability. This pressure, coupled with the refined design techniques, has produced more highly stressed components which require a high degree of dimensional and mechanical property consistency.
Improved consistency in carburised and hardened components can be obtained by reducing the spread in heat treatment response during the hardening operation. This spread in response is, in turn, a direct function of the material characteristics and the carburising parameters.
Reduction of the variation in heat treatment response will yield a product with a clearly defined and reliable strength gradient and microstructure, two key factors in determining part performance.
In addition, a third factor, dimensional control, which is particularly important for carburised gears and shafts, is directly related to the reduction in variability of the response to heat treatment.
Reducing this variation, primarily by controlling hardenability and quenching, will significantly limit the distortion variation common to carburising and hardening. For components such as carburised gears, reduced distortion translates to quieter operation and longer gear lives, often without resorting to more costly finishing processes, such as grinding after hardening.

THE CARBURISE AND QUENCH PROCESS

Carburising is a thermochemical heat treatment, with enrichment of the surface region with carbon. The aim of this treatment is to give the part optimal properties at the surface, in the case and in the core.
The process can be divided into three basic steps :
Step 1 : **carburising** : enrichment of the case with carbon to a certain depth **(C-profile)**.
Step 2 : **quenching** : transformation of the material to its hardened state **(HRc-profile)**.
Step 3 : **tempering** : lowering the peak stresses at the surface **(surface hardness)**.

A graphic representation of the result of step 1 and 2 is given in fig.1.

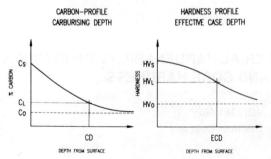

Fig 1: Result of the carburising and quenching step.

The carbon-profile is characterised by the carbon value at the surface (C_s), the carbon value of the base material (C_o) and the carburising depth (CD) at which a presetted carbon limit percentage (C_L) is reached. The hardness profile is characterised by the hardness value at the surface (HRC_s), the hardness value in the core (HRC_o) and the effective case depth (ECD) at which a presetted hardness limit value (HRC_L) is reached.

Generally the carburise and quench process can be represented in a process diagram, shown in fig.2.

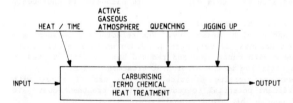

Fig 2: Process diagram of gas carburising

The **output** (part in the "hard"-state) is the result of the action of the **process parameters** (heat/time, active gaseous atmosphere, fast cooling, jigging up) on the **input** (part in the "soft"-state).
The variation in the output is the result of the variation, both in the process parameters and in the input. Special attention is to be given at the influence of the input. This is treated in detail in later sections.
The **capability** of the process is expressed in terms of the relation between average and spread in the output and the tolerance field in the specification.

The global action of the process parameters is :
   Heat/time : the amount of energy, as heat, given to the system in function of the time, let the heat treater establish a certain **temperature-time profile**.
   Temperature has a major influence on the diffusion of the carbon in the steel. Equation 1 gives the relation between the diffusion coefficient for carbon in gamma iron and temperature

$$D = 16.2 \exp \left(\frac{-137800}{RT}\right) \text{ mm2/sec} \tag{1}$$

   with : R = 8.314 J/K.mol
          T = Temperature in degrees Kelvin.
The influence of time is given by equation 2, for the simple one-dimensional diffusion (Fick's second Law)

$$C_x = C_o + (C_s - C_o) \left[1 - \text{erf} \left(\frac{X}{2\sqrt{D.t}}\right) \right] \tag{2}$$

   with : C_x: carbon percentage at depth X
          C_o: carbon percentage of the core material
          C_s: carbon percentage at the surface

          X : depth from the surface
          D : diffusion coefficient mm2/sec
          t : time in sec.

Based on this equation one can say that carburising depth is **proportional** with the **square root of time**.

Active gaseous atmosphere : the gaseous atmosphere is formed by a carrier gas (eg. nitrogen/methanol) and an enriching gas (eg. natural gas/air), for regulation of the activity of the atmosphere. This activity together with temperature is of prime importance for the **carbon transfer** between the atmosphere and the steel surface. The most common way of expressing the activity of the atmosphere is the **carbon potential** (C_p). The higher the carbon potential, the faster is the carbon transfer and the carburising process.

Quenching : this is the fast and controlled cooling of the load during which the transformation in the hard martensitic structure takes place. The quench must be fast enough to produce the martensite structure (above the critical cooling rate for the material) but not too fast in order to avoid too much deformations.

The quenching is characterised by a given **intensity**, which can be described by a **H-value** as stated by Grossman. The major control factors are the uniformity of the quench volume and agitation, the nature of the quench fluid, the temperature of the fluid and the proper and consistent loading of the parts within the quench trays.

Jigging up : the way parts are loaded on the trays is of prime importance for the macro-deformation of the part. Load density is also important for the uniform circulation of the quench fluid through the load.

The **input** (part in the "soft"-state) to the carburised and quench process can be described as :

material : the material is characterised by a chemical composition (carbon, alloying elements, impurities), a hardenability (Jominy-curve), a microstructure with inclusions and segregations (banding).

Geometry : the part has a certain shape with a given massivity and many changes in section often sharp.

History : parts and material entering the process have a history which can be described as thermal, mechanical and internal stresses.

BASICS USED FOR CALCULATIONS

Hardenability

The hardenability of the steel is a measure for the ease with which the hardness is reached into the material. The most widely used test method to evaluate the hardenability is the **Jominy test**, a single end quench test. The result of this test is the so-called **Jominy curve** for the material tested. This curve is a plot of the measured Hrc-hardness in function of the distance to the quenched end. A typical curve with the action of carbon and most of the alloying elements is given in fig.3.

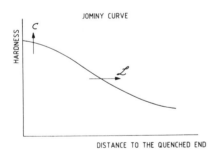

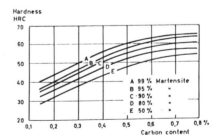

Fig 3: *Jominy curve with influence of carbon and alloying level*

Fig 4: *Hardness versus carbon content and percentage martensite in the micro structure*

During the years many methods have been proposed to calculate the Jominy curves from chemical composition and austenitic grain size. Even in recent years a lot of effort is still be done to increase the accuracy of the calculation methods for particular steel families. For this publication we used the **equations proposed by E. Just** for carburising steels. Two equations are used : one for carburising steels with carbon up to 0.25 % and one for steels with carbon from 0.25 % up to 0.60 %. Calculations for carbon higher than 0.60 % is not necessary for the purpose of this publication.

Equation for carburising steels with carbon up to 0.25 % :

$$J_{6..40} = 74\ (\%C)^{0.5} + 14\ (\%Cr) + 5.4\ (\%Ni) + 29\ (\%\ Mo) + 16\ (\%Mn) - 16.8\ E^{0.5} + 1.386\ E + 7\ \text{HRC} \qquad (3)$$

Equation for steels with carbon from 0.25 % up to 0.6 %

$$J_{6..40} = 102 \, (\%C)^{0.5} + 22 \, (\%Cr) + 21 \, (\%Mn) + 7 \, (\%Ni) + 33 \, (\%Mo) - 15.47 \, E^{0.5} + 1.102 \, E - 16 \, Hrc \qquad (4)$$

    with E   : distance (mm) to the quenched end

        (%X)   : percentage of element X
        J   : hardness HRC for distance E from the quenched
        6..40   end between 6 and 40 mm.

At distances below 6 mm the hardness is taken from the diagram HRC versus carbon percentage given in fig.4. This value is also the upper limit for all distances. The calculated values from equation 3 and 4 are not acceptable if they exceed the max. value for the upper curve of fig.4.

Quench rate

To calculate the quench severity and cooling rates in the part, several methods are developed and still are in development. To simplify the calculations we have used the **method of Grossman**. The quench severity can be described by the **Grossman H-value**, which depends on the quenching medium and its agitation. We also use the Grossman diagrams which relate diameters of round bars with distance to the quenched end given a certain H-value and for a specific location in the round bar. Fig.5 gives the monogram for the surface location and the near center location.

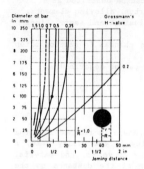 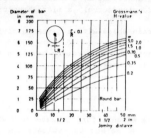

Fig 5: Relationship between H-value, representative diameter and Jominy distance for two specific locations in bars.

Relation between calculated Jominy curves and carbon limit (C_L)

For a given material a set of Jominy curves can be calculated, for all carbon % between core carbon content (C_o) and 0.6 % using equations 3 and 4 in combination with the upper curve in fig 4.

Fig.6 gives a calculated set for the DIN material 16 MnCr5, nominal analysis. The presetted hardness limit value, used for effective case depth determination, is clearly indicated. We use the limit value of 52.3 Hrc which corresponds with 550 HV.

It is easily seen that for increasing Jominy distances (decreasing cooling rates) higher carbon limits levels are needed to obtain the hardness limit value. It is easy to construct a diagram giving the **carbon limit value** in function of the Jominy distance. This is given in fig.6

The carbon limit value cannot be less than 0.33 %C, which is the minimum %C needed to obtain 52.3 Hrc in a full martensitic structure.

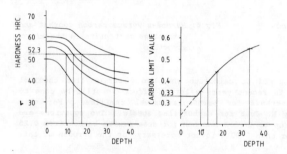

Fig 6: Calculated Jominy hardness curves for the average analysis of steel 16MnCr5 (left). Correlation of the carbon limit value (Cl) for the hardness limit value 52.3 Hrc and the distance from the quenched end (right).

A PRODUCTION EXAMPLE

In order to show the evidence of the influence of the material on the heat treatment result, an analysis is shown of a scrapped production batch. The part is a geared hollow shaft made of SAE 8620 material. Specification is core hardness minimum 30 Hrc and effective case depth to 510 HV : 0.38 - 0.63 mm.
Out of the batch, a large sample of 27 pieces was examined for core hardness and effective case depth. The data were statistically treated with the median-rank method. This is given in fig.7

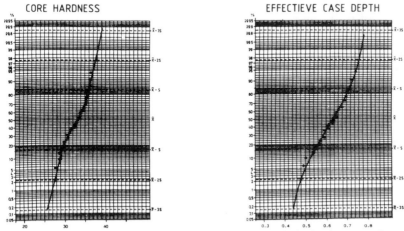

Fig 7: *Statistical analysis by use of the median-rank method of a large sample*

It is easily seen that there is an **S-shaped distribution** which indicates a mixing of two Gaussian distributions.
All pieces of the sample were further examined for chemical analysis. After this it was clear that there were two groups belonging to different smelts. (K4664 and K1391). Fig.8 gives the Jominy curves of the two smelts and the hardenability band of the material SAE 8620.

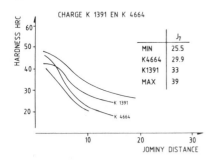

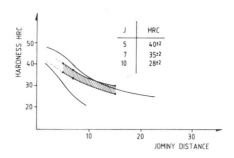

Fig 8: *Jominy curves of smelts K4664 and K 1391 in the Jominy band of steel SAE 8620.*

Fig 10: *New material specification for solving the capability problem*

A new statistical analysis with the median-rank method was carried out. The result is given in fig.9
It is easily seen that the samples of each smelt follow a Gaussian distribution and that the **smelt influence** is primarily a **shift in average core hardness and effective case depth**. The **standard variation** is relatively **smelt independent**. Table 1 gives all relevant data.
Based on these data it is impossible to guarantee the quality of all the pieces. In order to solve this problem, a new specification for the incoming material was accorded with the steel supplier / manufacturer.
The new specification for the hardenability band was :
   J = 5 mm : 40 ± 2 Hrc; J = 7 mm: 35 ± 2 Hrc; J = 15 mm: 25 ± 2 Hrc
This is graphically shown in fig.10

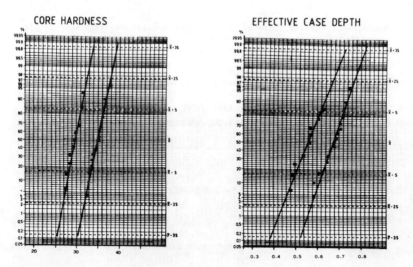

Fig 9: Statistical analysis by median rank method smelt separated.

Table 1	K 1391	K 4664
Jominy Hrc at 7 mm	33	28.5
Effective case depth		
average X	0.675	0.555
6 * standard dev.	0.30	0.34
X - 3 S	0.525	0.385
X + 3 S	0.825	0.725
Core hardness		
average X	34.9	29.9
6 * standard dev.	8.7	8.5
X - 3 S	30.5	25.5
X + 3 S	39.2	34

CORE HARDNESS : METHOD OF JOMINY EQUIVALENT DISTANCE (Jed)

Basics

- The hardness depends on the structure (% martensite) and the carbon percentage.
- The structure/hardness is determined by the cooling rate.
- The hardness at a given location in the part is equal to the hardness at the position in the Jominy curve with the same cooling rate. That cooling rate is the **Jominy equivalent cooling rate (Jec)** and the position in the Jominy probe is the **Jominy equivalent distance (Jed)**. This is graphically illustrated in fig.11

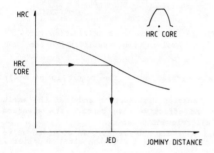

Fig 11: Definition of the Jominy equivalent distance (Jed)

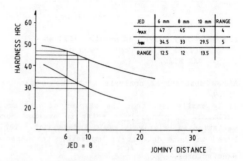

Fig 12: Graphic determination of the expected hardness interval for the steel 20MnCr5 with Jed = 8 ± 2 mm

Working procedure
- Take a small sample (5 pieces) in the first production batch (heat treatment procedure established).
- For each of the pieces measure the core hardness at the specified location.
- Determine the Jominy curve of the material used (Jominy test or calculated curve from chemical analysis).
- Determine for each part the Jominy equivalent cooling rate.
- Calculate average and standard deviation by an appropriate statistical method (Median-Rank method).
- Determine with the Jominy band for the material the core hardness interval which can be expected.

Application
Parts has to be carburised and quenched. The steel used is the DIN material 20 MnCr5. In the first heat treatment batch a small sample was analysed for core hardness and Jominy equivalent distance, which resulted in Jed = 8 ± 2 mm what can the heat treater guarantee over a long time period ? The results are given in fig.12

Analysing these results gives :
- Total expected range   : **17.5 HRc** (29.5 - 47)
- Range due to material   : **12 HRc** (33 - 45)
            quenching : **5.5 HRc**
- Procentual distribution :
            material (hardenability)   : **68.5 %**
            heat treatment shop (quench) : **31.5 %**

Material hardenability is by far the **major factor of influence.** Reducing the spread in core hardness is most efficiently done by reducing the specified Jominy band of the material. Due to the nature of the method used, the **core hardness is directly related to the base material hardenability.**

Example : steering pinion
The method is illustrated with an experiment done on a steering pinion (sample of 10 pieces). Part geometry, material, measuring locations and process are given in fig.13

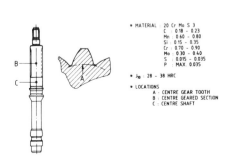

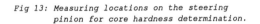

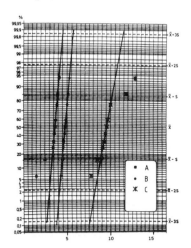

Fig 13: Measuring locations on the steering
        pinion for core hardness determination.

Fig 14: statistical evaluation
        by the median rank method
        for core hardness data of
        the steering pinion

The result of the statistical analysis for the 3 locations is given in fig.14, and fig.15 gives the relative position of the average Jed for each location on the Jominy curve.
Making all calculations gives the results in table 2.
In fig.14 it is seen that there is no perfect normal distribution for location A ʼand C and the standard deviation is smallest for location B. This effect is completely explained by the relative position on the Jominy curve. The **standard deviation is directly proportional to the hardness gradient at the intersection point.** At the intersection for A there is a remarkable difference in gradient at the left and at the right, from which it is easily understood that there can be a deviation from normal distribution at lower Jed. The long term statistical distribution depends on the relative position on the Jominy curve.

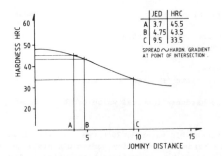

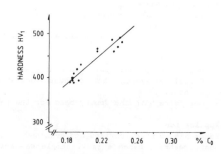

Fig 15: relative position of the Jed for the three
locations on the Jominy curve of the steel

Fig 16: regression analysis between core
hardness and carbon content of
the material

Table 2

	Position in the part		
	A	B	C
Average Jed	3.7	4.75	9.5
Stand dev. S Jed	.33	.30	.63
average HRc	45.5	43.5	33.5
average HV	454	432	329
Expected min HRc	42.9	41.1	27.8
" max HRc	47.1	46.8	40.8
" min HV	422	404	285
" max HV	472	470	400

Also a regression analysis was made for hardness HV for location B and the carbon percentage of the
material. The data are represented in fig.16 showing the very strong relationship.
From this, one can conclude that the spread on core hardness for a given location is mostly
determined by the variation in the chemical composition. **The real spread due to the quenching is
small.**
It is stated that for a given application (heat treatment cycle, quenching procedure and a
particular part) there is a 100 % correlation between core hardness and base material hardenability
expressed as the hardness on the Jominy curve at the Jominy equivalent distance for the location in
the part. It is easily seen that it is absolute necessary to **specify the location** where the core
hardness has to be measured.

EFFECTIVE CASE DEPTH : GRAPHIC 4-QUADRANT METHOD

Basics
The calculation of the hardness profile is possible when the cooling rate and the material
hardenability in each point are known.
As stated above, we use the formulas reported by E. Just for Jominy calculations and the monograms
and H-value of Grossmann to calculate the relation between part massivity (representative diameter),
quench severity (H) and distance on the Jominy curve. Determination of the H-value is possible with
the use of appropriate test probes. The carbon profile can be calculated with different methods or
can be measured on the test probe.

Calculation method : graphic 4-quadrant method.
In fig.17 is shown how the method works for constructing the hardness profile given a carbon
profile, a material and a quench severity.

1. Take a point on the carbon profile (depth a, %C = C).
2. Determine the Jominy distance which corresponds with the application. To do this, draw a
   horizontal line through the representative diameter ϕ and read at the intersection with the H-
   curve the Jominy distance (dist. b).
3. Calculate the Jominy curves for the base material and for the same material but with higher C-
   levels up to 0.6 %. Construct the set of Jominy curves.
4. For the Jominy distance sub 2°, draw a vertical line through the set of Jominy curves sub 3°.
5. At the intersection with the Jominy curve for C% carbon read the hardness value.
6. Construct one point on the hardness profile by the intersection of the horizontal line through
   the intersection point sub 5° and the vertical line through depth a.

7. Repeat the procedure till a sufficient number of points is available to draw the hardness profile.
8. For a given hardness limit value, determine the corresponding effective case depth (ECD).
9. For a depth equal to the ECD, read the corresponding carbon limit value (Cl) on the carbon profile.

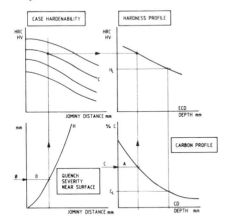

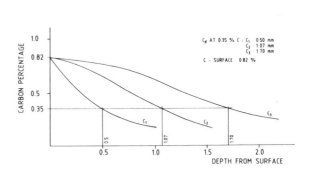

Fig 17: derivation of the hardness profile from Jominy curves, carbon profile and quenching severity

Fig 18: carbon profiles used in the simulations

The method was computerised on a personal computer so that the calculations can be made fast and easily.
With this model one can :

- predict the effective case depth for a given application.
- Calculate the influence of a modification in the process.
- Simulate and quantify process variations on material composition (hardenability), C-profile and quenching severity.

Calculations.

With the programm, calculations were made for the carburising steels listed in table 3.

Table 3	Chemical composition						Hardenability (Just)		
	C	Mn	Cr	Mo	Ni	Si Max.	J7	J15	J20
17 Cr3	0.14	0.40	0.60	–	–	–	30	20	–
	0.20	0.70	0.90			0.40	16	–	–
21NiCrMo2	0.17	0.65	0.4	0.15	0.40	–	26	16	–
	0.23	0.95	0.7	0.25	0.70	0.40	44	34	31
20MoCr4	0.17	0.70	0.3	0.40	–	–	30	20	17
	0.22	1.0	0.6	0.50	–	0.40	45	36	33
20CrMoS3	0.18	0.60	0.70	0.30	–	0.15	32	22	19
	0.23	0.80	0.90	0.40	–	0.35	45	35	32
16MnCr5	0.14	1.00	0.80	–	–	–	28	18	–
	0.19	1.30	1.10	–	–	0.40	41	32	29
22CrMoS35	0.19	0.70	0.70	0.40	–	–	37	27	24
	0.24	1.00	1.00	0.50	–	0.40	48	43	40
15CrNi6	0.14	0.40	1.40	–	1.40	–	34	25	21
	0.19	0.60	1.70	–	1.70	0.40	45	40	35
17CrNiMo6	0.15	0.40	1.40	0.25	1.40	–	41	34	30
	0.20	0.60	1.70	0.35	1.70	0.40	46	46	45

P : 0.025 %   S : 0.025 %   Cu : 0.1 %   V: 0.005 %   Al: 0.02 %

For each steel, 5 calculations were made regarding composition with respectively all elements at min. spec., 25 %, 50 %, 75 % of the spec. range and at max. spec.
For each severity, calculations are done for H = 0.1, 0.2, 0.4, 0.6 and 1.0

For representative diameter, calculations were made or diameter = 5, 10, 25, 50, 75 and 100 mm.
For the carbon profile three profiles are used as represented in fig.18 with a surface C % of 0.82 %
and carburising depths at 0.35 % C of 0.5, 1.07 and 1.7 mm.
In total **3600** calculations were made. For all calculations a hardness limit value of 52.3 HRc (=
550 HV 1) is used. It is impossible to give all calculated points in this article, so a selection
is made to give a representative idea of the role of the different factors.

Global Effect
To show the global effect fig.19 gives the calculated values of effective case depth for a H-value =
0.2 and with all steels at their nominal composition for the three carbon profiles and a hardness
limit of 52.3 HRc.

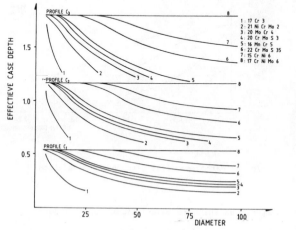
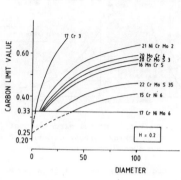

Fig 19: effective case depths calculated for all
        steels at nominal composition and the
        three carbon profiles for a quench
        severity of H = 0.2

Fig 20: carbon limit value for all steels at
        nominal composition for H = 0.2

From fig.19 it is easily seen that there is a **large difference in behaviour between different steel
families** : steels with lower base material hardenability, like 17Cr3, have a very strong **diameter
effect**, so that they are unadapted for carburising work on larger parts. Steels with very good base
material hardenability (like 17CrNiMo6) don't have a pronounced diameter effect even in quenching
conditions of low severity.
With fig.19 one can read up to which representative diameter the steel can be used for an accepted
difference in effective case depth. This effect is valid for each H-value but it becomes less
pronounced with higher H-values.
For each calculated value of the effective case depth, there is a corresponding value of the carbon
limit concentration. Based on these data, fig.20 is drawn which gives the curve of the carbon limit
value with the representative diameter for each steel at the nominal composition, a quench-severity
of H = 0.2 and a hardness limit value of 52.3 HRc.

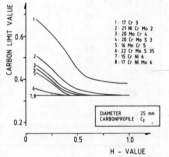
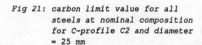
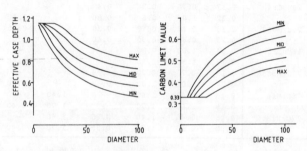

Fig 21: carbon limit value for all
        steels at nominal composition
        for C-profile C2 and diameter
        = 25 mm

Fig 22: effective case depth and carbon limit value for the
        hardenability range of the steel 16MnCr5 for carbon
        profile C2, H = 0.2

The curves are **independent of the carbon profile**. Again the diameter effect is clearly visible. The lower the curve the lesser is the diameter effect.
All curves intersect the Y-axis (d = 0) at a point of 0.25 %. The general form of these curves is given in equation 5.

$$C_L = 0.25 + f_c \cdot d^{(a-b \cdot d)} \text{ for } C_L \geq 0.33 \tag{5}$$

The coefficients a and b are primarily determined by the quench severity (H-value).
The coefficient f_c **is material specific** and is a measure for **the carburising power of the steel** : the larger f_c, the lesser the steel is suited for carburising work on larger parts.
f_c depends also on the quench severity H in a way that for higher H-values, f_c becomes smaller.
The f_c coefficient can be used as a basis for material classification with regard to their response to the carburising processes.
In order to show the **influence of the quench severity**, fig.21 is drawn for the carbon limit value with H-value for all steels at nominal composition, a representative diameter = 25 mm and a hardness limit value of 52.3 HRc. For more severe quenches (H-value higher), the carbon limit value drops rapidly, which means that for a given carbon profile the effective case depth will be higher. The difference between the different steels disappears more and more increasing the H-value. This effect is observed at all diameters in the way that it becomes more pronounced for larger diameters.

Effect of the material hardenability.
All figures discussed so far, refer to the steel with its nominal composition. For each steel, the hardenability varies from smelt to smelt, depending on the actual chemical analysis. In order to show this effect, fig.22 gives the effective case depth and carbon limit value with diameter for 16MnCr5 in the five material hardenability states, used in the calculations and H-value = 0.2, carbon profile 2 and hardness limit value of 52.3 HRc.
The effect of the smelt material hardenability is clearly visible. For diameters above 25 mm the difference between min. and max. composition is about 0.38 mm on effective case depth.
At smaller diameters the effect is less pronounced, due to the fact that the Jominy band is smaller for small Jominy distances.
Reducing this effect needs the use of the material 16MnCr5 with reduced hardenability band, which is possible with modern steel making processes.
The **material hardenability** effect exists for all the steels, but it is less pronounced for deeper hardening steels. The effect is also less pronounced when using a higher quench severity.
In order to show the relation between effective case depth and the base material hardenability, expressed by the HRc-hardness at 7 mm on the Jominy curve (J7), fig.23 is drawn for 3 steels, 15CrNi6, 16MnCr5 and 21NiCrMo2, for a diameter of 25 mm, H-value = 0.20 and carbon profile 2.

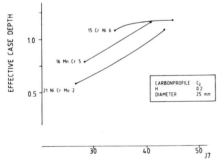

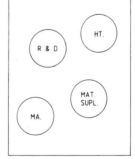

Fig 23: effective case depth related with base
material hardenability for the steels
21NiCrMo2, 16MnCr5 and 15CrNi6 for
carbon profile 2, H = 0.2 and diameter
= 25 mm

Fig 24: individual behaviour of the 4 spheres, of
interest

The strong positive relation between ECD and J7 is clearly visible for each material. For each steel the curve is different, meanly depending upon the alloying elements and their influence with carbon.
Because of this effect it can be advantageous to use steel with the same J7-value but a better response to carburising to obtain a deeper effective case depth, rather than to compensate the effect with a larger carburising time. (0.2 mm more on ECD requires a 10 to 20 % longer carburising time).

Relative significance between the different factors.
Good carburising practice (good controlled carbon potential, temperature and quenching) can give a process variation over long time periods of H-values within ± 0.05 and carbon profile within ± 0.05 % C at a given depth.

On the long term, the whole hardenability band of the material has to be taken into consideration. Based on these statements the following calculations were made for 16MnCr5 for diameter 25 and 50 mm, for H = 0.2 and 0.4 and carbon profile 2. The results are given in table **4**.

Table 4	d = 25			d= 50		
	ECD		Diff.	ECD		Diff.
Nominal	0.98			0.78		
H : 0.15 - 0.25	0.93	1.02	0.09	0.72	0.84	0.12
C2: - 0.05/+ 0.05	0.88	1.09	0.21	0.69	0.87	0.18
Anal.: min/max	0.80	1.15	0.35	0.61	0.96	0.35
All min/max	0.67	1.31	0.64	0.46	1.14	0.68
Nominal	1.15			1.04		
H : 0.35 - 0.45	0.14	1.16	0.02	0.99	1.10	0.11
C2: - 0.05/+ 0.05	1.05	1.28	0.23	0.93	1.16	0.23
Anal.: min/max	1.00	1.17	0.17	0.85	1.17	0.32
All min/max	0.87	1.32	0.45	0.71	1.31	0.60

Based on these results we can make a Pareto analysis of the total variation and it is easily seen that the **material hardenability is by far the most important factor**. About 50 to 60 % of the ECD-variation is explained by material hardenability variation. The second factor is the carbon profile and the third is the quenching severity.
The relative significance is dependent on the steel grade and the absolute value of H.

CONCLUSION

For good carburising practise, the heat treater disposes over good processes and equipment, which he runs with close control of all relevant process parameters. But this is not enough to meet the close tolerances asked for, because of the large influence of the base material and its hardenability on the heat treatment response (core hardness, effective case depth and deformations). This article has shown and quantified this material influence of the base material. This influence is not simple and also depends on quench severity and part massivity.
In order to **guarantee capability on the long term**, the heat treater cannot do this by its own. He has to collaborate with the other parties involved, namely the **R & D engineers** (material spec., heat treatment spec.), the **steel manufacturer** (steel processing with reduced hardenability spread) and the part **manufacturing engineers** (shaping processes, soft-hard relation on deformations). In the past all these spheres act like individuals with their own knowledge and optimalisation. It is clear that the individual behaviour is restricted within a certain reference frame, which is the real measure of the long term capability.
This is shown in fig.24. In modern practice, this is not sufficient. This means that we have to bring the spheres closer together.

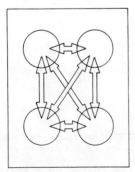

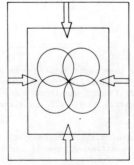

Fig 25: communication channels between all spheres of interest

Fig 26: clustering of the 4 spheres by communication with reduced reference space

To do so, we need to put some activation energy to the system. This means that we have to install good **communication channels** between **all** spheres.
This is shown in fig.25. In doing so, the spheres become closer together, even clustered, so that they act no longer like individuals but as a **cluster**.
This is shown in fig.26.

It is easily to see that in that situation **problems are really solved** and the cluster can act in a much narrower reference frame, which means that long term capability is much better and guaranteed. With some **creative thinking**, one can easily see in the contours of fig.26 the **four-leaved clover** as shown in fig.27. This is widely accepted as the symbol for a lucky and successful situation. It is also the end stage of what can be attended from a real **comakership**. This is the real **advanced heat treatment engineering**.

Fig 27: four-leaved clover : Success, Good Luck

<u>REFERENCES</u>

1) Jominy, E.E   : Hardenability test, Hardenability of alloy steels, ASM, 1939, p. 66

2) Jominy, W.E   : Hardenability test for carburising steels, Trans. ASM, 1938, p. 574

3) Grossmann M.A : Hardenability, Its relation to quenching, and some Quantitative Data, ASM
                   Cleveland 1939, p. 124

4) Just, E   : Formeln der Härtbarkeit, HTM 23, 1968, p. 85-100.

5) Wyss, V   : Kohlenstoff und Härteverlauf in der Einsatzhärtungsschicht verschieden legierter
                   Einsatzstähle, HTM 43, 1988, p. 27-35

6) Sponzilli, J.T: Product Quality Improvements by Effective Material and Process Control of
                   Carburised Components, Wolfson H.T. of Metals, 1990, p. 5-11

7) Lund, Th.   : Einfluß der Härtbarkeit des Grundwerkstoffs auf die Einhärtetiefe nach dem
                   Einsatzhärten, Ovako tech. Bericht 1/89

8) Vandewiele, B : Lange Termijn produktiebetrouwbaarheid in de warmtebehandeling, VWT-presentatie,
                   dec. 1990

Materials Science Forum Vols. 102 - 104 (1992) pp. 183-198

BENDING FATIGUE AND MICROSTRUCTURE OF GAS-CARBURIZED ALLOY STEELS

K.A. Erven*, D.K. Matlock and G. Krauss

Advanced Steel Processing and Products Research Center, Colorado School of Mines, Golden, CO 80401, USA

*now at The Timken Company, Canton, OH, USA

ABSTRACT

The bending fatigue performace of eight gas-carburized alloy steels was determined in specimens identically carburized at 927°C, quenched, and tempered at 150 C. The specimens were chemically polished prior to carburizing and tested in the as-carburized condition with a thin layer of surface oxidation. Relatively small variations in microstructure were measured by metallographic and x-ray techniques, and the endurance limits of the eight steels ranged from 1070 MPa to 1260 MPa. Manganese sulfide inclusions were shown to lower fatigue strength but no other systematic variations with microstructural parameters were identified. All fatigue cracks were initiated by intergranular surface cracks. The results are therefore considered to represent the upper end of the range of fatigue performance which can be expected from gas-carburized alloy steels in which bending fatigue cracks nucleate by cracking at prior austenite grain boundaries.

INTRODUCTION

A recent literature survey [1] of bending fatigue of carburized steels shows that measured fatigue limits range from 210 MPa (30 ksi) to 1950 MPa (210 ksi). There are many processing, testing, chemical composition, and microstructural variations responsible for this wide range of bending fatigue performance [2], but in many studies insufficient information is presented to separate the effects of the various factors on fatigue performance [1]. This study was designed to examine the effect of chemistry and microstructure on bending fatigue behavior of carburized steels. Carburized specimens from eight different chemical compositions, includng subsets of heats with systematic variations in sulfur and titanium content, were examined. All steels were carburized under identical processing conditions, and fatigue testing was performed on polished specimens with rounded corners. Therefore, processing variations were largely eliminated, and specimen variations such as surface roughness were minimized. As a result the effect of composition and microstructure on the fatigue performance of the eight carburized steels could be evaluated and compared.

EXPERIMENTAL PROCEDURE

Table 1 lists the chemical compositions of the eight steels. In addition to the elements listed, oxygen contents ranged from 8 to 11 ppm and nitrogen from 85 to 119 ppm. Sulfur content was systematically varied in the 8219 heats, from 0.006 to 0.029 wt pct, and titanium was varied in the 8719 heats, from 0.002 to 0.011 wt pct. The AISI/SAE 4320 and 8620 type steels were included in the test matrix in order to evaluate the fatigue performance of two widely used carburizing steels with different hardenabilities. All of the heats were ingot cast and rolled to billets between 140 and 152 mm in diameter.

TABLE 1 - Chemical Compositions of Test Heats (weight percent)

Steel Type	C	Mn	P	S	Si	Cr	Ni	Mo	Cu	Al	Ti
4320	.20	.61	.012	.018	.25	.56	1.77	.26	.19	.030	.003
8219	.20	1.40	.013	.006	.23	.61	.22	.21	.19	.024	.003
8219	.20	1.33	.011	.015	.21	.60	.31	.19	.24	.023	.002
8219	.20	1.43	.013	.029	.23	.62	.32	.18	.21	.030	.002
8620	.23	.85	.017	.022	.26	.57	.46	.21	.20	.026	.004
8719	.22	.87	.010	.023	.26	.53	.51	.21	.20	.023	.002
8719	.22	.89	.010	.019	.24	.55	.54	.21	.18	.027	.006
8719	.22	.87	.010	.020	.29	.55	.52	.20	.19	.029	.011

Figure 1 shows the geometry and dimensions of the specimens used for fatigue testing. The curved surface at the change in section thickness concentrates stress and simulates curvatures at the roots of gear teeth, and the rounded specimen edges reduce corner carbon build-up and associated variations in microstructure such as coarse carbide formation and excessive retained austenite [3]. The specimens were mechanically and chemically polished after machining. Fatigue testing of carburized specimens was conducted in a displacement-controlled fatigue machine operating at 31 to 32 Hz. Tension-tension loading with a load ratio (R) of 0.1 was applied.

Figure 2 shows the orientation in which the fatigue specimens were machined from the as-received bars. This orientation placed the long axes of elongated inclusions parallel to the curved location where maximum bending stress develops during testing.

Table 2 presents the processing schedule used to carburize the specimens. After the equalizing step, the specimens were quenched into oil at 66°C and tempered at 150°C.

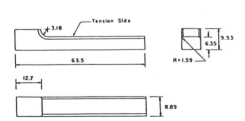

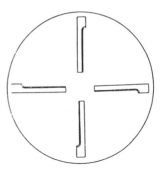

Fig. 1 - Schematic diagram of
specimen used to determine
fatigue behavior. Dimensions
in millimeters.

Fig. 2 - Specimen orientation
relative to cross sections of
as-received steel bars.

Table 2 - Carburizing Parameters

Description	Time (minutes)	Temperature °C	Temperature °F	Carbon Potential (percent)
Heat	18	921	1690	0
Carburize	135	927	1700	1.05
Diffuse	135	927	1700	0.82
Equalize	30	849	1560	0.82

Fracture surfaces were evaluated by scanning electron microscopy (SEM)
and metallographic specimens were prepared by standard polishing and etching
techniques. X-ray analysis was used to determine residual stresses and
retained austenite levels in the case microstructures of the carburized
specimens.

RESULTS

HARDNESS: Figure 3 shows the microhardness profiles measured from
carburized fatigue specimens and an 8620 test bar 16 mm in diameter. At any
given depth in the case of the carburized specimens hardness in the 8 heats
varied by 3 to 4 HRC points. Table 3 lists the near-surface and core hardness
of the various carburized specimens. The effective case depths, determined at
the distance from the surface where hardness was 513 HV (50 HRC), are also
listed. The interior case hardness of the 8620 test bar was lower than that
of the interior case hardnesses of the fatigue specimens. This observation
may be due to the relatively low hardenability of the 8620, slower cooling in
the bar and the resulting reduced transformation to martensite in the heavier
bar section relative to the smaller test specimens.

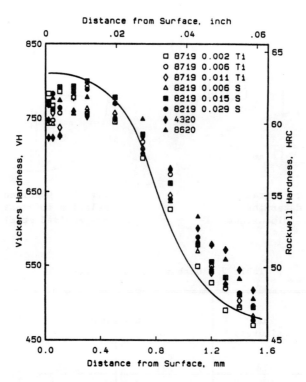

Fig. 3 - Microhardness profiles from test bar (solid curve) and fatigue specimens (data points).

Table 3 - Summary of Carburized Case Properties in the 8620 Test Bar and the Fatigue Samples

Sample	Hardness 0.02 mm from surface		Effective Case Depth		Core Hardness	
	HV	HRC	mm	inch	HV	HRC
Test Bar	809	64.3	1.22	0.048	431	43.7
4320	742	61.8	1.32	0.052	472	47.1
8219 - 0.006 S	767	62.8	1.38	0.054	458	46.0
8219 - 0.015 S	747	62.0	1.33	0.053	444	44.8
8219 - 0.029 S	723	61.1	1.53	0.060	451	45.4
8620	747	62.0	1.45	0.057	468	46.8
8719 - 0.002 Ti	783	63.4	1.25	0.049	460	46.2
8719 - 0.006 Ti	778	63.2	1.40	0.055	460	46.2
8719 - 0.011 Ti	747	62.0	1.35	0.053	460	46.2

MICROSTRUCTURE: As-polished surfaces were used to evaluate the
inclusion distribution and surface oxidation of the steel. Figure 4 shows
typical inclusion distributions in the 4320 and 8620 steels. Most of the
inclusions, as identified by SEM x-ray energy dispersive spectroscopy, were
manganese sulfide particles, and were present in amounts proportional to the
sulfur content of the steels. Few globular oxides were observed, consistent
with the low oxygen content of the steels.

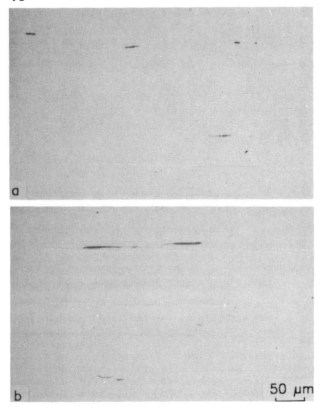

Fig. 4 - Micrographs of inclusions in steel types a) 4320 and
b) 8620, 200X.

Surface oxidation, typical of gas-carburized steels [4], was observed in
all specimens. Figure 5(a) shows a typical depth of surface oxidation, and
Figure 5(b) shows nontypical, extreme penetration of surface oxidation.
Surface oxidation has been found to occur in two zones: a shallow zone, about
5 μm deep, consisting of intragranular, chromium rich oxides, and a deeper
zone consisting of intergranular lamellar oxides rich in Mn and Si [5]. The
fatigue specimens in the present investigation were tested in the as-
carburized condition (no surface conditioning was done after carburizing) and
SEM examination of fracture specimens consistently showed a thin, 5 μm deep,
transgranular fracture zone, and occasional lamellar oxides along prior
austenite grain boundaries. Other investigations have shown that shallow
surface oxide zones have relatively little effect on fatigue performance,
provided that nonmartensitic surface structures do not form as a result of
alloying element depletion by alloy oxide formation [6,7].

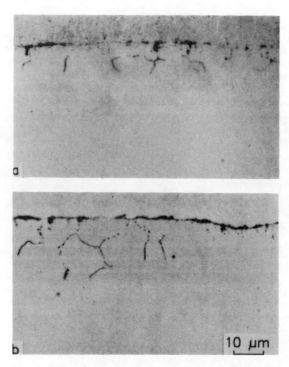

Fig. 5 - Surface oxidation of gas-carburized specimens of
a) 8719-low Ti steel, and b) 4320 steel.

Figure 6 and 7 show examples of the case and core microstructures,
respectively. The case microstructures consist of mixtures of plate
martensite and retained austenite and the core microstructures consist
exclusively of lath martensite. The size distribution of the martensite
crystals is controlled by the grain size of the parent austenite, and the
retained austenite content is determined by chemistry, with compositions which
give lower M_s temperatures producing higher amounts of austenite at a given
level of carbon. Table 4 lists austenite grain sizes measured in the cores
and surface retained austenite contents of carburized specimens from the eight
heats of steel. Retained austenite content increased to maximum values
between 31 and 39 pct at a distance of about 0.06 mm from the surface of the
specimens.

The surface residual stresses of all of the carburized specimens were
compressive, and were between 100 and 200 MPa at the surface. The compressive
stresses increased to subsurface maxima of 120 MPa for the 8620 steel and
between 220 and 300 MPa for the other steels.

FATIGUE PERFORMANCE: The fatigue performance of the eight carburized
steels was evaluated by determining the number of cycles of stress required to
cause failure at various stress levels. The applied maximum stress at which
specimens did not fail after 2 million cycles was used to establish the
fatigue limits of the various sets of carburized specimens. Figures 8, 9, and
10 show S-N curves for carburized 4320, 8620, and 8719/high Ti steels,
respectively. The open circles in the S-N plots represent data obtained from
specimens which SEM showed had fatigue cracks initiated at poorly rounded
specimen corners. The discontinuous shape changes at these corner sections
were felt to raise stress relative to smoothly rounded corners, and therefore
these data points were disregarded in estimating fatigue limits.

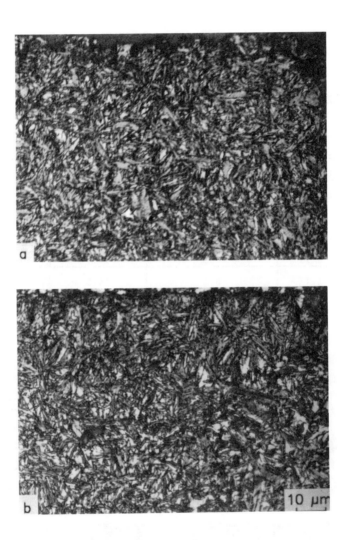

Fig. 6 - Representative micrographs of case microstructures from steel types a) 8219 with low sulfur, and b) 4320, 1000X.

Table 4 - Austenite Content and Grain Sizes of Test Heats

Steel Type	ASTM Grain Size	Grain Size (microns)	Percent Retained Austenite
4320	8.4	17.5	26.6
8219 - 0.006 S	7.8	22.0	22.7
8219 - 0.015 S	7.3	25.5	21.7
8219 - 0.029 S	7.1	27.0	19.9
8620	8.9	14.5	24.1
8719 - 0.002 Ti	7.9	20.5	19.7
8719 - 0.006 Ti	8.5	16.5	19.0
8719 - 0.011 Ti	8.5	16.5	19.8

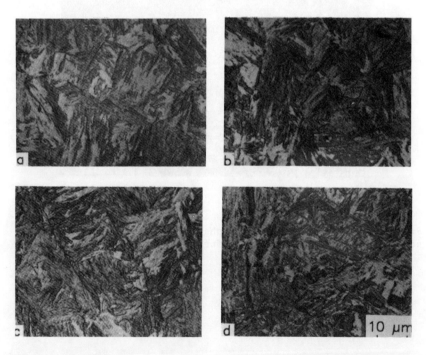

Fig. 7 - Representative micrographs of core microstructures from steel types a) 8719 with high titanium, b) 8219 with medium sulfur, c) 4320, and d) 8620, 500X.

Table 5 lists the fatigue limits determined for the carburized specimens
from the eight heats of steel. The range of measured fatigue limits for the
eight steels is relatively small, about 190 MPa (28 ksi). Thus the effects of
chemistry and microstructure are relatively small within the limits identified
in this investigation. Another general observation is that the reported
fatigue limits range well above many of those reported previously [1]. These
results show that good fatigue performance can be obtained in specimens
direct-quenched after carburizing and with thin layers of surface oxidation.

Table 5 - Endurance Limits of Test Heats from
Modified and Original S-N Curves

Steel Type	Endurance Limit	
	MPa	ksi
4320	1230	178
8219 - 0.006 S	1260	183
8219 - 0.015 S	1200	174
8219 - 0.029 S	1070	155
8620	1100	160
8719 - 0.002 Ti	1170	170
8719 - 0.006 Ti	1170	170
8719 - 0.011 Ti	1170	170

A major factor which contributed to variations in fatigue performance in
this investigation was sulfur content. Increasing sulfur content translated
to increases in MnS inclusions which in turn translate into lower fatigue
limits. This effect is clearly shown in the results for the 8219 steel series
with systematic variations in sulfur content. Higher sulfur contents
increased the probability that elongated MnS particles would contribute to
fatigue crack initiation at lower applied cyclic stresses [8].

FRACTOGRAPHY: Figure 11 shows macrographs of the fracture surfaces of
carburized 8219 specimens with three levels of sulfur and a carburized 4320
specimen. The case fracture zone is largely intergranular, typical of the
fracture associated with phosphorus segregation and cementite formation at
prior austenite grain boundaries in high carbon steels [9]. The core fracture
is ductile, and characterized by fibrous fracture associated with elongated
MnS inclusion particles. The core fractures of the 4320 steel with 0.018 pct
S and the 8219 steel with 0.015 wt pct S look very similar. As sulfur content
decreases, the aligned, fibrous fracture appearance of the core diminishes.
The MnS particles have no influence on the overload case fracture surfaces
which are dominated by intergranular fracture. The fatigue crack initiation
and growth sites are located on the rounded-corner side of the specimen, but
are too small to be readily identified in the macrographs of Figure 11.

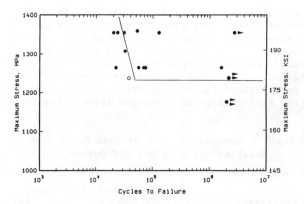

Fig. 8 - Fatigue data and S-N curve for steel type 4320.

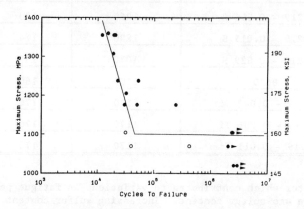

Fig. 9 - Fatigue data and S-N curve for steel type 8620.

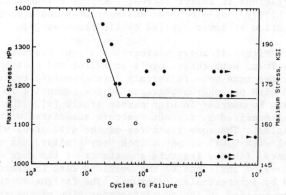

Fig. 10 - Fatigue data and S-N curve for steel type 8719 with 0.011 weight percent titanium.

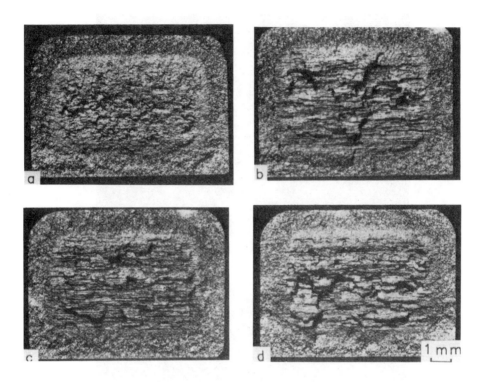

Fig. 11 - Representative fracture surfaces from steel type 8219
with a) low, b) medium, and c) high sulfur, and d) steel type
4320, 7.5 X.

Figures 12, 13, and 14 show details of fatigue fracture initiation and
growth. Figure 12(a) shows the typical progression of bending fatigue
fractures observed in the carburized steels of this investigation. Fracture
initiates by intergranular fracture, propagates by transgranular fracture, and
then becomes unstable, reverting to the intergranular overload fracture of
high carbon hardened steels. Eventually the overload crack enters the
specimen core and assumes the ductile, fibrous topology shown in Figure 11.
The flat transgranular region of stable fatigue crack growth is quite small
because of the low fracture toughness of the high carbon martensitic case [9].

Figures 12(b), 13, and 14 show higher magnification views of the
intergranular crack initiation and the transition to the transgranular crack
growth morphology. The 5 μm surface oxidized zone is visible in most of the
fractographs, but the intergranular crack zone is always significantly deeper.
Intergranular cracking was always associated with initiation, even when
inclusion particles were associated with initiation, as shown in Figure 14.
The presence of large inclusions close to the highly stressed surfaces of
fatigue specimens is believed to concentrate stress and lower the macroscopic
applied stress for intergranular fatigue nucleation [8].

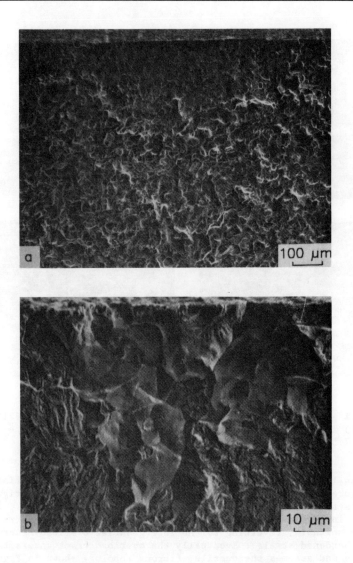

Fig. 12 - Fractographs of crack initiation sites in a specimen
of steel type 8219 with 0.006 weight percent sulfur at a) 100 X
and b) 1000 X.

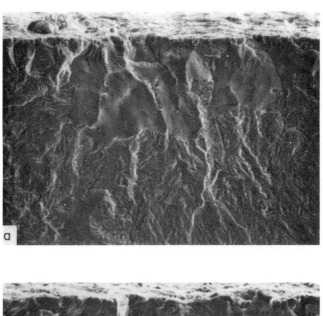

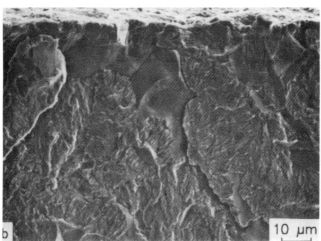

Fig. 13 - Intergranular crack initiation sites in carburized 4320 steel.
1000X.

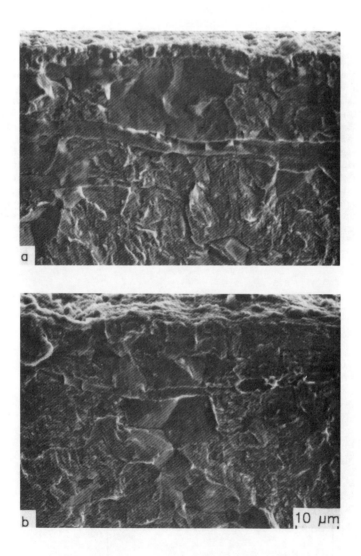

Fig. 14 - Fatigue crack initiation sites associated with inclusions in specimens of steel type 8620 at a) 1500 X and b) 2000 X.

DISCUSSION

Pacheco [7] recently showed that the bending fatigue of 8719 carburized
steel improved directly with decreasing austenitic grain size, and that quite
high levels of endurance limits, above 1400 MPa (203 ksi) could be achieved in
specimens with very fine austenite grain sizes. In order to compare the
results of this investigation with that of Pacheco's, endurance limits for all
of the carburized steels are plotted as a function of grain size in Figure 15.
Pacheco's results are identified by A, B, C, and D, and the results of this
investigation are identified by other letters as shown. Clearly the results
of this investigation do not directly correlate with grain size in the range
measured in this investigation, and other factors must be contributing to
differences in fatigue performance. One of the factors which has been
identified in this investigation is inclusion content. Nevertheless, the
range of fatigue limits for all of the alloys examined here is quite small,
and reflects the effect of small changes in grain size, chemistry, residual
stresses, and retained austenite on fatigue performance for constant
processing conditions.

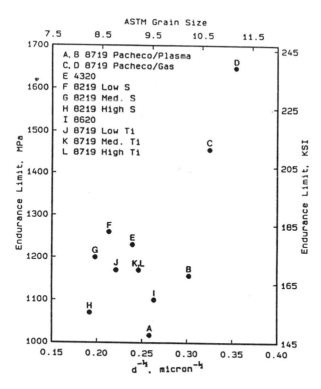

Fig. 15 - Endurance limit versus austenite grain size from
Pacheco study and current study.

The endurance limits identified by points C and D in Figure 15 were obtained from very fine-grained specimens with relatively low retained austenite contents. Fatigue initiation in these specimens occurred by transgranular fracture, not the intergranular mode noted exclusively in the specimens of this investigation. It appears that the carburizing of the steels in this study did not produce grain sizes fine enough to effect the transition from intergranular to transgranular crack initiation.

SUMMARY

Despite the intergranular mode of fatigue crack initiation, the endurance limits measured in this investigation are quite good, relative to those reported in a number of previous studies [1]. The endurance limits show little effects of nominal compositional and microstructural variations. The variations included austenitic grain sizes between ASTM No. 7 and 9, surface retained austenite contents between 19 and 27 vol pct, sulfur contents between 0.006 and 0.029 wt pct, surface residual stresses between 100 and 200 Mpa, and small differences in the contents of Ni, Mo, and Cr. Thus the results of this study define the range of high-cycle fatigue performance which would be expected in carburized microstructures typical of fine-grained alloy steels, direct quenched after carburizing and cyclically loaded without removal of a thin layer of surface oxidation.

ACKNOWLEDGEMENTS: The authors gratefully acknowledge the help of a number of individuals and corporations. Steve Donelson, Colorado School of Mines, for machining the fatigue specimens, Jon L. Dossett of Midland Metal Treating, Inc. for carburizing the specimens, The Timken Company for providing the steels and x-ray analysis, and Caterpillar Inc., for residual stress and retained austenite measurements. This research was supported by the Advanced Steel Processing and Products Research Center at the Colorado School of Mines, an NSF Industry/University Cooperative Research Center.

REFERENCES

1.   Cohen, R.E., Haagensen, P.J., Matlock, D.K., and Krauss, G.: SAE Technical Paper Number 910140, 1991, 11 pages.
2.   Parrish, G.: The Influence of Microstructure on the Properties of Case-Carburized Components, ASM International, Materials Park, Ohio, 1980.
3.   Jones, K.D., and Krauss, G.: Heat Treatment '79, The Metals Society, London, England, 1980, 188.
4.   Chatterjee-Fischer, R.: Met. Trans. A, 1978, 9A, 1553.
5.   Van Thyne, C., and Krauss, G.: in Carburizing: Processing and Performance, ASM International, 1989, 333.
6.   Cameron, T.B., Diesburg, D.E., and Kim, C.: Journal of Metals, 1983, 35, 37.
7.   Pacheco, J.L., and Krauss, G.: in Carburizing: Processing and Performance, ASM International, 1989, 227.
8.   Erven, K.A., Matlock, D.K., and Krauss, G.: J. Heat Treating, 9, 1991.
9.   Krauss, G.: in Case Hardened Steels: Microstructure and Residual Stress Effects, TMS, Warrendale, PA, 1984, 33.

Materials Science Forum Vols. 102 - 104 (1992) pp. 199-210

INFLUENCE OF SURFACE MICROSTRUCTURE ON BENDING FATIGUE STRENGTH OF CARBURIZED STEELS

A. Melander and S. Preston

Swedish Institute for Metals Research, Stockholm, Sweden

ABSTRACT

An investigation has been undertaken to assess the influence of the near surface microstructure, after gas carburising, on the bending fatigue strength of two standard carburising steel grades i.e.SS2127 and SS2511. The first steel has a lower nickel content and increased amounts of manganese and chromium compared to the second steel. For the SS2127 steel, non-martensitic transformation products were observed at the surface after carburising. This is shown to be due to a reduction in surface hardenability, caused by the depletion of certain alloying elements from the steel matrix which have a higher affinity for oxygen than iron. The maximum depth of internally oxidised grain boundary was also found to vary between the steels which were examined.

A difference in the bending fatigue strength was obtained for the two steel types. An explanation is given in terms of the different surface microstructures and the crack-tip stress intensity of cracks starting from the surface. It is proposed that the fatigue limit is controlled by the growth or permanent arrest of cracks from the oxidised grain boundaries.

INTRODUCTION

Ever increasing demands of carburised steel components are being required. This is achieved by optimising factors such as the hardenability, microstructure, residual stresses and steel cleaness. However, the presence of an internally oxidised layer at the steel surface is one factor which may result in a deterioration of fatigue properties. Removal of this layer by grinding will cause an improvement but at additional expense.

Internal oxidation arises because of the presence of water vapour and carbon dioxide in the carburising atmosphere. Normally the atmosphere is reducing to iron, but is oxidising to those elements with a greater affinity to oxygen, e.g. Si, Mn and Cr. The oxides are usually at the austenite grain boundaries. Both the quantity and type of oxide, as well as the depth of oxidation, are dependent upon the element concentration within the steel composition and also, the reaction temperature and time. Chromium oxides form nearest to the surface and manganese oxides to a greater depth, although not as deep as silicon oxides,[1].

The effect of Si, Cr and Mn on the depth of oxidation has recently been studied [2]. This work found a linear relationship between the depth and the composition parameter (10Si+Mn+Cr) which indicates a large influence of Si. Since an oxidised boundary may initiate fatigue failure [2,3], then it is reasonable to assume that improved fatigue properties can be obtained by decreasing the depth of oxidation, i.e. Si content. Another effect of internal oxidation is the depletion of alloying elements from the surrounding matrix. This may lead to a local decrease in hardenability, presence of non-martensitic transformation phases and tensile residual stresses at the carburised surface, [4].

In this investigation a detailed study using large scale heats has been undertaken to assess the influence of the internally oxidised layer on the fatigue properties of two steel types i.e. SS2127 and SS2511. A model has been proposed to relate the conditions existing at the carburised surface to the fatigue properties.

MATERIALS AND HEAT TREATMENT.

Three standard production carburising steel heats have been studied in this work of which two were the steel grade SS2127 and the third was the steel grade SS2511. Compositions of the heats are given in Table.1 below. All the steels were produced by electric arc melting, ladle refined and then, the cast ingots were subsequently hot worked to a bar of final diameter ~40mm.

STEEL	C	Si	Mn	P	S	Cr	Mo	Ni	Al
A	.16	.23	1.18	.007	.045	.98	.03	.15	.020
B	.17	.11	1.32	.016	.029	1.25	.07	.19	.031
C	.15	.26	1.02	.008	.039	.74	.06	.90	.026

Table.1. Compositions of the three steels in wt%.

Steels A and B (SS2127) were selected primarily on the basis of the difference in the silicon content. However, it may also be seen that steel B has higher levels of Mn and Cr which may also influence the internal oxidation behaviour. Steel C was chosen because of the increased Ni content. This element should not oxidise and therefore, surface hardenability should be maintained.

From each material, 25 rotating bend fatigue bars of diameter 10mm were machined from the centre of the as-received bars. The fatigue bars were ground and mechanically polished to a 1μm finish. All the test bars were treated in the same batch and carburised to have a surface carbon content of ~0.8wt% and a case depth of 1mm. After carburising, all these samples were then tempered at 180°C for 1 hour. Details of the test bar geometry and the carburising cycle have been reported elsewhere,[5].

EXPERIMENTAL METHOD.

Rotating bend fatigue testing was carried out for each steel, in the as-carburised condition, at a frequency of 3000rpm in laboratory air. The fatigue limit, defined at 10^7 cycles, was obtained using the stair-case method. Testing and evaluation were in accordance with ASTM standard E468-85 and ASTM STP 588. A selection of the fatigue fracture surfaces were examined using a JEOL JSM 50A scanning electron microscope (SEM).

For each carburised steel, a transverse section was taken from a fatigue bar and prepared by standard metallographic techniques for optical microscopy. These samples were etched in a 2vol% Nital solution for examination of the hardness and microstructural gradients below the carburised surface. Microhardness profiles were measured through the case, with a test load of 1kg, and the effective case depth was determined at a hardness of 550HV. A linear intercept method was used to determine the austenite grain size from these samples after etching in picral.

Using these polished and etched sections, the near surface microstructure was examined with the SEM. The concentrations of Si, Mn and Cr were measured in this region, at intervals of 1μm, down to a depth of 40μm. This analysis was made at a distance away from the oxidised grain boundaries using an Electron Probe Microanalyser, ARL-SEMQ.

The amount of retained austenite at the carburised surface was
determined by X-ray diffraction using Co radiation.   In order to
eliminate any effect of, for example, internal oxidation or
decarburisation, a surface layer of   20um was removed by
electropolishing in a perchloric/acetic acid solution.   The
intensity of the X-rays diffracted from the (200),(211)
martensite/ferrite and (200),(220) austenite peaks was
evaluated.   An integrated intensity, direct comparison method
was used to determine the austenite content,[6].   For each
steel, residual stresses were determined using the d versus $\sin^2 \psi$
method at depths of 0,10,20, 40,80 and 200µm below the
carburised surface.   Two measurements were taken from a fatigue
bar in the longitudinal direction at each depth.   These
measurements were made by X-ray diffraction using Cr radiation
and the (211) plane from martensite/ferrite.   The material was
removed to the required depth by electropolishing in a chromic
acid solution at a temperature of 50°C.

EXPERIMENTAL RESULTS

Microprobe analysis of the elements Si, Mn and Cr to a depth of
40µm is shown in figure 1.   These results indicate that element
depletion has occurred to a depth of 30-40µm.   A much larger
decrease in the Mn and Cr content within the surface region may
be seen for steels A and B compared to steel C.

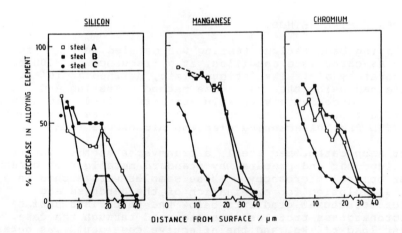

Fig.1.Decrease in the Si, Mn and Cr contents to a depth of
       40µm below the carburised surface for steels A, B and C.

Microstructures perpendicular to the carburised surface for
steels A,B and C are shown in figure 2.   The SS2127 steels had a
mixture of bainite and pearlite at the surface whereas steel
SS2511 had a mixture of martensite and retained austenite.   This

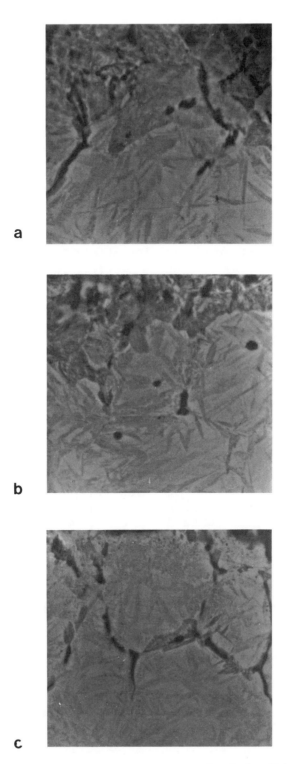

Fig.2. Surface microstructures for (a)steel A, (b)steel B and
(c)steel C.   (Magn. x2000)

latter steel did exhibit small areas of non-martensitic phases
but only close to the oxidised boundary. Microhardness
measurements taken a depth of 5-10μm revealed that, for steels A
and B, the value was 500HV and for steel C it was 670HV.

A summary of the carburised parameters of each steel is given in
Table.2.

STEEL	Mean grain size μm	Retain. austen. %	Effect. case depth mm	Hardness (HV1) Case(.2mm)Core(5mm)	
A	16	25	0.98	680	343
B	16	21	1.00	675	383
C	21	25	0.95	720	367

Table.2. Carburised parameters for steels A,B and C.

The reason for the lower austenite content in steel B may be due
to differences in the amount of surface material removed by
electropolishing. Although the values are similar for the two
steel types, it was found that the depth of austenite was
greater for steels A and B compared to steel C. There was no
evidence of any abnormal grain growth in any steel, but it was
noticed that steel C had a slightly larger mean grain size.
Hardness profiles through the carburised case are shown in
figure 3. It may be seen that steel C has a higher surface
hardness which would agree with the difference in the depth of
retained austenite. Steels A and B have a similar case hardness
but, steel A has a lower core hardness which is consistent with
the lower levels of alloying elements. In all three steels the
the core microstructure consisted of a mixture of martensite and
and bainite (<10%).

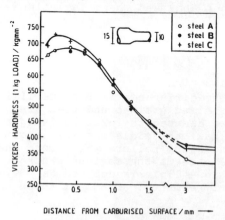

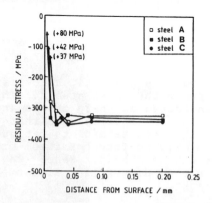

Fig.3.Hardness profiles through
     the carburised case for
     steels A,B and C

Fig.4.Residual stress profiles
     to a depth of 0.2mm for
     steels A,B and C

Figure 4 shows the residual stress profiles in the longitudinal direction for the three steels. It may be seen that tensile residual stresses exist at the surface and for depths greater than 10µm, compressive stresses are present. This figure indicates that the difference in residual stresses is rather small between the three steels.

Table.3 shows the mean stress amplitude and 95% confidence limits of the fatigue limit for each steel. These results show that the lower Si containing steel has the highest fatigue limit and also, that steel grade SS2127 has a higher value than the grade SS2511. This result reflects the differences in surface microstructure and depth of internal oxidation as discussed in the following sections.

STEEL	Mean fatigue limit MPa	95% Conf. limits for 10^7 cycles. MPa
A	991	971 - 1011
B	1065	1056 - 1074
C	890	868 - 912

Table.3. Rotating bend fatigue properties for steels A,B and C.

Examination of the fatigue fracture surfaces revealed that initiation had occurred at the surface and is most likely due to the oxidised layer. The fatigue crack had initially propagated in a transgranular mode which gradually became a mixed intergranular/transgranular mode. A slightly greater amount of intergranular failure was observed for the SS2127 steels. In the final fracture area a ductile dimpled fracture was present. A small number of failures (i.e <2%) were found to have been caused by inclusion clusters. When analysed these inclusions were found to consist of Ca, Al, Si and/or Mn and S.

MODEL FOR FATIGUE FAILURE.

A model is presented to explain the difference in fatigue limit between the three steels. The oxidised grain boundaries are considered as sharp cracks and are assumed to initiate fatigue failure. It is assumed that the maximum oxidised crack length is the most harmful to fatigue and also, failure will occur when the driving force of these cracks exceeds the fatigue crack growth threshold. Hence, the fatigue limit is defined as corresponding to the situation for permanent arrest of these cracks.

Based on the Kitagawa diagram [7], it may be shown that for the oxidised grain boundary cracks in a martensitic microstructure, the transition from short to long crack behaviour occurs for crack lengths >10µm. This value is also reasonable for a high strength bainitic microstructure. Hence, linear elastic fracture mechanics have been used to model long crack behaviour. It is concerned with the stress intensity of cracks propagating from the oxide boundaries and their relation to the crack growth threshold.

The stress intensity of a crack is sensitive to the exact stress distribution over the crack faces. Below the carburised surface a combination of changing applied and residual stresses exist and therefore, it is necessary to calculate the stress intensity with respect to the locally changing stress field. A method developed for a central crack exposed to a locally changing stress field [8] has been adopted in this work to evaluate the stress intensity of a surface crack. The stress intensity over the crack length can be written as

$$K = \sqrt{\pi \cdot a} \cdot \frac{2}{\pi} \cdot \int_0^a \frac{(\sigma_y^{bend} + \sigma_y^{res})}{\sqrt{a^2 - y^2}} \cdot dy \qquad \text{------(1)}$$

where a is the crack length and y is the point at which the stress is acting. Stress distribution due to the applied load consists of a linearly varying axial bending stress which is zero at the specimen axis, and a stress magnification at the notch radius. Using expressions given in the literature [9,10], the total axial stress as a function of the distance from the specimen surface (X) is given by the following expressions;

$$\sigma^{bend} = \frac{R-X}{R} \cdot \sigma_a \cdot \sqrt{\frac{r}{r+4X}} \qquad \text{for } 0 < X < \frac{r}{4} \cdot (K_T^2 - 1) \qquad \text{-----(2)}$$

$$\sigma^{bend} = \frac{R-X}{R} \cdot \frac{\sigma_a}{K_T^2} \qquad \text{for } \frac{r}{4} \cdot (K_T^2 - 1) < X < R \qquad \text{-----(3)}$$

where R (=5mm) is the specimen radius, r (=2.1mm) is the notch radius and σ_a is the axial stress at the notch surface.

The residual stress pattern has been represented by three constant stress levels within three different intervals from the surface as given in Table.4. Values σ_1^{res} and σ_2^{res} were based on experimental measurements, but σ_3^{res} was calculated by assuming force equlibrium across the test bar. It is appreciated that this residual stress profile is a simplified representation.

In this model we only consider the value of K_{max}, because fatigue crack growth occurs for positive stress intensities. Values of K_{max}, calculated using equation (1), are plotted in figure 5 against the crack length (a) for steel A loaded at the experimentally determined fatigue limit.

	STEEL A	STEEL B	STEEL C
Surface residual stress, σ_1^{res}	+80MPa	+80MPa	+46MPa
Residual stress depth, X_1	7um	7um	3um
Case residual stress, σ_2^{res}	-300MPa	-300MPa	-300MPa
Residual stress depth, X_{cd}	1mm	1mm	1mm
Core residual stress, σ_3^{res}	+169MPa	+169MPa	+169MPa

Table.4.Values for the residual stresses and distances used
in the calculations.

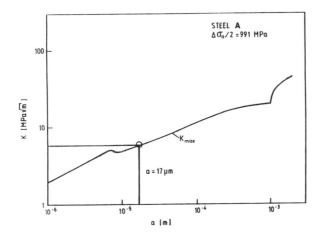

Fig.5 Positive part of stress intensity range versus crack
depth for steel A at the experimentally determined
fatigue limit.

This curve increases initially at a steady rate with crack length
until it makes a sudden drop, which is caused by the change from
σ_1^{res} to σ_2^{res}. A further increase in K_{max} then occurs, but at a
gradually decreasing rate due to the decreasing bending stress.
At the case depth the value of K_{max} suddenly increases because of
the change to tensile from compressive residual stress. The ring
on the curve corresponds to the depth of oxidised boundary and
these boundaries coincide with a K_{max} value of 5.7MPa$\sqrt{m}$. This
value is equivalent to the threshold stress intensity, K_{max}^{th}, if
the crack growth threshold controls the fatigue limit.

The same procedure discussed above was carried out for steels B
and C and the corresponding values of K_{max}^{th} were 5.6 and 4.5MPa$\sqrt{m}$
respectively. It is not surprising that steels A and B should
have a similar value in view of the fact that they had similar
surface microstructures, although different depths of oxidation.
Furthermore, the lower value for steel C may be expected based

On values reported elsewhere for martensite and bainite microstructures, [11]. For a 1%C-Cr bearing steel the threshold stress intensity of a martensitic structure was $\sim$3MPa$\sqrt{m}$ and $\sim$8MPa$\sqrt{m}$ in the case of a bainitic microstructure.

It is possible from this work to estimate the effect of different oxide depths on the change in the fatigue limit. Consider steel C which had an oxide depth of 17μm, but as a result of a lower silicon level this value was reduced to 12μm. The value of K_{max} at this new crack length is 3.9MPa$\sqrt{m}$, [12]. By approximating the bending stress as being constant over the crack length, an estimate of the increase in the fatigue limit can be made using this difference in K_{max} from the following expression

$$\Delta K_{max} = \Delta \sigma_a . \sqrt{\pi a}$$

---------(4)

In this work the experimentally determined fatigue limit value was 890MPa and the estimated increase from equation (4) is 98MPa. Therefore, if this steel had contained a lower silicon content of 0.1wt%, then we would have expected the fatigue limit to be higher at 990MPa.

DISCUSSION

The main function of silicon is to act as a deoxidising agent during steelmaking. In carburising steel compositions the amount of silicon is typically between 0.2 and 0.3wt%. At these levels no particularly adverse effect upon the mechanical properties has been observed. This fact, together with the relative cheapness, has meant that there was no reason to investigate whether this silicon content is the optimum. With modern steelmaking practice it is quite easy to reduce the silicon level in the steel composition.

It would now appear that the intenally oxidised layer can have a harmful effect upon the properties. The depth of oxidised grain boundaries which act as a source of fatigue failure, are strongly related to the silicon content,[2]. This factor as well as the presence of non-martensitic transformation products, must be considered when assessing the effect of the oxidised layer. The problem of internal oxidation is well understood [13] and arises because of the presence of
    (a)water vapour and carbon dioxide in the carburising
         atmosphere and
    (b)elements in solid solution having a greater affinity for
         oxygen than iron.
Three elements in the steel composition that are of particular concern are Si, Mn and Cr.

In [2] it was shown that as the silicon content decreased and hence the depth of oxidation, there was an increase in the fatigue limit. This is consistent with the present work. However, two observations in the present investigation are in disagreement with this reference. Firstly, non-martensitic transformation products were found at the carburised surface.

An explanation for their presence may be due to a more oxidising
atmosphere. The second anomaly was that for the same depth of
oxidation a higher nickel containing steel had a lower fatigue
strength. This could be explained in terms of the different
surface microstructures, i.e. a difference in the fatigue
threshold value.

The presence of non-martensitic phases is attributed to a local
reduction in hardenability caused by a decrease in elements
(e.g.Mn and Cr) from solid solution because of oxidation.
Another effect associated with these phases is the presence of
tensile residual stresses at the surface. It would appear that
both Mn and Cr have a dual role of oxide formation and
influencing the surface hardenability. In this work, the depth
of depletion of these two elements for the two steel types was
the same, i.e. 40μm. Although the amount of depletion was lower
for steel C which had a lower concentration of these elements.
The presence of Ni in this steel which should not oxidise, has
maintained the surface hardenability such that martensite
existed at the carburised surface. Molybdenum is another
element that should prevent any decrease in the surface
hardenability. It was stated in [14], that by adding nitrogen
to the steel surface at the end of the carburising cycle was a
more effective method of maintaining surface hardenability than
raising the carbon potential. However, this treatment will not
influence the internal oxides which have already formed.

CONCLUSIONS

1.The effect of reducing the silicon content to ~0.1wt% (Steels A
  and B) was a decrease in the depth of internal oxidation and a
  7% increase in the fatigue limit. Steel grade SS2127 had a
  higher fatigue limit than SS2511 at the same Si content and
  depth of oxidation.

2.For the SS2127 steels the surface microstructure consisted of
  pearlite and bainite whereas SS2511 was predominantly
  martensite and retained austenite. Depletion of Mn and Cr
  from the the matrix was found at depths up to 30-40μm and was
  worse for the SS2127 steels.

3.Critical fatigue cracks were expected to start propagating
  from the oxidised boundaries. The growth or permanent arrest
  of these cracks was assumed to control the fatigue limit.
  Stress intensities for these cracks have been calculated.
  The threshold was found to be 5.6 MPa$\sqrt{m}$ for the steel SS2127
  and 4.5MPa$\sqrt{m}$ for steel SS2511. In steel SS2127 where the
  threshold is constant, a shorter oxide depth allows for an
  increased fatigue limit.

ACKNOWLEDGEMENT

The work reported in this paper has been financed by the Swedish National Board for Technical Development and the Swedish Steel and Engineering Industry. The provision of materials and use of heat treatment facilities by the industry is gratefully acknowledged.

REFERENCES.

1) Chatterjee-Fischer, R.:Met. Trans. A., 9A, Nov., 1978, p1553.

2) Namiki, K. and Isokawa, K.:Trans. ISIJ., 26, Nov., 1986, p642.

3) Preston, S.:Mat. Sci. Techn., 7, 2, 1991, p105.

4) Hildenwall, B. and Ericsson, T.:J. Heat. Treat., 1, 3, 1980, p3.

5) Preston, S.:J. Heat Treat., 8, 2, 1990, p93.

6) Cullity, B. D.:'Elements of X-ray Diffraction', p391, Addison -Wesley, 1959.

7) Yamada, K, Kim, M. G. and Kunio, T.:'The Behaviour of Short Fatigue Cracks', Eds. Miller, K.J. and De Los Rios, E. R., Mech. Eng. publ., London, 1986, p261.

8) Albrecht, P. and Yamada, K.:ASCE Structural Division Journal, 103, 1977, p377.

9) Neuber, H.:'Theory of Notches', transl. F.A. Raven, Edwards Ann Arbor, MI, 1946.

10)Hammonds, M. M, Smith, R. A. and Miller, K. J.:'Fat. Eng. Mat. Struc.', 2, 1979, p139.

11)Beswick, J.:Met. Trans. A., 20A, Oct., 1989, p1961.

12)Preston, S. and Melander, A.:Swedish Institute for Metals Research, 1989, Report No. IM-2491.

13)Parrish, G.;Heat Treat. Metals., (part.1), 1, 1976, p6.

14)Gunnarsson, S.:Met.Treat.& Drop Forging, 30, 213, 1963, p219.

Materials Science Forum Vols. 102 - 104 (1992) pp. 211-222

CASE HARDENING IN THE NEXT CENTRUY

F. Kühn (a) and B. Lineberg (b)

(a) LOI ESSEN Industrieofenanlagen GmbH, D-4300 Essen 1, FRG
(b) Lineberg Consult AG, Sweden

ABSTRACT

Almost 40 years of experience from heat-treating of transmission components for cars and trucks is the base for demands towards the future methods and equipments. The end customers ask for reliable, efficient, safe, comfortable and "low cost" vehicles. The societies call for environmental positive solutions. All the shareholders want profit. The competition conditions will limit R&D resources, reduce open-minded exchange of experience and make small production units impossible.

The heat-treatment has a very big influence on the total cost - up to 20% - in spite of its own cost being only 6 to 9% of the manufacturing costs.

Optimizing heat-treatment, choice of material, design and co-operation between research laboratories, furnace manufactures and the production personal, will be a very important "tool" to get sufficient total efficiency for survival.

Bearing these facts in mind the authors give a survey of possible solution and point out areas where R&D is important to solve problems, which today are hindering practical and cost-efficient methods. Understanding of the importance of the total concept and avoiding suboptimizing on all decision levels will be shown as a competition weapon.

The authors are a combination of a heat-treater and a furnace manufacturer. The paper is partly based on enquiries made among colleagues in Europe and a study of papers presented in the period 1980 -1989.

INTRODUCTION

Since 1950, I have been working on the heat treatment of gearbox
components for automobiles and commercial vehicles. The experience
acquired during this period has been extrapolated to the requirements
of future processes and equipment. Feasible, practical solutions have
been developed from my ideas by my co-author. Reference is also made
to the development work which still remains necessary.

These requirements set high standards as demands, are often in con-
flict. Users demand reliable, efficient, safe, comfortable and low-
cost vehicles. Society calls for technologies and for plants compatible with environmental concerns. Employees strive for well-paid and
interesting jobs at factories where conditions of work are safe and
healthy. Shareholders are interested in profits.

At the same time, competition limits research and development
resources, restrict an open-minded exchange of views and experiences
between engineers and makes small production units impossible.

In this competitive environment, heat-treatment has a major influence
on total cost - which may be as high as 20 % - although the heat-
treatment operation itself only accounts for 5 to 9 % of the manufacturing cost.

Under these conditions, the optimizing of heat-treatment operations,
components design and choice of material and cooperation between
researchers, furnace manufacturers and automobile production
engineers will be critical to enhance productivity and efficiency
sufficiently for survival on a competitive market.

Against this backdrop, the paper reviews new trends and pinpoints
areas in which research and development work is required to overcome
problems and limitations which still prevent the application of more
practicable and more costefficient techniques. It shows that it is
crucial to adopt an overall approach avoiding suboptimal solutions,
to optimize the total value of the produced product.

Optimization is desirable in the following areas:

1. Safety
2. Environmental compatibility
3. Quality
4. Capacity/flexibility
5. Economy
   - Investments
   - Running costs
   - Second hand value
   - Influence on total concept

1. SAFETY

Safety of human beings and industrial plant is imperative. However,
accidents do occur and experience shows that, in 99 % of all cases,
they are due to human error.

Control systems are becoming increasingly standardized in many countries; this is resulting in wider experience and also in greater benefits for plant manufacturers who are able to incorporate the necessary measures at the development and engineering design stage.

HOWEVER, THESE CONTROL SYSTEMS MUST BE TREATED AS A MINIMUM REQUIREMENT.

The quantity of toxic and/or combustible substances in use in the hardening shops must be reduced. Even though experience shows that the standards applied today are already so high that it is now safer to work in a hardening shop than to smoke or drive a car.

However, when new methods and equipments are introduced, the following priorities must be kept in mind:

1. Safety
2. Environmental protection
3. Quality
4. Capacity
5. Costs

IN THAT ORDER!

A measure which enhances quality or reduces costs must never be introduced if it brings with it a safety or environmental protection risk.

Priority must be given to methods which entail less risk, e. g. vacuum instead of explosive gas in inlet and outlet chambers. Between these chambers, an atmosphere can be chosen which ensures the best quality of carburization as proposed below.

Experienced operating and maintenance personnel with a well-developed sense of responsibility are of course also essential. This high level of training and responsibility is important from the safety angle and to ensure pride in a job well done, which is a guarantee of successful results.

2. ENVIRONMENTAL COMPATIBILTY

The subject of environmental compatibilty involves four areas:

2.1 Waste gases

2.2 Waste water

2.3 Cooling fluids used in the hardening process

2.4 Refractory materials when the brickwork is renewed or the furnace dismantled

2.1 Waste gases

Waste gases are generated when fuel is used for heating purposes and from the carburization gas.

When gas-fired jacket-mounted radiant heating tubes are used, very low NO_x contents of 200 ppm can be achieved nowadays.

The only technique available at present to cut the CO_2 values is a reduction in the energy used for heating.

Further developments will enable all the waste gases originating from the carburization gas to be mixed with the combustion air. This in turn would permit the total elimination of waste gas.

2.2   Waste Water

Waste water is generated when water is used to clean or wash the components and trays before and after the heat treatment, and also from the water used for cooling purposes.

The volume of cooling water can be substantially reduced if the engineering design permits air cooling. The waste water from washing machines can be avoided by a combination of indirect heating to about 450 °C with simultaneous evacuation and steam treatment. (Fig. 1 (1) The volume of oil and water then needed is much lower than when a conventional washing process is used. It should be possible to return some of the oil used for precleaning to the machining facility, while the oil required for subsequent cleaning could be returned to the oil quenching bath. This would also enable tempering to be effected without waste gases and firing problems.

2.3   Quench oil consumption and waste gas can also be reduced by using oil grades with a high evaporation temperature and high resistance to ageing.

2.4   When choosing refractory materials, greater attention will have to be given in future to the aspect of their recycling or disposal after use. The suppliers of refractories must be asked whether recycling is possible. This is still doubtful in the case of fibre materials.

3.   QUALITY, OPERATION, MAINTENANCE

3.1   Quality

Quality is derived from the use of chosen steel and thereto related optimized methods, reliable plant and correct and responsible operation. Quality assurance (Q5 system), total quality management (TQM system) and company-wide quality assurance (Lit. Neue Züricher) can result in systems which permit the necessary monitoring with suitable documentation.

For quality assurance during heat treatment, practically all the data relating to materials, components, time sequences, furnace tempera- tures, gas compositions etc. can nowadays be measured and controlled. Only the quenching rate during production still cannot be measured, although it is just as important as the other factors.

The CAHRTQUENCH model (3) does, however, represent an important step forward to enable the hardening process to be included in quality assurance in conjunction with gas carburization (Fig. 2). This model has for the first time permitted extensive optimization of the over- all process. Other methods are, however, still awaited and remain necessary.

Experience shows that the temperature uniformity in general is greater in the case of continuous plants than in furnaces with batch operation because of the static relationship between the components and the energy source. The heating methods will have to be improved to enable the full benefits of batch-type furnaces.

3.2 Operation and Maintenance

Operation and maintenance also have a bearing on quality. Here, the concept of company-wide quality assurance must be used. This term denotes in particular data gathering and logistics, together with training and further training in quality assurance and quality management. At the same time, these measures will enable costs to be reduced.

4. CAPACITY/FLEXIBILITY

Various ways of improving the heat treatment capacity and flexibility are available today. In continuous plants (Fig. 3) the combined use of twin-track pusher-type furnaces and single-track furnaces of the same type provides the first tep towards greater flexibility. A second stage has been reached with the combination of rotary hearth furnaces (Fig.4)(4). A third stage will be represented by the use of the CAHRTQUENCH model when several cooling baths are operated.

The multi-chamber continuous discharge furnace (Fig. 5) with a linear push direction requiring minimal installation space and offering extremely short passage times, constitutes a complete innovation (internal LOI publication). Considerations of capacity and flexibility are closely linked here with the concept of economy.

5. PLANT COSTS, LEAD TIMES, PLANT UTILIZATION

5.1 Plant costs

The purchase price together with the current value of the operating costs is often a decisive factor in choosing the supplier. In future, however, quality and the overall cost structure will become increasingly important considerations, as outlined in section 7.4. For the final customer only the total concept is of interest. The aspect of cost optimization and cost distribution is illustrated by the example of a single track pusher-type furnace (see Figure 6) where the difference between the today used square section is compared with cylindrical section.

5.2 Lead times

The time required for enquiries, quotation and decisions must be shortened.

A standard (batch-type) furnace with no design modifications is available today in 6 months, while delivery times of 6 to 9 months are possible for straight forward, standard facilities built on the modular principle. 12 months or even more are needed for the supply of "tailor-made" complex continuous furnaces, to which must be added 3 to 4 months required for installations and starting the equipment.

This state of affairs will not be acceptable in the future. New methods are needed.

5.3 Plant utilization

Furnaces are used today for between 10 and 20 years for identical production processes in hardening shops.

Progress will in furture be made towards flexible use; this heading will include the place of operation. This means that the plant must be designed in such a way that it can be dismantled without too much difficulty and rebuilt elsewhere. This will result in a higher second hand value and, at the same time, make new methods and equipments a more economical proposition.

6.   OPTIMIZATION OF MATERIALS AND WORKPIECES

The interaction of all factors such as engineering design and development, procurement (materials and equipment items), total pro- duction costs, sales and guarantee etc. will determine a company's prospects of survival in face of the competition. Material analyses and the hardening characteristics which can be calculated from them, will help to maximize material properties and hence to increase the working stress resistance.

The CAHARTQUENCH program enables workpiece dimensions which permit efficient heat treatment to be determined already at the design stage.

If the machining and heat treatment of the materials are determined with reference to the distortation caused by hardening in such a way that this distortion can be calculated in advance and partly compen- sated, higher hardening plant costs will become economically viable. Fundamental research on the reduction of distortion shows (Lit.) that substantial progress is possible in this area.

Hard shaving, grinding and/or lapping after the hardening process is an expensive solution to this problem.

Shot Peening to increase the stresses in the surface is used mainly to compensate the influence of oxygen diffusion and the associated intergranular oxidation which reduces the properties of alloying elements.

However, the cost of this technique is too high and the degree of uniformity and controllability are low.

Often the cost involved will be equivalent to that of the entire case-hardening process. Ways must be found of avoiding the influence of oxygen in the process itself.

The factors outlined above clearly show that the overall costs - i. e. material costs and the cost of attaining the necessary working stress resistance for the individual component - must be taken into consideration to assess or select a heat treatment facility.

7.   PROSPECTS AND OVERALL PROFIT

7.1   Measures

The following measures are awaited:

- optimization of machinability;

- control of cooling;

- increase in temperature uniformity;

- quality assurance;

- reduction of the need for operating and maintenance personnel;

- reduction of lead times;

- flexible plant;

- heat treatment on production lines;

- optimization of materials and workpieces;

- review of overall profit;

7.2   Atmospheres and Heating

The four tables set out below illustrate experience to date with
different types of atmospheres. The tables make only cursory mention
of plasma processing, mainly because considerable problems still
exist here in respect of distortion and plant costs. However, these
will certainly be solved in due course and the illustrations (Fig. 7
- 9) show the advantages which can be obtained.

The THOR technique   which makes use of fast heating
(often by induction) to a temperature of about 1000 °C in a vacuum,
is uneconomical at present, because of:

-     uneven carbon distribution in the batch;

-     the inadequate quenching effect for workpieces with a wall
      thickness in excess of 20 mm;

-     sealing problems in the vacuum;

-     the need to adapt fixtures and gas supply pipes to each
      workpiece;

-     very expensive equipments.

Once these problems have been solved, it will become possible to use
the method for inline production.

The problems of carbon distribution and quenching have also still not been definitively solved when resistance heating by the ECM technique is used   . Technical breakthroughs can, however, be expected in this area.

7.3 Proposed technique

The following process sequence appears promising for the future in the light of the above observations:

7.3.1   Rinsing in water (250 °C) without a solvent in order to remove chips and solid particles. If centrifuging is possible. This rinsing operation may be excluded.

7.3.2   Pre-heating in a vacuum to approx. 500 °C

-      The residual oil is brought into the gas phase and can be returned to the gear cutting machine;

-      slow heating in order to reduce stresses and distortion;

-      the lower oxygen content in the material will protect the alloying elements and increase the design-dependent strength; shot peening will cease to be necessary;

-      no risk of explosion or poisoning.

7.3.3   Heating

Heating from 500 °C to about 900 - 950 °C should proceed in two stages. This would permit optimal temperature compensation in the batch. To avoid uneven carburization, the C-potential in the atmosphere must not be too high. Vacuum/plasma treatment is recommended in the first stage, as this will largely eliminate oxygen diffusion and increase uniformity. The atmosphere used in the second stage may consist of nitrogen based atmosphere.

In some cases indirect heating by means of radient tubes can be replaced in simple cases by plasma heating. This technique lends itself well to chamber furnaces and ensures a high degree of flexibility of special interests for commercial heat-treaters. Such a batch could however, be a part in a continuous furnace installation.

7.3.4   Carburization

The atmosphere, temperature and time are the three factors which jointly influence the results of carburization in every case. Since all the methods are governed by the same natural laws, the basis also remains the same. A combination of the advantages of the methods outlined above will also give optimal results.

To enable the highest possible C-potential to be employed, it is preferable to separate carburization from a further diffusion zone; this is possible when the multi-chamber pusher-type furnace described in section 4 is used.

Table 5 illustrates a prospected atmosphere which enables the
carburization, safety, quality and costs to be optimized. However,
some development work still remains necessary.

It would not be unreasonable to expect this atmosphere to become the
standard for gas-based carburization in ten years time.

7.3.5   Diffusion

The surface layer carbon content and the case depth should be set
with a specific carbon content in the diffucion zone which is
separated from the other stages.

7.3.6   Cooling (quenching)

Close attention must be paid to cooling. Wherever possible, a short
dwell time in a vacuum must be allowed in order to eliminate the
diffused hydrogen when remaining hydrogen influences the properties
of the components. Prior to quenching an overpressure by introducing
nitrogen avoids fire hazards and reduces the evaporation of the
quenching medium. At the end of the quenching cycle, the cooling
pattern might be controlled by means of a vacuum. Uniform cooling
reduces distortion and permits a more favourable action of the
alloying element.

Controlled quenching is thus required. Some development work remains
necessary to achieve this objective.

7.3.7   Tempering

To avoid the need for subsequent washing operations, tempering in a
vacuum is desirable. Most of the oil residue from the cooling process
can be returned to the quench bath afterwards.

7.4 Overall profit

The aspect of overall profit will be examined by reference to the
example of the manufacture of a gearbox. The total costs are assumed
to be 100 %.

These costs are shown in Figure 13

The heat treatment costs are only 3 - 5 % of the total.

On the other hand, heat treatment itself influences the principal
cost centres in manufacturing and external sourcing. These factors
are as follows:

1.   Influenced by annealing
      Machining   20 - 25 %

2.   Influenced by case hardening
2.1 Material   abt. 15 %
2.2 Finishing + assembly   abt. 20 %

In overall terms, 50 - 60 % of total costs are influenced. An
improvement of 10 % would therefore give a cost reduction of 5 - 6 %.
However, if the heat treatment results in an increase in strength,
i. e. a more uniform, optimal power transmission in all the components, equivalent to 10 % thus eliminate the need to develop a new
gearbox, this figure of 10 % can be set against the total production
costs of 100 %. This would give a potential cost saving of about 5 %
plus 10 %, i. e. about 15 % in all.

If this cost saving is set against the overall heat treatment cost,
it is readily apparent that a reduction equivalent to about four
times the actual heat treatment costs could be achieved. As the
actual capital investment costs for heat treatment representing about
1/3 of the total heat-treating costs, it becomes apparant that there
is a considerable margin for an increase in capital costs while still
obtaining a satisfactory total profit.

CONCLUSION

Carburizing is a several hundred years old method. Metallographists
made it a science around hundred years ago. Demands and
gascarburizing made it a massproduction method after the second world
war through continuous furnaces. During the last 10 to 15 years the
new developments have been few.

Resently however improved measuring methods, deep and widely spread
knowledge and computer techniques make it possible to ask for
improved results - safety, environmental demands, properties,
efficiency and flexibilty - and introduction of corresponding new
methods.

We have shown some important solutions and fields where development
are necessary.
More are to come like
o   ready-made furnace instead of the tailor made
o   "plug in" systems for 100 % pre-maintenance
o   "in-line" production of volume components and increased amount
       heat-treated by specialists outside own company
o   methods for controlled properties and distortion
o   almost completely automatic handling and control of
       installations

It is the author's opinion that improvements are in the long run
gained by choise of optimized steels and heat-treating methods.

Improvements like shot-peening are excellent to "save" an existing
component and therefore of great value. The method can however NOT be
process-controlled and is thereto very expensive. It should therefore
not be seen as future or final solution.

We have focused the importance of the "total concept" by an example
for cost calculation.

IV. NITRIDING AND NITROCARBURIZING

Materials Science Forum Vols. 102 - 104 (1992) pp. 223-228
Copyright Trans Tech Publications, Switzerland

MODEL DESCRIPTION OF IRON-CARBONITRIDE COMPOUND-LAYER FORMATION DURING GASEOUS AND SALT-BATH NITROCARBURIZING

M.A.J. Somers and E.J. Mittemeijer

Delft University of Technology, Laboratory of Metallurgy, Rotterdamseweg 137,
NL-2628 AL Delft, The Netherlands

ABSTRACT

In the present paper nitride and carbonitride compound-layer formation on pure ferritic iron is described for both gas and salt bath environments. Special attention is devoted to the development of porosity in the compound layer and to the uptake of carbon from the nitrocarburizing medium. It will be shown that two distinct mechanisms for the incorporation of carbon from the nitrocarburizing medium within the layer can be operative. Significant uptake of carbon occurs in the initial stage of nitrocarburizing in both gas and salt-bath media. This mechanism is responsible for a carbon-enrichment in the bottom part of the compound layer. The other mechanism is observed for gaseous nitrocarburizing only and becomes operative after the formation of channels by coalescence of pores: it leads to the development of a carbon-enriched zone, starting from the channel walls, in the porous part of the compound layer. On the basis of the present findings a wide variety of compound-layer microstructures may be understood.

I. INTRODUCTION

Nitriding and nitrocarburizing are the most versatile surface treatments of ferritic steel workpieces. These thermochemical treatments involve the reaction of nitrogen and carbon with the workpieces at temperatures below 843 K (570 °C). In practice, the nitrogen and carbon containing media can be a gas mixture, a salt bath or a plasma . Although the processes have been applied succesfully for many years, the microstructural development of the surface layer is largely unclear, in particular during nitrocarburizing. Fascinating phenomena that are comprehended insufficiently, comprise the development of porosity and the incorporation of carbon in the layer composed of iron(carbo)nitrides.

The present paper is concerned with the evolution of the microstructure of the compound layer during gaseous nitriding and during nitrocarburizing in a gas as well as in a salt bath and can be regarded as a synopsis of recent research [1-4].

The experiments providing the basis for the present work were performed with pure iron (except when stated otherwise) and, therefore, nitrogen and carbon supplied by the nitrocarburizing agent are responsible for the effects observed. The treatment times were shorter, equal to and longer than those in commercial treatments, in order to investigate the initial stage of compound-layer formation as well as the equilibrium stage strived for; in practice the process is terminated at an intermediate stage.

II. GASEOUS NITRIDING; PORE FORMATION

According to the (metastable) binary Fe-N phase diagram [5] only the nitrides γ'-Fe_4N_{1-x} and ε-Fe_2N_{1-y} can occur at usual nitriding temperatures and only a limited amount of nitrogen can be dissolved interstitially in the ferrite matrix (max. 0.4 at.% at 863 K). The surface region affected by nitriding, the case, can be subdivided in a surface-adjacent compound layer consisting of γ' and ε nitrides and a diffusion zone underneath, where, at the nitriding temperature, N is dissolved

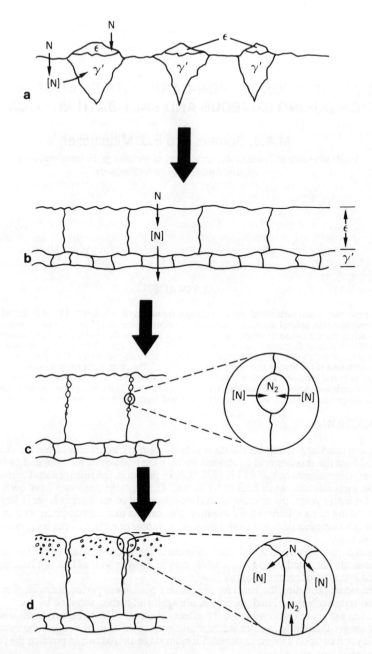

Fig.1:
Compound-layer formation during gaseous nitriding of pure iron:
a. dual phase ε(top)/γ'(bottom) nuclei at the surface;
b. isolating ε/γ' layer; nitrogen transport necessary for layer growth (into the substrate) is only
   possible by N-diffusion through the layer;
c. development of N_2 voids (pores) at grain boundaries in ε;
d. channels form by coalescence of pores; intragranular pore formation occurs too.

in ferrite. The diffusion zone is responsible for improved fatigue behaviour [6] but can also be important for enhanced wear resistance [7]. The compound layer has good wear and anti-corrosion properties [8]. At temperatures below the nitriding temperature precipitation of nitrogen as γ'-Fe_4N_{1-x} and/or α''-$Fe_{16}N_2$ [9,10] occurs in the diffusion zone, due to reduced equilibrium solubility of nitrogen in the ferrite lattice.

The formation of the compound layer is presented schematically in Fig.1a/b. After an incubation time, being the net result of the competition between nitrogen supply from the gas phase (by dissociation of ammonia) and nitrogen removal by diffusion into the substrate [11], γ' nucleates at the surface. Initial growth of these nuclei occurs by supply of nitrogen through adjacent ferrite, because diffusion of nitrogen in ferrite proceeds considerably faster than in γ'. On top of the γ' nuclei ε nitride develops. Lateral growth of such dual phase nuclei eventually leads to the establishment of an isolating compound layer composed of an ε sublayer at the surface and a γ' sublayer adjacent to the substrate. Continued growth of both compound layer and diffusion zone can only proceed by N diffusion through the nitride layer.

Frequently, the compound layer is porous, especially in the surface-adjacent part. The development of pores can be understood on the basis of thermodynamics [12]. The partial pressure of nitrogen gas (N_2) in equilibrium with ε nitride amounts to about 25 GPa (250,000 atm) at the nitriding temperature. The recombination of N atoms to N_2 is manifested as pores (voids) (Fig.1c). Pore formation is observed especially in the near surface region, because most nitrogen is present there (thus enhancing the thermodynamic driving force for N_2 development) and because it is the oldest part of the layer. It occurs preferably at energetically favourable sites in the microstructure, such as grain and twin boundaries. After prolonged nitriding, porosity also develops intragranularly (Fig.1d). Coalescence of individual pores leads to the development of channels that connect the interior of the compound layer with the outer atmosphere (at the surface), thus imposing the external nitriding conditions also at the channel walls in the near surface part [13]. In the bottom part of the channels, both recombination of nitrogen atoms at channel walls as well as continued coalescence of individual pores, deeper in the layer, with the channels create a local source of N_2. Hence, since refreshment of the gas mixture deep in the channels is less efficient than near the surface, the dissociation of ammonia at channel walls is hindered at larger depths.

III. GASEOUS NITROCARBURIZING; CARBON UPTAKE VIA CHANNELS

The model experiments were carried out at 843 K in a gas mixture containing only 3 vol.% CO in addition to 53 vol.% NH_3 and 44 vol.% H_2.

Initially, the formation of the compound layer is analogous to that observed for gaseous nitriding; the carbon content in the layer does not exceed 0.7 at.% C, about the maximum amount soluble in γ'. After the development of channels by coalescence of individual pores, a striking effect due to the presence of carbon in the gas mixture occurs. At some distance to the outer surface carbon is incorporated in the compound layer through the channel walls (Fig.2a).

Preferential uptake of carbon through the channel walls at some distance to the outer surface, points at different nitrocarburizing conditions within the channels as compared to those at the outer surface. The relative abundance of nitrogen gas in the channels, due to the ongoing recombination of nitrogen atoms and coalescence of pores with the channels (see section II), shifts the dissociation equilibrium of ammonia ($2NH_3 \Leftrightarrow N_2 + 3H_2$) to the left. On the contrary, the dissociation of CO is not hindered by the presence of N_2. Close to the surface, the gas mixture within the channels is refreshed relatively rapidly (N_2 removal; $NH_3/CO/H_2$ supply) as compared to the gas mixture at larger depth. Accordingly, in the bottom part of the channels a lower nitriding potential is present and more carbon can be incorporated there in ε carbonitride than near the surface. An alternative explanation has been proposed for the occurrence of the carbon accumulation enveloping the channels [14]. For several reasons this proposal does not provide a satisfactory explanation for the observations [15].

On prolonged nitrocarburizing the carbon enriched areas grow and cementite (θ-Fe_3C) nucleates at the channel walls (Fig.2b). Eventually, a continuous cementite sublayer develops, that is sandwiched between two ε carbonitride sublayers; a γ' sublayer still being existent between this $\varepsilon/\theta/\varepsilon$ sandwich and the iron substrate. A detailed description of the further microstructural evolution within the compound layer has been provided elsewhere [1,3].

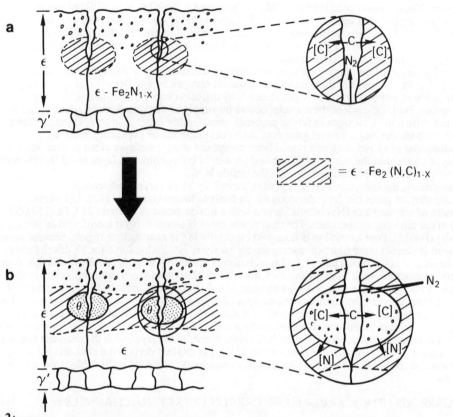

Fig.2:
Carbon uptake via channels during gaseous nitrocarburizing at 843 K:
a. preferential carbon uptake through channel walls;
b. development of laterally continuous carbon-enriched zone in ε and nucleation of cementite at the
   channel walls; small nitrogen solubility in cementite causes a fine dispersion of pores in θ on
   ε→θ conversion [1,2].

IV. SALT-BATH NITROCARBURIZING; PREFERENTIAL INITIAL CARBON UPTAKE

The present description of compound-layer formaton during salt bath nitrocarburizing has been
derived from experiments with pure iron and Fe-C alloys, treated according to the TF1 (Degussa)
proces at 853 K for times ranging from 5 minutes up to 4 hours.

The succession of phases forming at the outer surface during salt-bath nitrocarburizing is
determined to a large extent by different absorption kinetics for nitrogen and carbon. Initially, a
strong uptake of carbon occurs. Gradually, the uptake of nitrogen becomes more important and
eventually governs further evolution of the compound layer.

Firstly, cementite nucleates at the surface, immediately followed by the formation of ε
carbonitride, containing a relatively low nitrogen content (ca. 15 at.%) and a relatively high carbon
content (>5 at.%) (Fig.3a). The total amount of interstitially dissolved atoms in ε carbonitride at
the surface increases gradually. When the nitrogen content at the surface amounts to 19 - 20 at.%
N, γ (carbo)nitride nucleates. For nitrogen contents exceeding 20 at.% N, ε nucleates again at the
surface (Fig.3b). Now the amount of nitrogen offered is sufficient to stabilize ε without carbon.
Continued treatment leads to almost complete disappearance of the γ sublayer, that previously
became sandwiched between the two ε sublayers. A largely monophase ε layer results, that
exhibits a pronounced carbon accumulation in the substrate-adjacent area (Fig.3c).

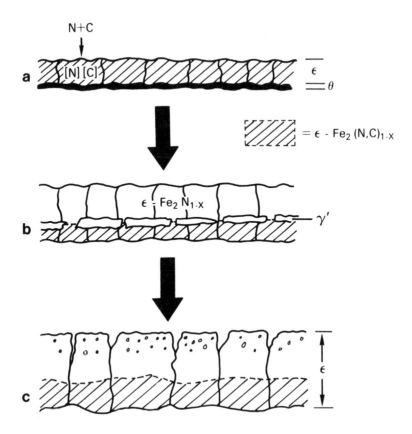

Fig.3:
Compound-layer formation during salt-bath nitrocarburizing (TF1 proces):
a. initially, cementite nucleates at the surface, immediately followed by nucleation of ε carbonitride;
b. gradually higher nitrogen and lower carbon contents at the surface lead to the phase sequence
   (top to bottom): ε (high N)/γ/ε (low N;'high' C);
c. γ' sublayer disappears by coalescence of the sandwiching ε layers; channels are formed by
   corrosive attack of the salt bath; N$_2$ pores form intragranularly.

The morphological appearance of the porosity in the compound layers obtained in the salt bath deviates considerably from that observed for gaseous nitrocarburizing (cf. section III). Apart from pore formation due to N$_2$ "precipitation", also severe porosity is observed that can be attributed to corrosive attack of the compound layer by components in the salt bath. The interpretation of corrosive attack as a main cause for channel formation was confirmed by (i) a coral-like appearance of the outer surface (ii) a decrease of specimen mass during nitrocarburizing and (iii) an increase of the amount of iron in the salt bath [4].

V. NITROCARBURIZING IN PRACTICE

Usually, in practical gaseous nitrocarburizing larger volume fractions of carbon containing gases are applied than in the model experiments discussed in section III. Then, similar to commercial salt-bath nitrocarburizing (section IV), the compound layer formed on iron substrates contains a carbon accumulation in the substrate-adjacent region. This is most likely caused by a mechanism comparable to that valid for salt-bath treatments, comprising initially strong carbon uptake and gradually increasing nitrogen uptake. Possibly, also in this case the first compound nucleating at the surface is cementite. In the porous region of the compound layer a (second) carbon enrichment can occur, which originates from preferred carbon uptake though the channel walls (section III). The two carbon-enriched areas do not have to be connected, because they have different origins.

The presence of cementite (or other carbides) in steels also provides a source of carbon during compound-layer formation. Carbides are transformed into (carbo)nitrides. For plain carbon steel, for example, cementite is converted preferentially into ε carbonitride, because (i) ε (as compared to γ) can contain relatively high carbon contents in combination with relatively low nitrogen contents and (ii) ε and θ are crystallographically alike [16]. Consequently, the compound layers, formed during nitrocarburizing of plain carbon steels, contain a larger fraction of ε carbonitride the larger the amount of cementite in the substrate. Because of the cementite→carbonitride conversion, the distribution of ε in the compound layer is closely related to the distribution of pearlite in the steel substrate [4]. The carbon released from this conversion is incorporated in ε carbonitride throughout the compound layer. Additionally, carbon is also released by the conversion of carbides to (carbo)nitrides in the diffusion zone (underneath the compound layer). This carbon is partly taken up in the compound layer (see [17] and discussion in [1]). At the outer surface of the compound layer the nitrocarburizing conditions usually impose as the equilibrium phase a nitride or a carbonitride with a small amount of carbon. Hence, in the surface-adjacent part of the compound layer, decarburization occurs. This effect also contributes to the carbon enrichment near the substrate.

VI. CONCLUSIONS

The microstructural evolution of the compound layer during ferritic nitrocarburizing is determined to a large extent by the kinetics of nitrogen and carbon uptake.

Pore formation occurs during both gaseous and salt-bath nitrocarburizing as a consequence of the thermodynamic metastability (with respect to nitrogen gas at 1 atm) of the (carbo)nitrides, leading to the precipitation of nitrogen gas (N_2). Such pores grow, coalesce and form channels that connect the outer surface and the interior of the compound layer. In addition, during salt-bath nitrocarburizing channels are also formed by corrosive attack of components in the salt bath.

The channels in the compound layer play an important role during further microstructural evolution. During gaseous nitrocarburizing nitrogen is removed and carbon is preferentially incorporated in the layer through channel walls at some distance from the outer surface.

A carbon accumulation in the substrate-adjacent part of compound layers on pure iron is caused by an initially strong carbon uptake and occurs for both gaseous and salt-bath nitrocarburizing. On nitrocarburizing steels, carbon from the substrate contributes to this local carbon enrichment.

REFERENCES

1) M.A.J. Somers, E.J. Mittemeijer- Surf. Engng., 1987, 3, 123.
2) M.A.J. Somers, E.J. Mittemeijer- Härterei-Tech. Mitt., 1987, 42, 321.
3) M.A.J. Somers, E.J. Mittemeijer- Proc. "Heat Treatment '87", The Metals Society, London, 1988, 197.
4) M.A.J. Somers, P.F. Colijn, W.G. Sloof, E.J. Mittemeijer- Z. Metallkde., 1990, 81, 33.
5) O. Kubachewski- "Iron- binary phase diagrams", Springer- Verlag, Berlin, 1982, 67.
6) E.J. Mittemeijer-Härterei-Tech. Mitt., 1984, 39, 1.
7) F. Hoffmann, P. Mayr-AWT Seminar "Nitrieren und Nitrocarburieren", 1984, Berlin.
8) T. Bell- Heat Treat. Met., 1975, 2, 39.
9) L.J. Dijkstra- J. Metals, 1949, 1, 252.
10) K.H. Jack- Proc. Roy. Soc., 1951, 208A, 216.
11) H.C.F. Rozendaal, E.J. Mittemeijer, P.F. Colijn, P.J. van der Schaaf-Metall. Trans., 1983, 14A, 395.
12) B. Prenosil- Härterei-Tech. Mitt., 1973, 28, 157.
13) M.A.J. Somers, E.J. Mittemeijer-Metall. Trans., 1990, 21A, 189.
14) J. Slycke, L. Sproge- Surf. Engng., 1989, 5, 125.
15) M.A.J. Somers, E.J. Mittemeijer- Proc. Conf. "Nitrieren und Nitrocarburieren", Darmstadt 1991,
    Eds. E.J. Mittemeijer and J. Grosch, Arbeitsgemeinschaft Wärmebehandlung und Werkstoff-Technik, 24.
16) E.J. Mittemeijer, W.T.M. Straver, P.J. van der Schaaf, J.A. van der Hoeven- Scripta Metall., 1980, 14, 1189.
17) F. Hoffmann, Härterei-Tech. Mitt., 1981, 36, 255.

Materials Science Forum Vols. 102 - 104 (1992) pp. 229-242

CONTROL OF THE COMPOUND LAYER STRUCTURE IN GASEOUS NITROCARBURISING

L. Sproge (a) and J. Slycke (b)

(a) AGA Innovation AB, Lidingö, Sweden
(b) SKF Engineering & Research Centre BV, Nieuwegein, The Netherlands

Abstract

The various reactions occurring in nitrocarburising atmospheres are described and discussed based on thermodynamic principles such as equilibria and reaction kinetics. Utilizing a model for nitriding and nitrocarburising systems is shown how to design a furnace atmosphere with specific properties, i.e with different nitriding and carburising properties. Furthermore the influence of different parameters, such as the nitrogen- and carbon activities, on the constitution of the compound layer is demonstrated. The analysis of the nitrocarburising atmosphere and the control of the process are also described and discussed.

Introduction

The reason to why the nitrocarburising process has been subjected to a such a great interest could be found from substantial improvements of several technological properties like resistance to adhesive wear, fatigue resistance and corrosion resistance. Given that gaseous nitrocarburising can be performed in conventional furnaces, can be considered a low cost operation and very small dimensional changes are induced, only helps to emphasize this interest even more. In this paper we focus on the gaseous nitrocarburising in contrast to other methods such as salt bath, fluid bed and plasma methods. The reactions in the nitrocarburising system referred to as gas exchange reactions and mass transferring reactions will be described together with some mechanisms that control them.
This knowledge is the base for how to control the process and how to choose the correct atmosphere for specific needs.

Chemistry of gaseous nitrocarburising

<u>Gas Exchange Reactions</u>:

Nitriding is usually performed at 500 to 550°C in an atmosphere based on ammonia with or without additions of diluting gases like nitrogen and hydrogen. The "nitriding power" of the atmosphere is monitored and controlled either by measuring the content of residual ammonia or, better, the "nitriding potential":

$$N = \frac{p_{NH3}}{(p_{H2})^{1.5}} \, , \tag{1}$$

where p_{NH3} and p_{H2} denotes the partial pressures of ammonia and hydrogen in the atmosphere. The nitriding power of the atmosphere can also be determined as the thermodynamic equivalent to the nitriding potential, the nitrogen activity, which will be discussed in detail later in this paper.

Nitricarburising is normally performed at a temperature of 550 to 600°C in the same base atmosphere as for nitriding, but with additions of one or more constituents containing carbon, oxygen and/or hydrogen (e.g. endothermic or exothermic gas, carbon dioxide, methanol etc.). Besides a "nitriding power", nitrocarburising atmospheres also exhibit a "carburising power", preferable expressed as the carbon activity, which also will be discussed later in this paper.

The content of ammonia in the in-going gas mixture is thermodynamically unstable which means that there is a strong driving force for decomposition of ammonia according to the following reaction:

$$NH_3 \rightarrow \tfrac{1}{2} N_2 + 3/2 \ H_2 \ . \tag{2}$$

However, the decomposition of ammonia is a slow process. Similar to all gaseous exchange reactions at heat treating temperatures, this reaction also requires a catalysing surface in order to reduce the activation energy for the rate controlling step. In this case it is most probably the recombination of two nitrogen atoms at the catalysing surface, to a nitrogen molecule which may be compared to the case when the reaction occurs on an iron surface [1,2]. The surfaces normally present in a heat treating furnace are not catalysing the ammonia decomposition reaction sufficiently to bring the reaction close to its equilibrium during the available time (the mean residence time of the gas in the furnace). This fact makes it possible to maintain a high nitrogen activity, which is hundreds or even thousands times that of nitrogen gas and thereby makes the nitriding reaction possible in nitriding, nitrocarburising and carbonitriding in ammonia-

containing atmospheres. Therefore, in nitrocarburising only a
fraction of the added ammonia will decompose and produce hydrogen
and nitrogen gas, the remainder will act as the source for the
nitriding reaction.

When carbon- and oxygen-containing species are present in the
nitrocarburising atmosphere, the so-called water gas shift
reaction can take place[3]:

$$H_2 + CO_2 \leftrightarrow H_2O + CO .\tag{3}$$

This occurs in order to adjust the contents of hydrogen, carbon
dioxide, water vapour and carbon monoxide of the ingoing gas
mixture towards the equilibrium composition. This reaction is
fairly close to its equilibrium in well functioning nitrocarburising furnaces and it can be concluded that the kinetics for
this reaction under such conditions is satisfactory. The relation
between the partial pressures of these gases in the furnace
atmosphere is determined by the equilibrium equation for the
water gas shift reaction:

$$K_3 = \frac{p_{H2O} \cdot p_{CO}}{p_{H2} \cdot p_{CO2}} ,\tag{4}$$

where K_3 is the equilibrium constant for reaction 3.

A third, less well-known, gas exchange reaction in active atmospheres containing ammonia and carbon monoxide is the reaction
where hydrogen cyanide is formed:

$$NH_3 + CO \leftrightarrow HCN + H_2O .\tag{5}$$

This reaction is also believed to be close to equilibrium in well
functioning nitrocarburising furnaces but probably less so in
equipment where thin and non-uniform compound layers are obtained. The amount of hydrogen cyanide formed is small, in the
range of 0.05 to 1% (by volume), but is nevertheless believed to
be of great importance for the kinetics of the nitrocarburising
and carbonitriding processes [4,5]. The equilibrium equation for
the hydrogen cyanide reaction becomes:

$$K_5 = \frac{p_{HCN} \cdot p_{H2O}}{p_{NH3} \cdot p_{CO}} ,\tag{6}$$

where K_5 denotes the equilibrium constant for reaction 5. In the
case where the amount of hydrogen cyanide is close to equilibrium
with high contents of ammonia, this component also exhibits a
very high nitrogen activity. This means that hydrogen cyanide can
transfer nitrogen to a steel surface in parallel to the classical

nitriding reaction (Reaction 7 below). Two prerequisites are needed for hydrogen cyanide to become a nitriding component of importance: Firstly, since hydrogen cyanide is formed inside the furnace, the catalytic properties of the surfaces available in the furnace must be such that the kinetics for reaction 3 and 5 are promoted; Secondly, the low content of hydrogen cyanide makes it essential that an efficient circulation is maintained inside the furnace for the transport of hydrogen cyanide from the catalysing surfaces to the steel surface. These catalytic properties of the furnace lining play an important role in giving the composition of the furnace atmosphere. If the various surfaces are not active enough in catalysing the desired gas exchange reactions, insufficient amounts of carbon monoxide and hydrogen cyanide are produced to ensure a satisfactory compound layer growth rate.

Mass Transferring Reactions:

The classical nitriding reaction where ammonia transfers nitrogen from the gas to the steel surface has already been mentioned:

$$NH_3 \leftrightarrow \underline{N} + 3/2\ H_2\ , \hspace{4cm} (7)$$

were $\underline{N}$ denotes nitrogen in solution in iron. The nitrogen activity a_N, obtained via this reaction can be determined according to the equilibrium equation:

$$a_N = K_7 \cdot \frac{p_{NH3}}{(p_{H2})^{1.5}}\ , \hspace{4cm} (8)$$

where K_7 is the equilibrium constant for reaction 7. In this paper the nitrogen activity will be related to that of pure nitrogen gas at a pressure of one atmosphere (1.013 bar). In this standard state the nitrogen activity will be unity. Nitrogen transfer via reaction 8 is fairly slow. The kinetics of this reaction have been studied by Grabke [1,2], who found that the rate controlling steps were the dehydrogenation of the ammonia molecule at the surface of iron. Once the nitrogen atom is free from hydrogen it rapidly dissolves into the steel surface.

As indicated above, the hydrogen cyanide formed according to Reaction 5, can also transfer nitrogen and carbon to the steel surface. This occurs according to the following reaction:

$$HCN \leftrightarrow \underline{N} + \underline{C} + \tfrac{1}{2}\ H_2\ . \hspace{4cm} (9)$$

The nitrogen activity induced by this reaction can be found by employing the equilibrium equation:

$$a_N = K_9 \cdot \frac{P_{HCN}}{a_C \cdot (P_{H2})^{\frac{1}{2}}} , \tag{10}$$

where K_9 is the equilibrium constant for reaction 9. In this equation it is assumed that the carbon activity (a_C) is controlled by another reaction and that it is known. The activity values of Equations 8 and 10 will be the same if the assumption that the gas exchange reactions (Reactions 3 and 5) are at equilibrium is valid.

The reaction that determines the carbon activity of an atmosphere containing both carbon monoxide and hydrogen is:

$$CO + H_2 \leftrightarrow \underline{C} + H_2O . \tag{11}$$

The carbon activity can thus be written:

$$a_C = K_{11} \cdot \frac{P_{CO} \cdot P_{H2}}{P_{H2O}} , \tag{12}$$

where K_{11} is the equilibrium constant for this rapid carbon transferring reaction. The standard state for carbon used in this work is pure graphite.

Besides the mass transferring reactions discussed so far, there is one more surface reaction which may transfer nitrogen between the atmosphere and the steel surface:

$$N_2 \leftrightarrow 2 \underline{N} . \tag{13}$$

However, the nitriding potential of molecular nitrogen gas is very low. The reaction rate of this reaction is also slow and thus the influence of it can normally be disregarded when discussing nitrocarburising. However, the nitrogen activity of the compound layer is very high which means that the reverse of Reaction (13) may occur on the outer surface of nitrocarburised parts. The high nitrogen activity may also build up a high nitrogen gas pressure in a cavity. This equilibrium gas pressure can be found by employing the equilibrium equation:

$$P_{N2} = K_{13} \cdot a_N^2 , \tag{14}$$

where K_{13} is the equilibrium constant of Reaction 13. The equilibrium pressure of this reaction will be extremely high and may cause nucleation and growth of pores, mainly in the grain bound-

aries, in the compound layer upon nitrocarburising. This is
analogous to the porosity, sometimes observed in the outer part
of the case after (incorrect) carbonitriding [6].

Atmosphere composition

The composition of the furnace atmosphere could be calculated by
modeling the nitrocarburising system. Such a model has been
developed by Slycke [7] and published by Slycke and Sproge [8].
By giving the ingoing gas mixture, gas flow, furnace temperature
and furnace interior surface area, the resulting atmosphere
composition together with the nitrogen- and carbon activity could
be calculated.

By adjusting the nitrogen- and carbon activities of the nitrocarburising atmosphere to suitable levels the constitution of the
resulting compound layer can be effectively controlled.

For atmospheres based on ammonia and carbon dioxide, this adjustment could be brought about by additions of hydrogen and/or
carbon monoxide. Figure 1 shows how the addition of hydrogen
reduces the calculated nitrogen activity and increases the carbon
activity of such a base atmosphere. By altering the ammonia
content in the in-going gas mixture the nitrogen- and the carbon
activities will respond as shown in Figure 2.

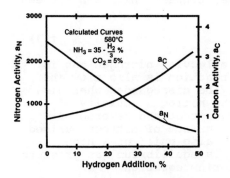

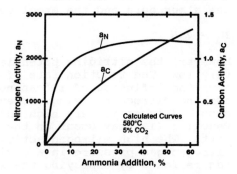

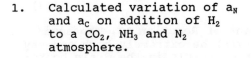

1. Calculated variation of a_N
    and a_C on addition of H_2
    to a CO_2, NH_3 and N_2
    atmosphere.

2. Calculated variation of
    a_N and a_C vs. NH_3 content
    in the in-going gas
    mixture in a CO_2, NH_3 and
    N_2 atmosphere.

If mainly the carbon activity is to be adjusted addition of
carbon monoxide could be utilized, see Figure 3. Only minor
adjustments of the nitrogen and the carbon activities are possible by altering the carbon dioxide content in the in-going gas
mixture, see Figure 4. Note, however, the sharp increase in
carbon activity, from none to very small contents of carbon
dioxide.

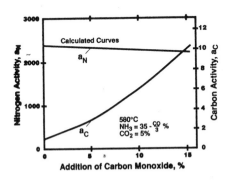

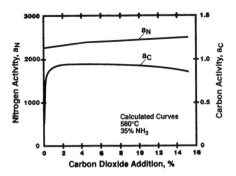

3. Calculated variation of a_N
    and a_C on addition of CO
    to a CO_2, NH_3 and N_2
    atmosphere.

4. Calculated variation of
    a_N and a_C vs. CO_2 content,
    in the in-going gas
    mixture in a CO_2, NH_3 and
    N2 atmosphere.

Compound layer constitution versus atmosphere composition

Figure 5 and 6 show a number of scanning electron micrographs
(back scatter mode) showing the compound layer micro-structure of
unalloyed low carbon steel (SS Steel 1147 corresponding to Din St
14 and SS Steel 1311 corresponding to USt 37-2), nitrocarburised
under various conditions. For these Steels all the compound
layers exhibit the same sequence of phases as outlined in detail
in an earlier paper [4]. The outer part of the compound layer
consists of epsilon phase (ϵ) with a high nitrogen and low carbon
content. At intermediate depths, a two-phase structure consisting
of epsilon and gamma prime phases ($\epsilon + \gamma'$) is present. The bottom
part of the compound layer again consists of only epsilon phase
(γ'), now with low nitrogen and high carbon contents. Similar
observations have also been made by Somers and Mittemeijer [9]
and Sommers et al. [10].

5. Compound layer structure of steel 1147 (Din St 14) after
    nitrocarburising in various types of atmospheres (580°C/2h):
    a. 35%NH$_3$+5.0%CO$_2$+N$_2$,   Ipsen RTQ-1 (a$_N$=2390, a$_C$=0.93)
    b. 33%NH$_3$+4.8%CO$_2$+5.0%CO+N$_2$,   Ipsen RTQ-1 (a$_N$=2317, a$_C$=2.95)
    c. 28%NH$_3$+4.0%CO$_2$+21%H$_2$+N$_2$, Ipsen RTQ-1 (a$_N$=1304, a$_C$=1.54)
    d. 60%NH$_3$+40%Endogas, 570°C,   Ipsen RTQ-4,(a$_N$=1693, a$_C$=23.5)

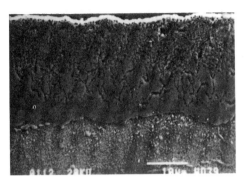
a

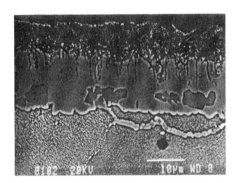
b

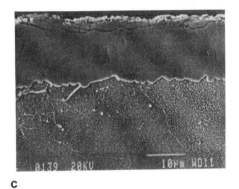
c

6.   Compound layer structure of steel 1311 (USt 37-2) after
     nitrocarburising in various types of atmospheres (580°C/2h),
     Ipsen RTQ-1:
     a. 35%NH$_3$+5.0%CO$_2$+N$_2$,   (a_N=2390, a_C=0.93)
     b. 35%NH$_3$+2.0%CO$_2$+5.0%CO+N$_2$,   (a_N=2270, a_C 4.86)
     c. 12%NH$_3$+3.0%CO$_2$+6.0%CO+10.0%H$_2$+N$_2$, (a_N=1236, a_C=4.87)

Figure 7 shows an iso-thermal section of a tentative Fe-N-C phase
diagram with nitrogen- and carbon contents given in weight
percent. In Figure 8, a calculated version of the phase diagram
is shown, now plotted versus the nitrogen- and carbon activities
[11].

In this diagram it is shown how the activities for nitrogen and carbon changes in the substrate surface through the compound layer as well as the diffusion zone after nitrocarburising.

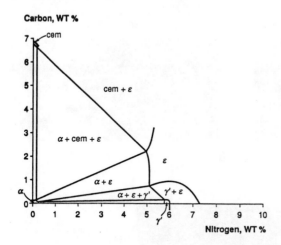

7. Tentative Fe-N-C phase diagram at 570-580°C.
Sizes of α, γ' and cementite fields are not in scale [11].

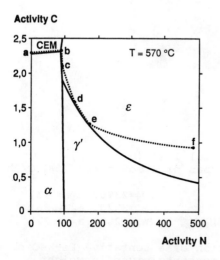

8. Calculated version of Fe-N-C phase diagram [11], plotted versus nitrogen- and carbon activity. In the phase diagram is the nitrogen- and the carbon activities shown for the substrate surface through the compound layer as well as the diffusion zone after nitrocarburising. The standard state for the nitrogen- and carbon activity respectively is: $a_N=1$ for pure N_2 at 1 atm and $a_C=1$ for pure graphite.

By tracing the activity gradients for nitrogen and carbon from
the inner part of the diffusion zone out to the outer part of the
compound layer a path similar to the one shown in the Fe-N-C
phase diagram in Figure 8 will appear (cf. also the micrographs
in Figures 5 and 6). At the inner part of the diffusion zone (a),
a_N is zero since no nitrogen is present and a_C is determined by
the equilibrium between cementite and ferrite. When moving
outwards, the nitrogen is found to be dissolved in ferrite and to
some extent also in cementite. At some point the saturation level
is reached (b) and cementite is transformed into epsilon phase
and the activities are determined by the equilibrium between
ferrite and epsilon phase. Further out in the diffusion zone the
ferrite and epsilon phase becomes richer in nitrogen and leaner
in carbon until the interface between the diffusion zone and the
compound layer (c) is reached. The inner part of the compound
layer consists of epsilon phase which has low nitrogen and high
carbon activity. The high solubility of carbon in epsilon phase
makes the activity value to drop rapidly when moving further out
in the compound layer and the gamma prime region is encountered
(d) and a two-phase zone with epsilon+gamma prime is observed.
The activities of nitrogen and carbon is in this inter-layer controlled by the equilibrium between epsilon and gamma-prime
phases. The volume fraction of gamma prime phase formed is
depending on the availability of carbon, both from the substrate
as well as through the inward carbon diffusion supplied by the
atmosphere.

Further out in the compound layer (e) the nitrogen activity
becomes higher and again a mono-phase epsilon layer is observed,
now with high nitrogen and low carbon activity. At the outer
surface the activity values approaches the ones of the nitrocarburising environment (f).

The phase relations in the Fe-N-C system and the sequential
mechanism for the formation of the compound layer are discussed
in more detail in the References [4, 9-11].

In Figure 5a a highly porous compound layer is shown. This sample
was nitrocarburised in a synthetic atmosphere with ammonia,
carbon dioxide and nitrogen gas in a small sealed quench furnace
(Ipsen RTQ-1). With the addition of 5% carbon monoxide, no actual
change in the compound layer morphology could be observed for the
steel SS 1147 (Figure 5b), although the carbon activity was
estimated to have increased by a factor three. With the addition
of about 20% hydrogen, the nitrogen activity was reduced to about
one half of its original value and the carbon activity increased
slightly. As can be seen in Figure 5c, the porosity under these
circumstances is more or less fully suppressed. In Figure 5d a
sample which was nitrocarburised in an ammonia and endogas
mixture in a sealed quench furnace (Ipsen RTQ-4) is shown for
comparison. There is no obvious difference in the morphology of
the compound layer micro-structure, except for the degree of
porosity, which is more limited in this sample. The estimated
nitrogen activity is also intermediate to that of the other
samples, but the carbon activity is dramatically higher.

In Figure 6 the same phenomena as discussed above are shown for
steel SS 1311, but the composition of the atmosphere, i.e the
estimated nitrogen- and carbon activities, differs slightly.
The sample in Figure 6a is nitrocarburised in the same atmosphere
as for the sample in Figure 5a above. By increasing the carbon
activity by approximately a factor five together with a slight
reduction of the nitrogen activity see Figure 6b, a reduced
fraction of gamma prime phase is observed.

In Figure 6c the gamma prime phase is almost totally suppressed.
In addition the porosity is practically fully avoided. In this
case the atmosphere was composed from 12% ammonia, 3% carbon
dioxide, 6% carbon monoxide and 10% hydrogen with the remainder
nitrogen.

Control of the nitrocarburising atmosphere

In order to monitor and control the nitriding and carburising
power of the nitrocarburising atmosphere it is necessary to
analyze the composition of the atmosphere analogous to atmospheres for e.g nitriding and carburising.

The nitrogen activity is derived from the analysis of the ammonia
and the hydrogen content in the atmosphere by utilizing equilibrium equation 8. A more comprehensive nitrogen activity analysis
also involves the analysis of HCN and utilizing equilibrium
equation 10, (knowing the carbon activity). At good process
conditions these two equilibrium equations should yield the same
nitrogen activity value.

The carbon activity could be evaluated by measuring the content
of carbon monoxide, hydrogen and water vapour, utilizing equilibrium equation 12.

It is obvious that the monitoring of an atmosphere for nitrocarburising calls for the analysis of more gases compared with other
heat treating processes. There are also many practical difficulties involved, e.g high contents of water vapour together with
carbon dioxide and ammonia leading to precipitation of ammonium
carbonate clogging the gas analysis pipes and tubings. However,
heating of all piping and analysis equipment to a temperature
above the atmosphere dew point will help to avoid this problem.

The atmosphere gas analysis could be carried out by utilizing
infrared analysis (IR) of all gases but hydrogen which has to be
measured by heat conductive methods or by gas chromatography.It
is also possible to utilize gas chromatography for analysis of
all the gases. Another interesting analysis technique is Fourier
Transform InfraRed analysis (FTIR). With this technique it is
possible to analyze all gases but hydrogen (and nitrogen) in one
instrument. One major advantage is that this technique does not
need calibration (and calibration gases) in operation.

Together with gas analysis it would also be possible to utilize
some kind of in-situ probe measuring the development of the compound layer together with the diffusion zone. Such a probe
(nitriding sensor) is being developed at IWT in Germany [12].

The control of the nitrocarburising atmosphere regarding nitro-
gen- and carbon activity based on gas analysis is performed
through adjustments of the ingoing gas mixture according to what
is outlined earlier in this paper.

Discussion

In a well functioning nitrocarburising equipment it is reasonable
to assume that the gas exchange reactions are at (or close to)
equilibrium. Under such conditions the surface reactions can
rapidly supply nitrogen (and carbon) to the surface and the
growth rate of the compound layer will be controlled by the
diffusion rate of nitrogen through the compound layer itself and
a parabolic growth of the compound layer can be expected. How
ever, if the catalytic properties of the furnace lining are not
sufficient for the gas exchange reactions, as sometimes experienced in new or re-bricked furnaces, these reactions will be
retarded. The consequence of this will be that the additional
path for transfer of nitrogen (and carbon) to the surface offered
by Reaction 9 is cut off. Although the nitrogen activity of the
classical nitriding (Reaction 7) is high, its reaction rate is
not sufficient to keep the surface saturated with nitrogen, which
will result in a reduced growth rate of the compound layer. In
such a case, the growth of the compound layer will be reaction
rate controlled and a linear growth rate can be expected.

The role of carbon in the nitrocarburising process is two-fold.
Firstly, the presence of carbon in the form of carbon monoxide
will facilitate the formation of hydrogen cyanide; as mentioned
above, this in its turn will open a new route for the transfer of
nitrogen (and carbon) to the steel surface. Secondly, the presence of carbon will promote the formation of the epsilon phase
instead of the gamma prime phase. The epsilon phase has very
large ranges of solubility for both nitrogen and carbon which
will promote the growth of the compound layer because steeper
concentration gradients can be maintained.

The optimum level of the nitrogen- and the carbon activities must
be chosen accordingly for each steel grade depending on the
properties desired; for example should the compound layer exhibit
a high or a low degree of porosity?

Studying the phase diagram (Figure 8) it could be concluded that
(for plain carbon or low-alloy steels) the carbon activity of the
nitrocarburising atmosphere probably not should exceed that of
cementite in order to obtain the fastest growth rate of the
compound layer together with a compound layer morphology containing epsilon phase to a major extent. A too high carbon activity
in the atmosphere would lead to precipitation of cementite at the

outermost surface effectively hindering or slowing down further growth of the compound layer and a too low carbon activity would result in a lower fraction of epsilon phase.

A low carbon activity will also imply that the formation of hydrogen cyanide will not be promoted to a sufficient degree ensuring the second nitrogen transfer reaction, see Reaction 9 [8].

It should also be pointed out that practical considerations referring to type of furnace etc. must be taken when composing the furnace atmosphere in order not encounter problems with sooting etc.

By utilizing gas analysis of ammonia, hydrogen, hydrogen cyanide, carbon monoxide and water vapour the nitrogen- and carbon activities of the nitrocarburising atmosphere could be monitored and controlled.

Gas analysis could also be utilized together with probes directly measuring the development of the nitrocarburising (compound layer thickness, compound layer morphology as well as the diffusion depth). However, this technique is not today fully available commercially. Gas analysis equipment for nitrocarburising atmospheres are today available but call for further development in order to lover cost and increase reliability.

The adjustment of the carbon activity of a nitrocarburising atmosphere, without changing the nitrogen activity, could for example be carried out by small additions of carbon monoxide to the base atmosphere. If hydrogen is added the nitrogen activity will decrease and the carbon activity increase. Many control concepts based on analysis of different gases and utilizing different gases as addition- and control gases are possible. Furnace equipment together with quality demands and type of steel have to be the base for the decision which technique to choose. Here the modeling [8] of the nitrocarbrising system would be a powerful tool.

References

1. Grabke, H.J.: Ber. Bunsengesell., 42(1968), pp.533-548.
2. Grabke, H.J.: Arch.Eisenhüttenwes. 46(1975), PP.75-81.
3. Wünning, J.: Z. Wirtschaft. Fert. 69(1974), pp.80-85.
4. Slycke, J., Sproge, L.: Surf. Engineering 5(1989), pp125-140.
5. Slycke, J.,Ericsson, T.:J. Heat treating 2(1981), pp.3-19 and pp.97-112.
6. Slycke, J., Ericsson, T.: Proc. "Heat Treament '81", pp.185-192, Metals Society, London 1981.
7. Slycke, J.: personal communication 1988-1989.
8. Slycke, J., Sproge, L.: Proc. "AWT-Tagung, Nitrieren und Nitrocarburieren", pp.9-23, AWT Arbeitsgemeinschaft Wärmebeh. und Werkstofftech., Darmstadt 1991.
9. Somers, M.A.J., Mittemeijer, E.J.: Surf. Engineering 3 (1987), pp.123-137.
10. Somers, M.A.J., Colijn, P.F., Sloof, W.G., Mittemeijer, E.J.: Z. Metallkde. 81(1990)1, pp.33-43.
11. Slycke, J., Sproge, L., Ågren, J.: Scan. J. Metallurgy 17(1988), pp.122-126.
12. Klümper-Westkamp, H., Mayr, P.:Proc. "AWT-Tagung, Nitrieren und Nitrocarburieren", pp.202-212, AWT Arbeitsgemeinschaft Wärmebeh. und Werkstofftech., Darmstadt 1991

Acknowledgements

The authors are indebted to AGA Innovation AB and to SKF Engineering & Research Centre BV for the possibility to publish this material. We also want to extend our gratitude to our colleagues at Atlas Copco MCT AB, Stockholm, where most of the trials were made and to Nils Lange at the Royal Institute for Technology, Stockholm, for his valuable contribution to the scanning electron microscopy studies.

Materials Science Forum Vols. 102 - 104 (1992) pp. 243-248

KINETICS OF COMPOUND LAYER GROWTH DURING NITROCARBURIZING

H. Du and J. Agren

Division of Physical Metallurgy, Royal Institute of Technology,
S-100 44 Stockholm, Sweden

ABSTRACT
Compound layer growth during nitrocarburizing is analyzed by means of a mathematical model taking into account the diffusion through the layers as well as the thermodynamic behaviour of the various phases. The calculations are compared with recent experimental information. Pore formation is discussed.

1. INTRODUCTION

The nitrocarburizing process has a number of advantages compared to traditional surface hardening processes, e.g. case hardening, but is physically more complex and consequently its fundamental mechanisms are not yet satisfactorily understood. During conventional case hardening carbon and sometimes nitrogen are added to the steel surface at a rather high temperature where the steel is austenitic and the surface layer is subsequently hardened by a rapid quench causing a a transformation into martensite. Nitrocarburizing is performed at a low temperature where ferrite rather than austenite is stable. The addition of nitrogen and carbon causes the surface layer to undergo a phase transformation and no rapid quench is necessary. The phase transformations can occur in many different ways and depend on temperature, time and constitution of the surrounding media in quite a complex way. For example, when adding only nitrogen to pure iron a glance at the binary Fe-N phase diagram, see Fig. 1 taken from Frisk (1), shows that γ' will be stable at intermediate nitrogen partial pressures and ε phase at higher pressures. When also adding sufficient amounts of carbon the ternary Fe-C-N phase diagram given by Slycke et al. (2) and by Du and Hillert (3), shows that γ' may be suppressed and only ε would form.

In order to completely understand the formation of the various phases during nitrocarburizing one should consider not only the equilibrium information represented by the phase diagram but also kinetic information. The present study was thus initiated in order to combine the thermodynamic and kinetic information into a self-consistent picture of the phase transformations during nitrocarburizing. Such a picture will be a basis for optimization and precise control of industrial nitrocarburizing.

2. MODELS OF DIFFUSIONAL GROWTH DURING NITROCARBURIZING

The model and the mathematical treatment were developed in detail in a recent report by Du and Ågren (4) and here we will only summarize its main features. Its most important assumptions are:

. Local equilibrium at phase interfaces
. Parabolic time dependence
. Diffusion coefficients independent of composition
. Coupling between C and N diffusion is neglected.

The first assumption implies that the growth rate is given by the rate of diffusion through the

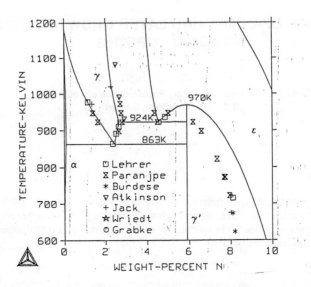

Fig. 1: The Fe-N phase diagram. From Frisk (1)

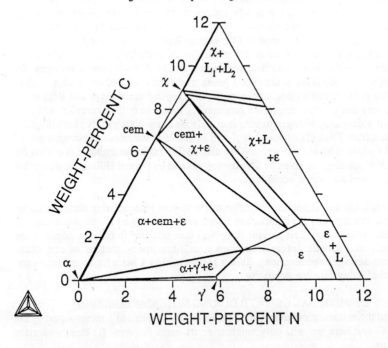

Fig. 2: Isothermal section of Fe-C-N phase diagram at 550 °C. From Du and Hillert (3).

various layers and the other assumptions allow us to find analytical solutions based on the error function as soon as the local equilibrium tielines characterizing the phase interfaces are established. In a ternary system at a given temperature and external pressure the Gibbs phase rule shows us that a tieline is fully characterized by giving the activity of one of the three components. The position of the tieline thus represents one unknown quantity, i.e. an activity. The migration rate of the phase interface represents another unknown and we thus have two unknowns for each moving phase interface. These may be fixed by considering the massbalance for both nitrogen and carbon. For example, for an ε phase growing into α we have

$$\frac{\upsilon^{\varepsilon/\alpha}}{V_m} (u_C^\varepsilon - u_C^\alpha) = J_C^\varepsilon - J_C^\alpha \tag{1}$$

$$\frac{\upsilon^{\varepsilon/\alpha}}{V_m} (u_N^\varepsilon - u_N^\alpha) = J_N^\varepsilon - J_N^\alpha \tag{2}$$

where $\upsilon^{\varepsilon/\alpha}$ is the migration rate of the ε/α phase interface, V_m the volume per mole of Fe atom, assumed independent of composition and the same in all phases, and u_C^ε, u_C^α and u_N^ε, u_N^α are the C and N contents on both sides of the phase interface respectively. J_C^ε, J_C^α and J_N^ε, J_N^α are the corresponding fluxes. The concentration variable u is related to the ordinary mole fraction x by the relation

$$u_C = x_C/x_{Fe} \tag{3}$$

and a similar relation holds for N.

In order to perform calculations we must have a thermodynamic description from which a tieline may be calculated when external pressure, temperature and an activity are given. Furthermore diffusivities in the various phases must be known. The thermodynamic parameters used was given in the recent assessment by Du and Hillert (3) and the diffusivities compiled from various sources are given in ref. 4.

The calculations were performed using the PARROT program developed by Jansson (5).

It should be emphasized that the present calculations are based on constant carbon and nitrogen activities at the steel surface. This is a reasonable assumption for long times but not for short times when the finite rate of the reactions at the steel surface becomes important. In a coming study we will apply numerical procedures to consider the phase transformations during nitrocarburizing and it will then be possible to take into account also the finite rate of the surface reactions.

3. RESULTS OF GROWTH CALCULATIONS
When solving the two fluxbalance equations we find that the solutions give rise to two different regimes representing different mechanisms of transformation. For low carbon activities (a_c less than around 0.5 with graphite as reference state) at the steel surface but sufficiently high nitrogen activities we find that close to the surface an ε layer will form and between it and the ferritic matrix a γ' layer forms. In accordance with our assumptions both layers grow parabolically, i.e. as $\sqrt{time}$. At higher carbon activities γ' will not form and we will only find the ε layer at the surface and α phase matrix below.

The results of a series of calculations are summarized in Figs. 3 and 4 depicting iso-growth rates curves for ε and γ'. Isoactivity lines for C and N are also included in the diagrams.

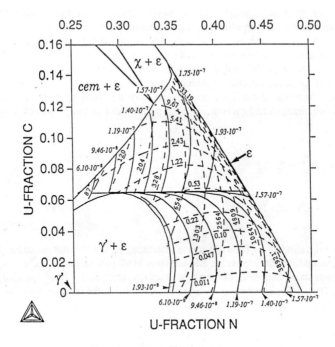

Fig. 3: Calculated iso-growth rate lines for the ε phase (solid lines) together with iso-activity lines of C and N (dashed lines) in ε phase at 575 °C. From ref. 4.

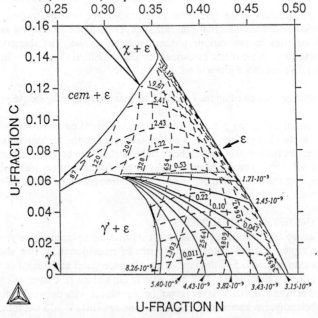

Fig. 4: Calculated iso-growth rate lines for the γ' phase (solid lines) together with iso-activity lines of C and N (dashed lines) in ε phase at 575 °C. The dotted line is the limit line for existence of γ' phase. From ref. 4.

Increased nitrogen activity yields a higher growth rate for ε and a lower for γ'. For the low carbon activity regime the calculations show that the growth rate of ε is more than one order of magnitude larger than that of γ'. When increasing the carbon activity under constant nitrogen activity the growth rate of ε will increase whereas that of γ' will decrease.

When comparing our calculations with experimental information on the thickness of the compound layers from Somers and Mittemeijer (6) and Wells and Bell (7) we do not find a perfect agreement between experiments and calculations. The calculated compound layer is thinner than the observed one. However, the discrepancy is less than a factor two which may be regarded as satisfactory when taking into account that the present calculations were based on several simplifying assumptions. The most uncertain parameters are the diffusivities for C in ε and C and N in γ'. In order to improve the predictability of the model one would thus need more accurate measurements on the diffusivities. Moreover, it should be emphasized that taking the finite rate of surface reactions into account would yield still lower growth rates of the compound layers. On the other hand, the appearance and disappearance of the γ' phase and even more complex reactions, which are sometimes observed, cannot be understood without taking variable surface conditions into account.

5. FORMATION OF PORES

The high nitrogen activities used in nitrocarburizing correspond to extremely high nitrogen partial pressures and there is a strong tendency for formation nitrogen-gas pores. Experimentally a high porosity is observed at high nitrogen activities. The pores are formed preferentially along grain boundaries, see for example ref. 8. From a pedagogical point of view it may be instructive to consider homogeneous nucleation of gas bubbles. The critical size of a gas bubble is given by

$$r_{cr} = 2\sigma/P_{N_2} \tag{4}$$

where r_{cr} is the critical radius, σ the surface tension assumed isotropic and P_{N_2} is the nitrogen partial pressure and is proportional to the nitrogen activity squared. The probability of homogeneous nucleation is proportional to

$$p = \exp(-\Delta G^*/kT) \tag{5}$$

where k is Boltzmanns constant, T the temperature and ΔG^* is given by

$$\Delta G^* = \frac{16\pi}{3} \frac{\sigma^3}{P_{N_2}^2} \tag{6}$$

The surface tension between solid and vapour is around 2 Jm^{-2} and the probability p is plotted as a function of the nitrogen activity in Fig. 5. As can be seen p starts to increase drastically at some critical activity value. It should be observed that the critical nitrogen activities correspond to very small critical sizes. In reality the pore formation will ocucur at grain boundaries and the probability for grain boundary nucleation is also included in Fig. 5, dashed line. However, in practice the poreformation may be controlled by the growth of the bubbles which will occur by means of nitrogen diffusion and vacancy diffusion, i.e. Fe self diffusion.

A more refined treatment taking into account the diffusional growth of the bubbles is under preparation (9).

SUMMARY

A mathematical model for the growth of the compound layers during nitrocarburizing is presented. The predictions of the model are compared with experimental results and the agreement is found satisfactory when considering the various approximations applied. However, it is concluded that more accurate values for the C and N diffusivities in the various phases are needed. The formation of pores is discussed. It is concluded that the nucleation of pores will occur at the very high nitrogen activities used in nitrocarburizing.

ACKNOWLEDGEMENTS

The authors express their thanks to Mr. Lars Sproge for many valuable discussions. This is a part of a project sponsored by the Swedish Board for Technical Development.

REFERENCES

1. K. Frisk: Calphad Vol. 15, (1991) p 83

2. J. Slycke, L. Sproge and J. Ågren: Scand. J. Metall. Vol 17 (1988) p 122

3. H. Du and M. Hillert: Trita-Mac 0435, Royal Inst. of Techn. 1990, S-100 44 Stockholm SWEDEN

4. H. Du and J. Ågren: Trita-Mac 0, Royal Inst. of Techn. 1990, S-100 44 Stockholm SWEDEN

5. B. Jansson; Trita-Mac 0234, Royal Inst. of Techn. 1984, S-100 44 Stockholm SWEDEN

6. M.A.J. Somers and E.J. Mittemeijer: Surface Engineering, Vol. 3 (1987) p 123 7. A. Wells and T. Bell: Heat Treatment of Metals, Vol. 2 (1983) p 39

8. L. Sproge and J. Slycke: In this conference

9. H. Du: to be published

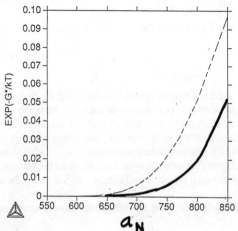

Fig. 5: Calculated probability for homogeneous nucleation of nitrogen gas bubbles (solid line) and grain boundary nucleation (dashed line). From ref. 9.

Materials Science Forum Vols. 102 - 104 (1992) pp. 249-258

NITRIDING OF SINTERED STEELS AND NITROCARBURIZING OF STEELS IN FLUIDIZED BED FURNACE

F. Cavalleri, G. Tosi (a) and P. Cavallotti (b)

(a) Dipartimento di Meccanica, Politecnico di Milano, Italy
(b) Dipartimento di Ch. Fisica Applicata, Politecnico di Milano, Italy

ABSTRACT

A large research program on nitriding and nitrocarburizing thermochemical treatments of steels was carried out in a fluidized bed furnace (Tab. 1,Fig. 1).
Sintered and compact steels of several compositions, classic and not were tested .
Our goal was :
- to get strong layers not porous
- to define optimum pre-oxidation time
- to reach optimum treatment temperature and atmosphere composition
- to set out the treatment cycle for all steels examined
- to reduce the processing time with respect to standard processes
The fluidized bed furnace is easy to handle flexible and with it we coud attain the proposed aims.
Here we do not consider handling and running problems,concentrating on treatments results [20,21,22,23].

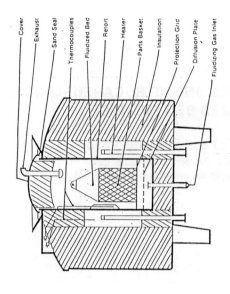

Cover, Exhaust, Sand Seal, Thermocouples, Fluidized Bed, Retort, Heater, Parts Basket, Insulation, Protection Grid, Diffusion Plate, Fluidizing Gas Inlet

FIG. 1 : FLUIDIZED BED FURNACE

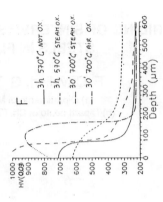

- 3h 570°C NOT OX.
- 3h 570°C STEAM OX.
- 30' 700°C STEAM OX.
- 30' 700°C AIR OX.

F

FIG. 2 : MICROHARDNESS CURVE OF F SPECIMEN

TAB. 1 : FURNACE SIZE AND OPERATION PARAMETERS

RETORT DIMENSION:	DIAM.	250 mm
	DEEP	500 mm
	VOL.	24 dm³

MAXIMUM TEMPERATURE	1250°C
INSTALLED POWER	30KW
FLUIDIZING GAS	15 m³ AT 25°C (1.4bar)
	2 m³ AT 1200°C "
Al₂O₃ BED WEIGTH	75 Kg
MAXIMUM LOAD	35 Kg

TAB. 2 : SINTERED STEELS

Sample		Density(g/cm³)
E	ASC 100-29+1% Cu	7
F	ASC 100-29+1% Cu	7.3
G	SC 100-26+2.5% Cu	6.3
H	ASC 100-29+2.5% Cu	7.1
Q	SC 100-26+2.5% Cu	6.9

TAB. 3 : COMPACT STEELS

C43	UNI 7847
35CrMn4	UNI 7847
39NiCrMo3	UNI 7874
38CrAlMo7	UNI 6120
18CrMo4	UNI 7846
X35CrMo05KU	UNI 2955
48MnV3	

TAB. 4: DIMENSIONAL CHANGES (mm)

C43	0.043
39NiCrMo3	0.021
38CrAlMo7	0.020
48MnV3	0.043
18CrMo4	0.017
35CrMn4	0.010
X35CrMo05KU	0.007

SINTERED STEELS

The material tested had density ranging from 6.3 to 7.3 g/cm and Cu content from 1 to 2.5% (Tab. II).
Atmosphere composition was 50% NH_3, 50% N_2 .
Two treatment temperatures were tested.
First temperature was 570°C, with or without pre-oxidation with steam; second, temperature was 700°C, pre-oxidation with water or with air in the same furnace.
Treatment time was 15' to 1 hour.
Fig.2 shows microhardness curves of F specimens for both temperatures: 570°C and 700°C.
In both treatments pre-oxidation influence is clear and microhardness values after treatment at 700°C is very high. A cross section of specimen H pre-oxidized with steam is presented in fig.3, showing secondary precipitation of γ' between the compound zone and diffusion zone.
Maximum microhardness value (900 HV0.03) is reached just at the boundary of the diffusion zone .
Decreasing the nitriding time, it is possible to observe an homogeneous precipitation (fig.4) and microhardness values exceeding 1100 HV0.03.
The x-ray diffraction pattern from this zone (fig.5) shows an α structure with minimum amount of γ'.
Pre-oxidizing with steam microhardness values can be as high as those of samples pre-oxiidized with air; in the first case the layer appears more thick; in the second case, where the layer appears continuous and compact, coherent precipitation is observed where microhardness has maximum values.
Pre-oxidation with air in the same furnace is a very important feature also in view of industrial utilization.

COMPACT STEELS

Nitrocarburizing was made on many compact steels standard and not standard (Tab.III).
Temperature was in the range 570°C-700°C; and atmosphere composition was 47% NH_3, 3% C_3H_8, 50% N_2.
Microhardness values, as well as layer thickness, are

FIG. 3 : S.E.M.   MICROGRAPH   OF
                  H SPECIMEN :
      SECONDARY PRECIPITATION

FIG. 4 : S.E.M.   MICROGRAPH   OF
                  H SPECIMEN
       COHERENT PRECIPITATION

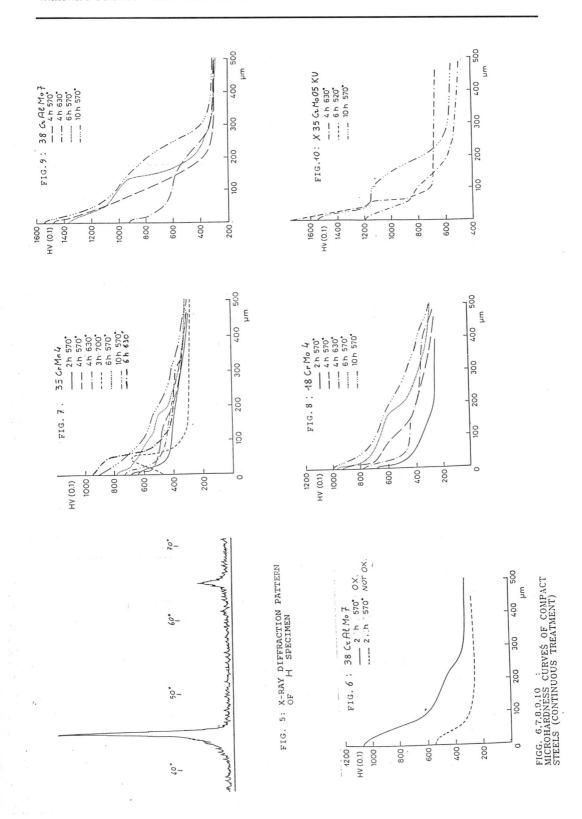

FIG. 9 : 38 Cr Al Mo 7
— 4 h 570°
—·— 4 h 630°
—··— 6 h 570°
········· 10 h 570°

FIG. 10 : X 35 Cr Mo 05 KU
—·— 4 h 630°
········· 6 h 520°
—··— 10 h 570°

FIG. 7 : 35 Cr Mn 4
——— 2 h 570°
— — 4 h 570°
—·— 4 h 630°
········· 3 h 700°
—··— 6 h 570°
—— 10 h 570°
—··— 6 h 630°

FIG. 8 : 18 Cr Mo 4
——— 2 h 570°
— — 4 h 570°
—·— 4 h 630°
········· 6 h 570°
—··— 10 h 570°

FIG. 5 : X-RAY DIFFRACTION PATTERN
OF H SPECIMEN

FIG. 6 : 38 Cr Al Mo 7
——— 2 h 570° OX.
—·— 2 h 570° NOT OX.

FIGG. 6,7,8,9,10 :
MICROHARDNESS CURVES OF COMPACT
STEELS (CONTINUOUS TREATMENT)

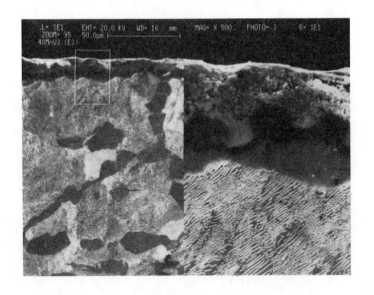

FIG. 11 : S.E.M.   MICROGRAPH   OF
        48MnV3   STEEL

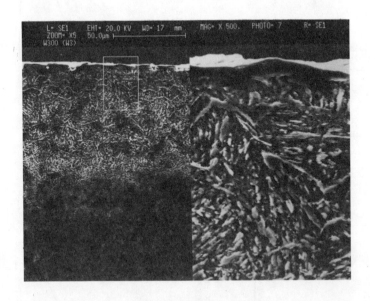

FIG. 12 : S.E.M.   MICROGRAPH   OF
        X35CrMo05   STEEL

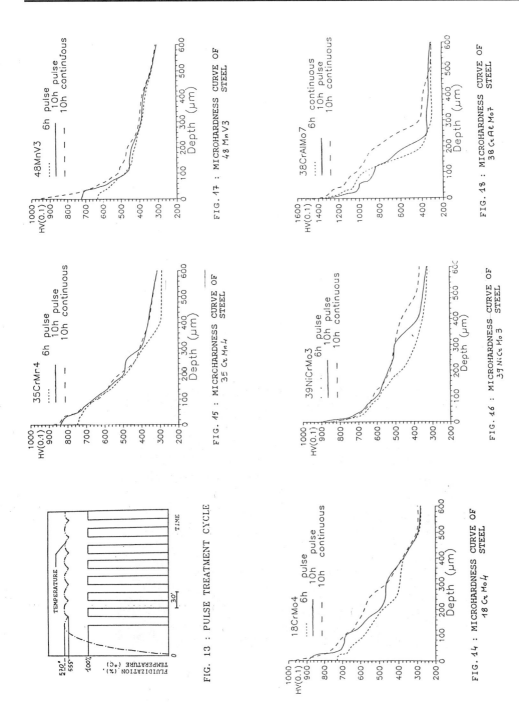

FIG. 17 : MICROHARDNESS CURVE OF
48 MnV3   STEEL

FIG. 18 : MICROHARDNESS CURVE OF
38 CrAℓMo7   STEEL

FIG. 45 : MICROHARDNESS CURVE OF
35 CrMn4   STEEL

FIG. 46 : MICROHARDNESS CURVE OF
39 NiCrMo3   STEEL

FIG. 13 : PULSE TREATMENT CYCLE

FIG. 14 : MICROHARDNESS CURVE OF
18 CrMo4   STEEL

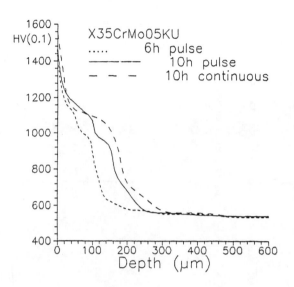

FIG. 19 : MICROHARDNESS CURVE OF
X 35 Cr Mo 05   STEEL

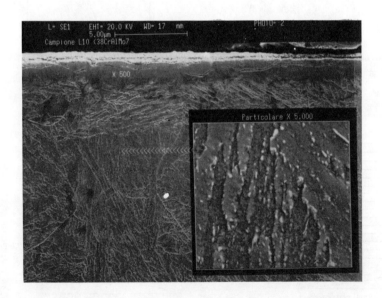

FIG. 20 : S.E.M.   MICROGRAPH   OF
38 Cr Al Mo 7 STEEL

strongly dependent on the pre-oxidation treatment (fig.6).
An optimum treatment temperature and time was found for
each steel often at variance with respect to standard
treatments.
E.g. 35CrMn4 steel shows an optimum microhardness
profile after treatment at 630°C for 4 hours (fig.7).
A layer depth of 200 µm was obteined after only 6 hours
treatment time as show in fig. 8,9,10.
Micrographs show layers compact,homogeneous and without
cracks (fig. 11,12).
Further researches were carried out nitrocarburizing steels
by pulse treatments with same atmosphere and treatment
times of 6 or 10 hours. In this case we had to reduce pre-
oxidation time to about a half.
Fig.13 show the pulse treatment cycle, fluidization for
15',stop for other 15' in sequence.
Results for the pulse treatments are presented in
fig.14,15,16,17,18,19.
Micrographs of 38CrAlMo7 steel are reported in fig.20.
Wear behaviour was tested with a pin on disk,Tin
coated;disk rotation speed was 2.8 m/s,load 2.5 N,time 1
hour.Samples weigth loss is negligible, showing very high
wear resistence in those conditions, and the dimensional
analysis are reported in table IV.

CONCLUSIONS

Several thermochemical treatments can be realized in
sequence in the same furnace with fluidized beds.
Oxidation of sintered and compact steels before nitriding
and nitrocarburizing proved to be effective increasing
microhardness values and layer thickness.In very short
times compact and homogeneous hard layers are
obteined,without cracking problems.
Nitriding and nitrocarburizing with same treatment times
and layer properties on steels,but with great savings in
gas consumpion,could be realized by the fluid pulse
technique.

REFERENCES

1-P.Sommer,KTruhlar-Harterei TM 37 (1982) N°2,58-63
2-P.Sommer-Z.Wirtsch-Fertig 79 (1984),302-306
3-P.Sommer-Gas Warme Int. 33 (1984), 303
4-P.Sommer-Wire World Int. 25 (1983),425
5-P.Sommer-Heat treatment of metals 4 (1985), 99-102
6-J.E.Japka-Metal Progress 123 (1983),1
7-P.Sury-Br.Corrs.J. 13 (1978),31
8-A.Hendry-Corros.Sci. 18 (1978),553
9-G.F.Bocchini-Metall.Ital. 67 (1975),589
10-T.Bell,N.Dunford-Heat Treatment 83 (1983),163
11-P.Cavallotti   D.Colombo   et   alt.-Metall.Ital.   75 (1983),703
12-P.Cavallotti-D.Colombo   et   alt.-   X onv.Naz. Trat.Term.   Salsomaggiore (1984),279
13-K.Szcrecinski,M.Wysiecki,W.Wawrzynski-J.Heat   Treating 3 (1984),249
14-M.Rosso,G.Scavino-Metall.Ital. 79 (1987),291
15-L.W.Crane,R.Mehra-Powder Met. 20 (1977),90
16-H.Ferguson-Metal Progress 107 (1975),81
17-S.Mridha,D.H.Jack-Metal Science 16 (1982),398
18-P.C.Van   Viggen,H.C.F.Rozendaal   et   alt.-J.Met.sci.   20 (1985),4561
19-M.Caprioglio,A.Mancuso-XXI   Conv.Naz.   A.I.M.   Bologna (1988)
20-F.Cavalleri,P.Cavallotti,G.Tosi   XXII Conv.Int.   A.I.M. Bologna (1988),565
21-F.Cavalleri,P.Cavallotti, G.Tosi   E.M. 88 Bologna 1988
22-F.Cavalleri,P.Cavallotti, G.Tosi   XII   Conv.Naz.Trat. Term.   Salsomaggiore (1989),123
23-M.Boniardi,F.Cavalleri,P.Cavallotti,G.Tosi   XXIII Conv.Naz. A.I.M. Ancona,(1990),377

Materials Science Forum Vols. 102 - 104 (1992) pp. 259-270
Copyright Trans Tech Publications, Switzerland

THE STRUCTURE AND PROPERTIES OF ION NITRIDED 410 STAINLESS

M.F. Danke (a) and F.J. Worzala (b)

(a) Stuttgart University, Germany
(b) Department of Materials Science and Engineering, University of Wisconsin, Madison, WI 53706, USA

ABSTRACT

The structure and resulting properties of ion nitrided AISI 410 stainless steel were examined using microhardness/depth correlations, optical microscopy, x-ray diffraction, scanning auger microscopy and pin-on-disc wear testing. The examined samples exhibited predominantly γ'- (Fe_4N) or ϵ- ($Fe_{2-3}N$) compound layers.

The microhardness/depth correlation of the stainless steel samples showed a definite ion nitrided case area. The hardness increase in this area was found to be of factor 4., the γ'-sample was superior to the ϵ- sample.

The structural change of the modified layer was confirmed by optical microscopy. Additionally, the compound layer on the ion nitrided samples was observed.

The composition of the surface material was examined by x-ray diffraction. A mixture of both iron nitrides, γ'- and ϵ-compound, was recorded in the samples. Only a predominance of one compound - correlated to the ion nitriding treatment - was found.

The Auger Line Scans of the stainless steel for the element nitrogen showed a distribution exclusively in the nitrided surface area. This correlated to the hardness/depth characteristic and showed the major effect of nitrogen on the properties of ion nitrided stainless steels. The higher nitrogen concentration was found to be in the ε-samples.

The wear tests showed primary impact of ion nitriding on the volume loss of the tested samples. Substantial improvement of wear resistance after ion nitriding in the unlubricated tests was observed. Almost no wear was observed in lubricated tests. The wear resistance of the surface layer in AISI 410 was lower than in the diffusion zone. A light microscopic examination of the wear tracks of AISI 410 showed a mixture of abrasive and adhesive wear in all examined unlubricated samples. Lubrication reduces the wear significantly.

INTRODUCTION

The modern use of metals in all areas of technical progress has increased the demand for high performance materials. Though the special properties of ceramics or other high tech materials have displaced ferrous alloys in some fields, steels are still used in most applications. Their properties have been improved by means of structural modification along with a better understanding of microstructural arrangements. Today there are a variety of different methods available to increase the properties of ferrous alloys according to their field of use.

Ion nitriding is a glow discharge surface modification technique, which is primarily used to increase the surface hardness of ferrous alloy steels. It is superior to conventional gas-nitriding processes due to shorter nitriding times, controlled case growth and no environmental problems.

The ion nitriding process is a chemical reaction of reactive ions in a glow-discharge gas. During ion nitriding, the specimen is the cathode (negative potential) of the d.c. glow-discharge. Impinging ions, usually a mixture of nitrogen and hydrogen, deliver enough energy to the specimen to heat it to the nitriding temperature. By selecting the composition of the gas, a γ'-layer with diffusion zone, an ε-layer with diffusion zone or just a diffusion zone can be achieved on the surface [1-8].

The following work presents the results of an examination of ion nitrided 410 stainless conducted with the PSII-group at the University of Wisconsin, Madison, USA. The Ion nitrided surface layers were examined by means of light microscopy, scanning auger microscopy and x-ray diffraction. The properties of the modified surfaces were compared by measuring microhardness and pin-on-disc wear.

EXPERIMENTAL CONDITIONS

The alloy used for this study was an AISI 410 stainless steel, which had been austenitized at 950° C, quenched in water and tempered at 625° C for 2 hours to achieve a tempered martensitic structure. To examine the results of the ion nitriding process on this steel, discs 6 mm thick and 25 mm in diameter were used.

The samples marked "A" were ion nitrided at 550° C for 5 hours in a gas mixture, which is designed to produce predominantly γ'-nitrides. This mixture contained 15 to 20 at.% N_2 and the balance argon gas.

The samples marked "B" were also ion nitrided at 550° C for 5 hours. This time an ε-nitride gas mixture was used containing 60-70 at. % N_2, 1-3 at. % CH_4 and the balance argon gas.

The hardness profiles of the ion nitrided cross section samples were obtaineded by Knoop microhardness measurements using 25g loads.

The surface phase structure was investigated by x-ray diffractometry.

Polished cross-section samples were etched and examined using light microscopy.

The nitrogen distribution in the ion nitrided samples was determined with a scanning auger microprobe.

The wear characteristics of ion nitrided AISI 410 were examined with a pin-on-disc wear tester using AISI 52100 rider balls (radius: 2.38mm) sliding on the surfaces with a constant speed of 2.64 m/min and a normal load of 1kg.

DISCUSSION

The hardness profiles of the ion nitrided samples were measured on ground and polished (1μm diamond paste) cross-section samples. These polished surfaces enabled the use of a low load micro-hardness tester. Each reading was taken with a 25g load and more than three readings were taken to determine the exact value.

The samples exhibited a sharp interface between the nitrided area and the base material. The hardness and the thickness of the modified γ'- surface area were higher than those of the ε-area - figure 1.

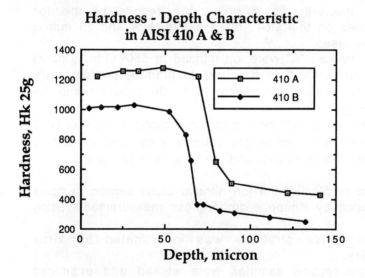

Figure 1 : Hardness Profile
of Stainless Steels AISI 410
AISI 410A : γ'-layer: 410B : ε-layer

The change in structure in the ion nitrided area was observed by light microscopic examination of etched cross section samples.

The samples AISI 410A and B show, after etching with 4%NITAL reagent, compound layers of even thickness. The interface between compound layer/diffusion zone and diffusion zone/base material are not sharply defined - figures 2A&B.

After etching with FeCl₃ a distinct interface between the ion
nitrided area and the base material was visible on the
stainless steel samples - figures 3A&B.

The thickness of both compound layers was about 5 μm,
whereas the modified layers were 76 and 65 μm, AISI 410A
and AISI 410B respectively.

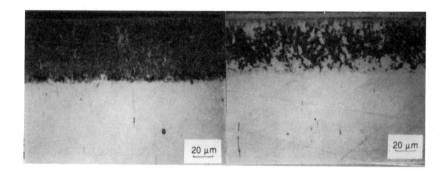

A)   B)

Figure 2 : Cross Section Micrographs of AISI 410A&B (NITAL)
A: γ'-Sample ; B: ε-Sample

A)   B)

Figure 3 : Cross Section Micrographs of AISI 410A&B (FeCl₃)
A: γ'-Sample ; B: ε-Sample

The surface phase structures of the ion nitrided samples were
examined by x-ray diffractometry - figures 4A-C.

The very pure base material - figure 4A - exhibited, after ion nitriding, the distinct peak pattern of γ'-Fe$_4$N, ϵ-Fe$_{2\text{-}3}$N, CrN and martensite (M) - figures 4B&C.

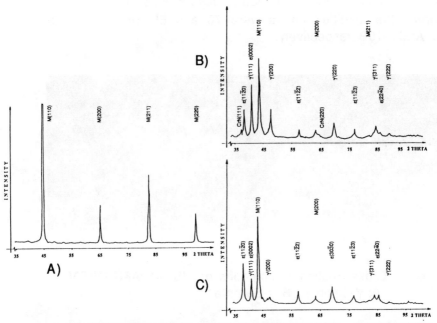

Figure 4 : X-Ray Diffraction Scans of AISI 410
A: AISI 410 untreated; B: AISI 410A:
C: AISI 410B

The results of x-ray diffractometry indicated, that neither of the ion nitrided specimens appeared to have an exclusive nitride layer but rather a mixture of γ'- and ϵ- ion nitrides. However, a correlation was visible to the ion nitriding treatment.

The concentration/depth profiles obtained by auger line scans showed nitrogen exclusively in the nitrided surface layer - figures 5A-6B. The concentration of nitrogen was higher in the B-sample than in the A-sample. Only the B-sample exhibited a surface peak of nitrogen. The nitrogen profiles showed a sharp boundary separating the nitrided region from the bulk.

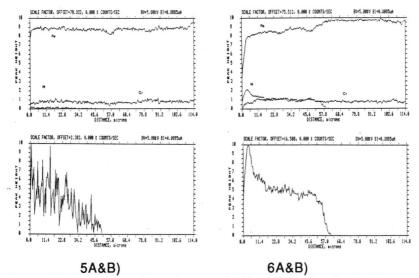

5A&B)   6A&B)

Figure 5A&B : Auger Line Scans of AISI 410A (Fe, N, Cr//N)
Figure 6A&B : Auger Line Scans of AISI 410B (Fe, N, Cr//N)

The wear testing was carried out on lubricated and unlubricated surfaces. All surfaces were polished to a mirror like finish. Only one specimen, the ε-nitride 410 stainless, was tested with the rough (1.5 μm) ion nitrided surface.

The high volume loss of the untreated base material in the unlubricated test was drastically reduced by lubrication - figure 7.

The unlubricated ion nitrided surface (SU) of the γ'-sample exhibited a higher wear than the lubricated, but still lower than the lubricated untreated material.

The unlubricated, rough surface of AISI 410B (ε-sample) changed the wear behavior from abrasive to adhesive, which is indicated by a negative volume loss. Material from the rider built up on the sample surface. The lubricated test exhibited no change in the wear characteristic.

A 12 μm layer was removed from the surface to expose the diffusion zone (DZ) for wear tests - figure 8.

The volume loss of the sample was again drastically reduced after ion nitriding.

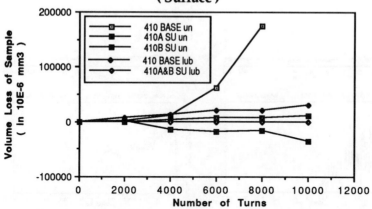

Figure 7 : Wear Characteristic of AISI 410 (Surface)

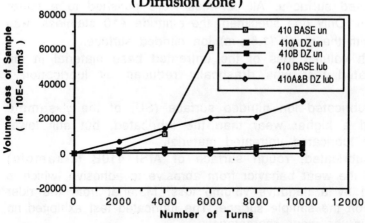

Figure 8 : Wear Characteristic of AISI 410 (Diffusion Zone)

Both ion nitrided samples exhibited about the same wear in the unlubricated tests. Both lubricated tests showed no measurable wear.

Figure 9 shows the x-ray scans of the γ'-sample (surface and diffusion zone). The amount of iron nitrides was reduced in the diffusion zone, whereas the amount of martensite was increased. However, almost no ε compound was detectable in the diffusion zone. There was still a large amount of CrN present.

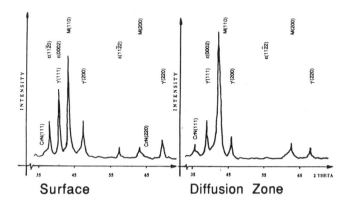

Surface   Diffusion Zone

Figure 9 : X-Ray Scans of AISI 410A
(Surface and Diffusion Zone)

Figure 10 shows the x-ray scan of the ε-sample (surface and diffusion zone). Similar characteristic changes can be shown, the amount of iron nitrides is again reduced, the amount of martensite is increased.

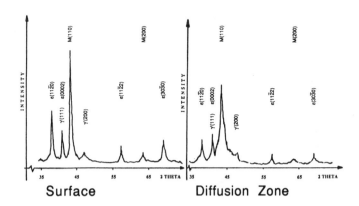

Surface   Diffusion Zone

Figure 10 : X-Ray Scans of AISI 410B
(Surface and Diffusion Zone)

In both cases only one compound (γ' or ε) was visible in the diffusion zone. This explains the improved wear characteristic of the diffusion zones. The brittle ε-phase is no longer present to cause cracks and material removal. It can be shown, that the wear is higher on the surface than in the diffusion zone. Yet there is no drastic difference in wear behavior between both ion nitrided samples (γ'- and ε- sample). The light microscopic examination of the wear tracks showed a mixture of both abrasive and adhesive wear in all unlubricated tests.

CONCLUSIONS

Evaluation of ion nitrided samples of AISI 410 was conducted by Knoop hardness measurement, x-ray diffractometry, optical microscopy, scanning auger microscopy and pin-on-disc wear testing.

(1) The samples showed a distinct compound layer on the surface, consisting of a mixture of both γ'- and ε- iron nitrides.
(2) The surface hardness of the samples was significantly increased by ion nitriding. The samples showed a very drastic hardness drop at the interface of the ion nitrided region to the base material.
(3) The samples showed a defined diffusion zone under the compound layer after etching with NITAL reagent or $FeCl_3$.
(4) Scanning auger microscopy exhibited the limitation of nitrogen to the surface area and a concentration drop to zero at the same depth, where the hardness dropped to the base level. The nitrogen concentration in the surface layer was higher in the ε-compound sample than in the γ'-compound sample.
(5) Ion nitriding improved the wear resistance of the samples substantially. The wear was higher on the surface than in the diffusion zone. Rough ion nitrided surfaces showed predominantly adhesive characteristics (lubricated and unlubricated tests). Unlubricated tests in the diffusion zone showed high amount of adhesive wear besides abrasive wear. Unlubricated tests on polished surfaces showed predominantly abrasive characteristics with some adhesive behavior. Lubricated tests on ion nitrided surfaces showed essentially no wear.

ACKNOWLEDGEMENTS

The author would like to thank the people associated with this project, Prof. F. J. Worzala (Material Science and Engineering Department) for helpful suggestions and responsible guidance through the project, and the PSII-Group (College of Engineering, University of Wisconsin, Madison) for providing analytical facilities and technical support.

REFERENCES

[1]   T. Spalvins, Tribological and Microstructural Characteristics of Ion Nitrided Steels, Metallurgical and Protective Coatings, Thin Solid Film, 1983, 108,157-163

[2]   Kovacs, William, "An Introduction to Ion Nitriding", Proceedings of an International Conference on Ion Nitriding, Cleveland, Ohio, USA, September 1986, ed. T. Spalvins, American Society of Metals - International, 9-12

[3]   T. Spalvins, Proceedings of an International Conference on Ion Nitriding, Cleveland, Ohio, USA, September 1986, American Society of Metals - International

[4]   T. Spalvins, "Plasma Assisted Surface Coating/ Modification Process: An Engineering Technology", Proceedings of an International Conference on Ion Nitriding, Cleveland, Ohio, USA, September 1986, American Society of Metals - International, 3-8

[5]   A. M. Staines, T. Bell, Technological Importance of Plasma-induced nitrided and Carburized Layers on Steel, Thin Solid Films, 86 (1981), 201-211

[6]   Sz. Farkas, E. Filip, Z. Kolozvary, Plasma Nitriding Improves Service Behavior of Textured Machine Components, Proceedings 18th. International Conference on Heat Treatment of Materials, Detroit, MI, May 1982, American Society for Metals, Metals Park, OH,.181-185

[7]   B.Edenhofer, Production Ion Nitriding, Source Book on Ion Nitriding, American Society for Metals, Metals Park, OH, 1977, 181-185

[8]   E. K. Jones, D. J. Surges, S. W. Martin, Glow Discharge Nitriding in Production, Source Book on Ion Nitriding, American Society for Metals, Metals Park, OH, 1977, 181-185

Materials Science Forum Vols. 102 - 104 (1992) pp. 271-278

A FRACTURE MECHANICS APPROACH TO TOUGHNESS CHARACTERIZATION OF NITRIDED LAYERS

R. Roberti (a), G.M. La Vecchia (b) and G. Colombo (c)

(a) Dipt. di Meccanica, Politecnico di Milano, Italy
(b) Dipt. di Meccanica, Università di Brescia, Italy
(c) Oerlikon Italiana S.p.A., Milano, Italy*

ABSTRACT

A method is proposed to assess fracture toughness of nitrided layers. Blunt notch fracture mechanics specimens are employed; specimen sides must be protected against nitriding, or lateral nitrided layers must be grinded away before loading. During loading crack advancement in the nitrided layer at the notch root is measured either by means of a travelling microscope, or specimens can be unloaded at evenly spaced intervals. Resistance to crack propagation of the nitrided layer is expressed by the relationship between the apparent stress intensity factor applied to the specimen and the amount of crack growth. Results on a series of nitrided layers show the method to be very promising.

INTRODUCTION

Mechanical characterization of surface layers in nitrided steels is usually carried out by means of Vickers hardness measurements; surface hardness is the most straightforward property that can be obtained but, for a better analysis of case hardness properties, Vickers hardness indentations are generated on a sectioned sample in order to outline the hardness profile from the outer surface throughout the unaffected base steel.
Actually hardness profile measurement is the most widespread method for mechanical characterization of nitrided layers, as it can be of use for both quality control of the nitriding process and selection of steel and/or nitriding treatment parameters. In fact, hardness profile is directly related to the nitrogen content distribution through the hardened case and provides a quantitative measure of the properties that are to be met, according to the hardness distribution required for a specific industrial application.
Almost since nitriding was introduced, numerous researchers attempted to measure the hardened case toughness, mostly by means of impact test on notched specimens; however, impact energy is mainly spent in propagating fracture

(°) now at LARO S.p.A., Sirtori (Como), Italy

through the base steel and the superimposed effect of the nitrided case is almost negligible. Moreover, sometimes nitrided layer toughness is related to the cracking occurring in correspondence of Vickers or Brinell indentations; the method, however, provides only a qualitative, and frequently uncertain, assessment of toughness [1].

Measurement of nitrided layer toughness is still of great interest, not only for a more reliable design of nitrided components, but also because toughness is expected to be particularly selective, and therefore able to distinguish among nitrided layers that yield a different in service behaviour, although they have equivalent hardness profile and/or case microstructure.

A method for the measurement of nitrided layer fracture toughness is proposed; results on a series of nitrided layers show the method to be very promising.

EXPERIMENTAL METHODS

Toughness testing

Blunt notch Charpy V type specimens have been employed; blunt notches have been used instead of fatigue precracks due to the impossibility of nitriding the steel in correspondence of a crack tip. Notch root radius has been varied for steel 1 in the range 0.065 - 0.7 mm; standard Charpy V specimens, with 0.25 mm notch root radius, have been employed for the other steels. Previous experiments have shown the results are independent of notch root radius [2]. Charpy V type specimens have been loaded in three point bending by means of a screw driven testing machine at a crosshead speed of 0.1 mm/min. At prefixed load levels the specimens have been unloaded and the amount of crack advancement measured. Fracture propagation in the nitrided layers always occurred at loads lower than the general yield load of the specimen.

Since the crack faces are very close each other, heat tinting resulted unsatisfactory to mark crack propagation; therefore, crack advancement has been measured by metallographic inspection of polished lateral surfaces. Before polishing, the lateral nitrided layers have been grinded as they affect crack propagation in correspondence of the specimen sides. Polishing before testing proved to be the easiest method to reveal cracks; on the contrary, if polishing was done on unloaded specimens, a metallographic etching had to be applied in order to outline the presence of the crack. The apparent stress intensity factor K_A has been calculated according to present fracture mechanics standards using the notch depth as the crack length. In most cases several specimens have been loaded at different load levels and the corresponding crack length measured; a single specimen technique has also proved to be successfull provided the crack length is measured by means of a travelling microscope or by metallographic observation of the unloaded specimens.

In some cases the lengths of the cracks in correspondence of Vickers indentation corners have been measured; the relatinship between indentation load and cumulative crack length has been used to evaluate fracture toughness, according to some of the relationships proposed in the literature.

Materials

Six casts of UNI 39 CrMoV 10 type steel have been used; they have been selected to evaluate the influence of impurity elements and microsegregation. The effect of these parameters on the nitriding behaviour, however, is not taken into account in the present paper.

Charpy V type specimens have been gas nitrided according to the following

TABLE I - Mechanical properties of the base steels and characteristics of the
nitrided layers.

Steel	σ_Y	σ_{UTS}	Elong.	Red. of area	KV	HV10	case depth (at 600HV)	white layer	ASTM grain size
	MPa	MPa	%	%	J		μm	μm	
1	879	1006	13	55	68	791	275	< 5	6/7
2	812	974	20	62	95	797	265	11	7
3	823	970	22	68	130	785	255	< 5	8
4	837	986	20	61	105	795	270	11	6/7
5	807	960	19	61	102	810	270	12	6/7
6	824	983	19	59	108	800	270	8	6/7

procedure: 1) activation stage: T = 530 °C, t = 10 h, H_2 = 24.5%; 2) diffusion
stage: T = 545 °C, t = 35 h, H_2 = 45%.
Case depth and white layer size are reported in Table I, along with mechanical
properties of the base steels. Micrographic appearance of the nitrided layers
is shown in Fig.1.

RESULTS AND DISCUSSION

Hardness profiles of the nitrided cases studied are reported in Fig.2. As
expected nitrided layers surface hardness is not affected by impurity level
and microsegregation of the steel and hardness profiles show a very low
scatter throughout the whole hardened case. Therefore, from the point of view
of hardness profile, all the nitrided steels investigated seem to be
coincident. Only minor differences can be observed from the comparison of the
microstructural appearances (Fig.1): nitrided cases in steels 2, 4, 5, and 6
are very similar, with nitrides precipitation on grain boundaries
perpendicular to the diffusion direction, steel 1 shows very fine and evenly
distributed nitride precipitation, while steel 3 compares with steels 2, 4, 5,
and 6 except grain boundary nitrides are almost negligible.
As reported above, the lateral hardened cases of Charpy V type specimens have
been removed by grinding in order to observe crack propagation at the root of
the nitrided notch. In some cases bending experiments have been carried out on
ungrinded specimens and the crack length has been evaluated at different
distances from the lateral surfaces. A set of results for specimen loaded up
to an apparent K_A applied of 120 MPa$\sqrt{m}$ is shown in Fig.3. Crack length is
higher in correspondence of the lateral nitrided surfaces up to a depth of
about 450 μm, while it holds constant thoughout the remaining thickness of the
specimen.
Therefore, if the nitrided case is removed from the specimen sides, it is
possible to measure crack length at the notch root on the new lateral surfaces
and this measure is representative of the whole crack front.

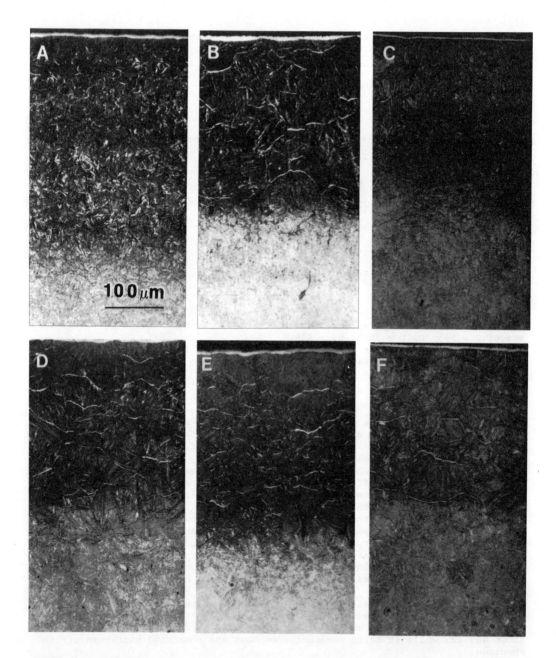

Fig.1 - Microstructural appearance of nitrided layers; A) steel 1, B) steel 2, C) steel 3, D) steel 4, E) steel 5, and F) steel 6.

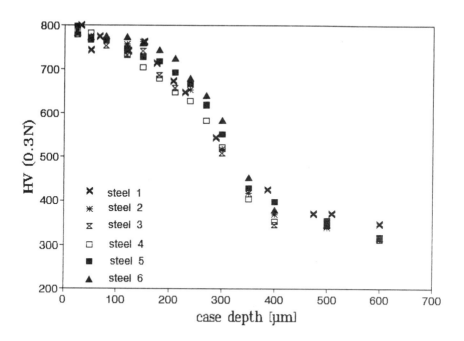

Fig.2 - Hardness profiles of the nitrided cases studied.

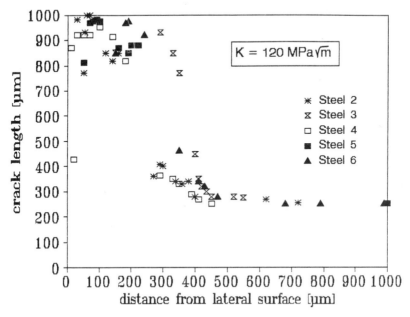

Fig 3 - Crack length in nitrided layers at sides of Charpy V type specimens.

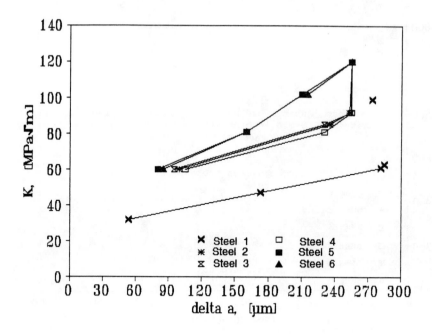

Fig. 4 - Resistance to crack propagation of nitrided layers studied.

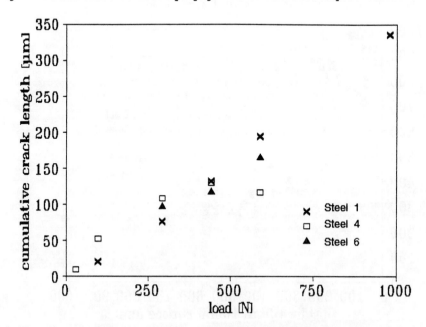

Fig. 5 - Crack length at Vickers indentation corners vs. indentation load.

The relationship between the apparent stress intensity factor applied to the
bend specimen and the crack length is reported in Fig.4. The experimental
points are fitted by almost linear lines that represent the increasing
fracture toughness of the nitrided steel within the hardened layer at
increasing depth from the notch root surface. These lines can be regarded as
the crack resistance curves of the different hardened cases tested, and show
different behaviours for the various nitrided steels.
The relationship between fracture resistance and the other hardness case
properties has not yet been established and needs a more comprehensive characterization of the nitrided layers in order to be fully investigated. At the
moment only for steel 1 it appears possible to envisage a relationship among
microstructural appearance, base steel toughness properties and nitrided case
fracture resistance.
Crack advancement within the hardened case proceeds at increasing load up to a
certain level and then stops notwithstanding load is further increased. Crack
propagation starts again only when fracture resistance of the base steel is
overwhelmed. The amount of growth prior to crack arrest seems to be directly
related to the nitrided case depth.
In Fig. 5 the cumulative length of cracks at the corners of Vickers
indentation are reported as a function of indentation load. For loads higher
than the ones reported in Fig. 5 a diffuse cracking occurred around the
hardness indentation and crack length could not be measured.
A linear relationship between crack length and indentation load is found, as
reported in the literature for other brittle materials [3-4]. The slope of the
lines interpolating the data in Fig. 5 can be used to extrapolate the fracture
toughness [5-6] and K_{IC} values of the order of 11-16 MPa$\sqrt{m}$ are obtained. These
values are lower than those obtained for the applied apparent stress intensity
factor in Fig. 4 and are likely to represent the fracture toughness measure of
the surface white layer of the nitride case.

CONCLUSIONS

The experimental reported in the paper results show that by loading blunt
notch nitrided Charpy V type specimens it is possible to obtain a resistance
curve where the crack advancement within the nitrided layer is a function of
the applied apparent stress intensity factor K_A. The rising curve can be
viewed as the resistance towards crack propagation offered by the subsequent
nitrided layers of decreasing hardness that are progressively encountered by
the crack tip in the nitrided case. Crack advancement undergoes an arrest when
hardness level is about 600 HV; if load is further increased the specimen
behaves like a precracked, non case hardened, three point bend sample.
K_A-crack advancement curves seem to be more selective in the characterization
of the mechanical properties of nitrided layers; in fact, various hardened
cases, although similar as for hardness profile, yielded different toughness
properties. Investigation about the relationship among microstructure, steel
chemical composition, microhomogeneity and toughness is still in progress; the
base steel toughness, however, seems to affect the nitrided layer toughness to
a great extent.
Toughness assessment by measurement of corner cracks at Vickers indentations
yielded $K_{\overline{I}C}$ values that seem to be relevant particularly to the external white
layer and less selective with respect to the toughness of the complete
hardened case.

REFERENCES

[1] Thelning, K.E.: Steel and heat treatment, Butterworths, London, 1975, 389.
[2] Roberti R., La Vecchia G.M, Colombo G.: Metall. It., 83, 1991, 29.
[3] Palmqvist S.: Jernkontorets Ann., 141, 1957, 300.
[4] Exner H.E.: Trans. Met. Soc. of AIME, 245, 1969, 667.
[5] Peters C.T.: J. Mat. Sci., 14, 1979, 1619.
[6] Ponton C.B., Rawlings R.D.: Mat. Sci. and Tech., 5, 1989, 865.

Materials Science Forum Vols. 102 - 104 (1992) pp. 279-284
Copyright Trans Tech Publications, Switzerland

INFLUENCE OF THERMOCHEMICAL TREATMENT ON CORROSION RESISTANCE OF STEELS

J. Straus and D. Hitczenko

Institute of Presision Mechanisms, Warszawa, ul. Duchnicka 3, Poland

ABSTRACT

In this paper attempts are made to explain the influence of heat treatment parameters (time and temperature of nitriding and oxidation processes) on corrosion resistance of impregnated oxynitride layers in a NaCl environment. The experiment was performed on a few selected constructional carbon and alloy steels. The advantageous influence of increase in thickness and porosity of oxynitride layer on corrosion resistance of these steels was found. Some technological recommendations had been formulated. The best results were obtained for the 10 steel grade [0.1% C], which resisted corrosion for 740 hours till the first indications of a red rust occurred.

INTRODUCTION

The first studies on utilization of impregnated oxynitride layers for corrosion resistance were conducted at the beginning of the eighties. At the Institute of Precision Mechanics in Warsaw [IMP] research on this topic started in 1985. This research is connected with technologies developed at the Institute; oxidation - ONS, controlled nitriding - NITREG and nitro-oxidation in a fluidized bed -

TABLE I
Parameters of the thermochemical processes

No	Process	Nitriding			Oxydizing		
		Temp.°C	Time	Atm.comp.	Temp.°C	Time	Atm.comp.
1	TERMOFLUID	560	2h	90%NH_3 + +10%H_2O	560	1min	H_2O
2	TERMOFLUID	560	2h	90%NH_3 + +10%H_2O	560	5min	H_2O
3	TERMOFLUID	620	2h	90%NH_3 + +10%H_2O	530	1h	H_2O
4	NITREG	570	4h	40NH_3 + +60%N_2	-	-	-
5	ONS	570	4h	40NH_3 + +60%N_2	540	1/2h	ONS
6	ONS	620	4h	NH_3	540	2h	ONS

•The 38HMJ steel was nitrided at the temperature 530°C

TERMOFLUID. This work is an attempt to systematize results obtained in IMP and will be continued.

EXPERIMENTAL PROCEDURES

Quenched and tempered specimens, made of steels; 45 (0.45% C),
40H (0.40% C, 1.0% Cr), 38 HMJ (0.38% C, 1.6% Cr, 0.3% Mo, 1.0%
Al) and 10 (0.1% C), were subjected to complex processes of thermochemical treatments, described in Table 1. Some of the
specimens were impregnated with the KO-1S preparation. The KO-
1S preparation and all the thermochemical treatments were
developed and performed in IMP. The corrosion resistance investigations were carried out in a salt spray chamber, according to the
ASTM B117 Standard. A 5% NaCl solution was applied. The corrosion resistance were determined by counting the number of
hours needed till the first indications of a red rust occurred. The
surface roughness, which resulted from the thermochemical treatment, and thickness of the diffusion layer were measured. The
changes in microstructure of the surface layer of the investigated
steels were examined.

EXPERIMENTAL RESULTS

The changes in microstructure of the oxynitride surface layer of the
investigated steels are shown in Figure 1 (an example of 45 steel
grade). Increase in temperature and duration of oxynitriding
processes result generally in a definite increment of thickness of
the surface layer which consists of two sub-layers: an upper one,
gray, oxide-like; and a bottom one, bright, nitride-like. The development of the oxide sub-layer, resulting from the change of oxydizing
parameters, is veryinteresting. This sub-layer increases at the cost
of the nitride one [Fig. 1a,b]. In the case of longer oxidizing times
oxygen migrates into the nitride sub-layer [Fig. 1c,d]. At the upper
part of the nitride sub-layer, a dark etched ,oxygen rich band is
created. This phenomenon is accompanied by increase of porosity
of the nitride sub-layer and occurrence of a specific zone of colum-
nar nitrides in the transition region. Increase in temperature and

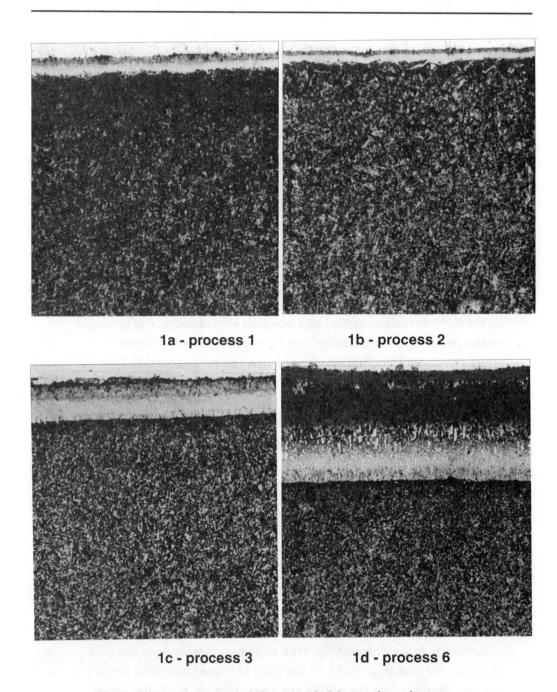

1a - process 1　　　　　　1b - process 2

1c - process 3　　　　　　1d - process 6

Fig.1　Microstructure of the oxynitride surface layers -
an example of 45 steel grade. Process number
according to the Table 1.

duration of thermochemical treatment results in the increase in porosity [Ra, Rz] of the specimen surface. This phenomenon was more significant in the case of plain carbon rather than alloy, constructional steels. The experimental results of corrosion resistance are shown in Figure 2. Investigations of rusted specimens revealed that pitting corrosion occurred. A definite connection was found between the corrosion resistance, and the thickness [and indirectly microstructure] and porosity of the created oxynitride layer on the surface. Definitely, the impregnation of oxynitride layer with KO-1S preparation resulted in advantageous corrosion resistance. More difficult to define, however also advantageous , was the influence of addition of oxidizing processes to nitriding ones. However oxidizing should be carried out carefully, so as not to thin the nitride sublayer out. Application of oxidation processes only, without nitriding, didn't provide any beneficial results. The obtained results can be applied to formulate some practical conclusions:

-in NaCl environment the best corrosion resistance was obtained when applying an nitriding process to the plain carbon steel at $620^{\circ}C$, and to the alloy steels at lower temperatures in the range 560 - $570^{\circ}C$,

-anticorrosion nitriding processes should result in the oxynitride layers of thickness at least 17 micrometers,

-the surface roughness (characterized by Ra coefficient) about 1.5 micrometer seems to be beneficial.

The authors are grateful to Prof. J. Tacikowski and Prof. Z. Rogalski for their help and carrying out different variants of thermochemical treatment of specimens.

CORROSION RESISTANCE

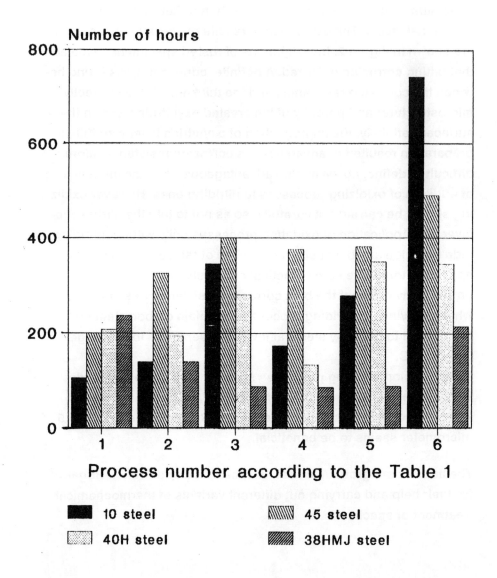

Fig.2 Corrosion resistance; time needed for the first indication of a red rust vs. type of the process

Materials Science Forum Vols. 102 - 104 (1992) pp. 285-300
Copyright Trans Tech Publications, Switzerland

MODIFIED PROCESS PARAMETER IN NITROCARBURIZING IMPROVE COMPONENT PROPERTIES AND GIVE GREATER ECONOMY

G. Wahl

Hanau

A B S T R A C T

Two-step treatment broadens the already wide range of application and improves the economics of nitrocarburizing.

The process as described can be performed where 2 or more nitrocarburizing baths are available. Salt baths are excellent for carrying out the two-step nitrocarburizing sequence.

TENOPLUS$^{(R)}$ treatment is advantageous in that it improves component properties and economy and that component geometrics and dimensions remain within the range of the conventional treatment.

1. INTRODUCTION

Nitrocarburizing is a thermochemical surface engineering process which, because of its process characteristics and broad range of application, undoubtedly has a long and successful career ahead of it. The numerous process variants available to users today are an indication of the importance nitrocarburizing in modern heat treating technology.
In addition to the oxidative post treatment, the use of higher temperatures has aroused great interest in recent years. Higher treating temperatures produce much thicker nitride layers or, in other words, permit a distinct reduction in treating time. In many cases the plant capacity ·can be increased and economics improved.

2. NITROCARBURIZING AT HIGHER TEMPERATURES

As shown in Slide 1, raising the treating temperature from 580°C to 630°C doubles the thickness of the compound layer without the treating time having to be increased.

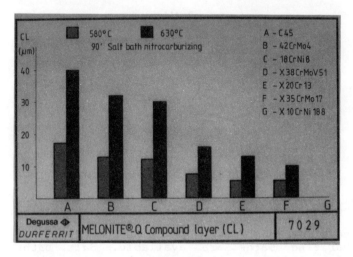

Slide 1

In addition to producing a thicker compound layer, high temperature treatment forms an intermediate layer on unalloyed and low alloyed materials which, depending on the cooling speed after nitrocarburizing, consists either of carbon/nitrogen bainite or carbon/nitrogen martensite with a high percentage of austenite. At the phase boundary between compound and diffusion layers the intermediate layer, which is initially austenitic, grows into the diffusion layer on a time and temperature related basis. The higher the alloying content the higher the temperature at which austenite transformation takes place. As an example, layers on unalloyed and low alloyed steels after 90 minutes nitrocarburizing at 630°C

plus oxidative cooling are shown in Slide 2.
The thickness of the intermediate layer fluctuates between
5 and 15 µm depending on the kind of steel.

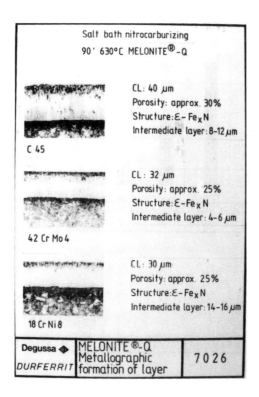

Slide 2

Numerous references to improved properties such as better
corrosion resistance and higher fatigue strength are to be
found in technical literature. To obtain further information
on the corrosion resistance of various alloyed steels a number of investigations were carried out by us. We established
that nitrocarburizing at 630°C produced no increase in
corrosion resistance on unalloyed and low alloyed steels
over the same treatment at 580°C, although much thicker
layers were obtained. On high chrome steels, however, the
higher temperature has a much more favourable effect.

Remarkable increases in resistance can be obtained in the
salt water spray test.

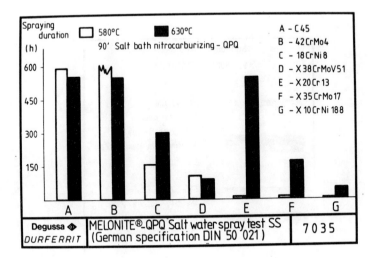

Slide 3

3. TENOPLUS Two-step nitrocarburizing

As there was no difference in the resistance of unalloyed from
that of low alloyed steel qualities obtained at the higher
treating temperature, we were curious to know whether a combination of nitriding temperatures would influence the properties and offer any benefits. In the tests the parts were first
preheated in air to 350°C or preoxidized and then nitrocarburized at 630°C, thereafter at 580°C, intermediate cooled in
an oxidative cooling bath at 380°C and then put into water.
Where a high level of corrosion resistance was required the
parts were mechanically processed and post oxidized in an AB 1
cooling bath.

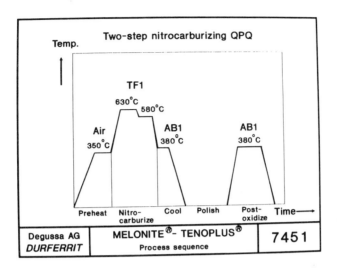

Slide 4

Reversing the process sequence by nitrocarburizing first at
580°C and then 630°C produced poorer results with regard to
the corrosion resistance at equal layer thickness. Therefore
we concentrated our work on the sequence shown in this slide.

3.1 Thickness of compound layer

To refresh your memories, the layer thicknesses obtainable by
the various treating parameter are shown again.

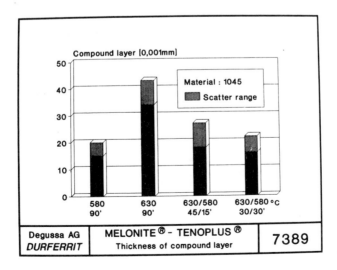

Slide 5

After 90 minutes at 580°C the compound layer on material 1045 is usually 15-20 µm thick' and after 90 minutes at 630°C 40 µm. Using the two-step process, after 45 minutes at 630°C and 30 minutes at 580°C the layers are similar to those obtained in normal treatment, that is to say 16-22 µm.

3.2 Surface roughness

Good polishability is essential for the high quality manu-facture of hydraulic parts, rods for vibration absorbers and gas springs as well as components running against soft coun-ter partners. In this connection we also investigated the in-fluence of the various treating parameter on the roughness of ground samples.

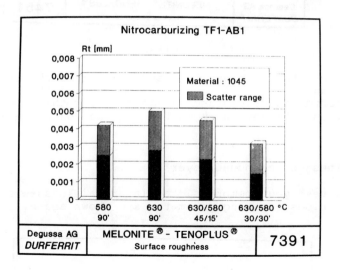

Slide 6

It is an established fact that an increase in roughness is to be expected during nitrocarburizing, independent of the pro-cess used. In our example, the ground samples had surface roughnesses of R_t = 1-1.5 µm prior to treatment. Nitrocarburizing by the different process variants caused an increase in roughness to R_t = 3-4 µm. Therefore the change in surface roughness during two-step treatment was within the usual range.

3.3 Growth in diameter

We have also investigated the changes in dimension and geometry caused by nitrocarburizing. With regard to the diameter, approx. one half to two thirds of the compound layer thickness on hardened and tempered or normalized parts of material 1045 can be taken as the average after nitro-carburizing.

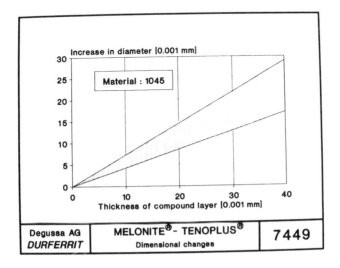

Slide 7

This means that, in the case of a compound layer 20 μm thick, the increase in diameter will be around 8-15 μm and with a layer 40 μm thick approx. 20-30 μm.

3.4 Geometrical changes

With regard to geometrical change, the first results to hand are those obtained on treated cylindrical rods 8 mm in diameter and 250 mm long. In the "as supplied" condition these rods had a maximum run-out of less than 0.03 mm.

A statistical evaluation of batches of 100 samples revealed that the geometrical change after 90 minutes nitrocarburizing at 580°C is similar to that occurring in two-step treatment for 30 minutes at 630°C + 30 minutes at 580°C.

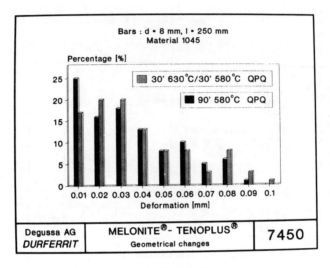

Slide 8

4. COMPONENT PROPERTIES AFTER TENOPLUS two-step nitrocarburizing

4.1 Wear resistance

Measurings were carried out on an Amsler wear test machine to obtain information on the wear resistance to be expected. The load applied was 50 N at a speed of 200 r.p.m. The wear or running partners consisted of rollers of identical steel which had been subjected to identical treatment.

Slide 10 shows the time-related wear measured on grease-free samples under dry running conditions.

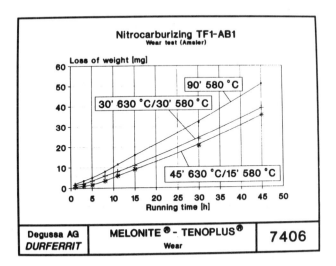

Nitrocarburizing TF1-AB1
Wear test (Amsler)

Loss of weight [mg]

90' 580 °C

30' 630 °C/30' 580 °C

45' 630 °C/15' 580 °C

Running time [h]

| Degussa AG DURFERRIT | MELONITE ® - TENOPLUS ® Wear | 7406 |

Slide 9

As you can see from the run of the wear curves, the weight loss experienced by layers approx. 20-28μm thick during two-step nitrocarburizing tends to be lower. This of course is to be regarded with the reservations usual in assessing wear experiments and their reliability.

4.2 Fatigue Strength

The nitrogen absorbed by the diffusion layer increases the rotating bending fatigue strength and the rolling fatigue strength of components. It is well documented in technical literature that, by nitrocarburizing at higher temperatures, an increase in fatigue strength compared with the normal treatment is achieved. In this connection I would like to make reference to the Works of Pakrasi. Our investigations on notched rotating samples confirm this tendency.

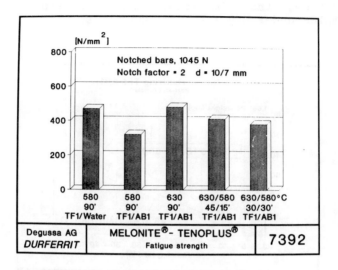

Slide 10

The slower rate of cooling in the oxidative AB 1 salt bath has an effect on the fatigue strength of unalloyed steels, as indicated in the two left-hand columns. If parts are nitrocarburized at 630°C for 90 minutes the fatigue strength after salt bath cooling is similar to that of water-cooled samples after normal treatment.

Depending on the treating time at 630°C, the values of samples treated by the two-step process are also higher than those of samples treated normally for 90 minutes at 580°C and cooled in the oxidative salt bath.

4.3 Corrosion resistance

Nitrocarburizing plus oxidation is being employed in many branches of industry where highly corrosion resistant components are required. In many cases nitrocarburizing plus post oxidation is used as an alternative to surface treatment processes such as chrome or nickel plating and other galvanic coatings.

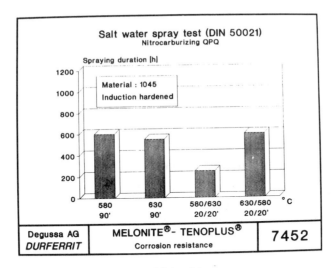

Slide 11

The numerous applications of MELONITE QPQ treated parts encouraged us to carry out tests with two-step nitriding. Results obtained by various process variants are given again in the slide now showing. As you can see, nitrocarburizing at higher temperatures does not give any improvement in corrosion resistance, in spite of the much thicker layers over normal treated samples. We have also tried out a combination of higher and lower nitrocarburizing temperatures.

If parts of equal layer thickness are first treated at the lower temperature and then at the higher, the corrosion resistance is much poorer than if treatment is done in the reverse order.

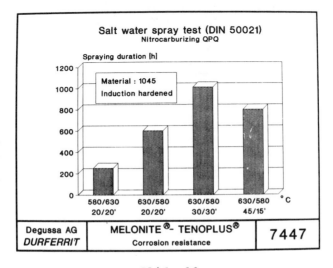

Slide 12

The results of corrosion tests obtained on parts treated by the TENOPLUS two-step nitrocarburizing process are summarized in this slide. The resistance of 1000 hours obtained in a salt spray test on the samples nitrocarburized for 30 minutes at 630°C plus 30 minutes at 580°C are indeed remarkable.

5. COST ASPECTS AND PLANT CAPACITY

In an examination of the economics, nitrocarburizing compares most favourably with other surface engineering processes. Comparisons of cost carried out by various users showed significant savings.

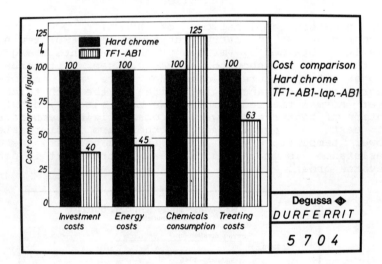

Slide 13

Comparing the costs with those of hard chrome plated components serves as an example. Investment and energy costs had
a particular influence on the economics. A saving of 37 %
over hard chrome plating was possible with the QPQ process.

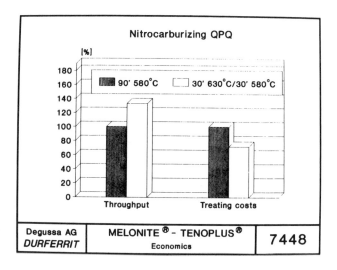

Slide 14

On this favourable basis it is possible to make further
savings and increase plant capacity with two-step nitrocarburizing. Comparing the normal treatment with two-step
nitrocarburizing, the throughput can be raised by approximately one third and the costs reduced by approx. 20-25 %,
assuming layers of similar thickness.

6.   DISCUSSION OF THE RESULTS

The investigations carried out again confirm the positive
influence of nitrocarburizing plus oxidative post treatment
of the compound layer. By using various treating temperatures
the plant capacity and economy can be increased.

We have of course tried to find out the reason for the two-step treatment having a favourable effect on the component properties. Glow discharge spectroscopy investigations carried out at IWT in Bremen do not, however, yet permit any definite conclusions to be drawn. It can, however, be seen that two-step treatment causes changes in the carbon and nitrogen behaviour of the compound layer particularly in the lower region thereof. In other words, a distinct enrichment with nitrogen is achieved in the lower third of the layer in the direction of the diffusion layer during the 580°C stage.

7. APPLICATIONS

Two-step nitrocarburizing broadens the range of application with improved component properties and reduced treating costs. As you can see from the test results, the loadability of the components is increased as well as the resistance to corrosion and wear.

These results also coincide with the experience gained by us in connection with nitrocarburizing at higher temperatures. We see fields of application in the automotive and motor industries, for the treatment of hydraulic parts as well as for applications requiring an economical heat treatment to obtain components with good wear and corrosion resistance and increased fatigue strength.

The process can be performed in manually operated or automated salt bath plants.

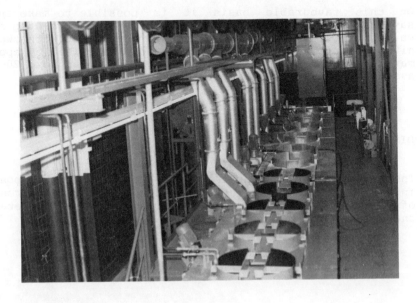

Slide 15

To represent the almost unlimited range of applications of nitrocarburizing I would like to refer to the treatment of golf club heads, as carried out in the USA in large numbers.

Slide 16

References:

1. Kunst, H.   Eigenschaften von Proben und Bauteilen nach Badnitrieren und Abkühlen in einem Salzbad. Härterei-Techn. Mitt. 33 (1978)1, S. 21-27.

2. Wahl, G.   Anwendung der Salzbadnitrocarburierung bei kombinierter Verschleiß- und Korrosionsbeanspruchung Z. f. wirtsch. Fertig. 77 (1982) 10, S. 501-506

3. Wahl, G.   Salzbadnitrocarburieren nach dem QPQ-Verfahren VDI-Z 126 (1984) 21, S. 811-818

4. Ohsawa, M.   Der heutige Stand des Nitrierens im Fahrzeugbau in Japan. Härterei-Techn. Mitt. 34 (1979) 2, S. 3-10.

5. Taylor, E.   QPQ Salt Treatment that prevents Corrosion Metal Progress (1983), Heft 7, S. 21-25

6. Rauch, B.   Metallische Korrosionsschutzschichten für Stahl, Untersuchung von Verfahren zur Substitution. Dissertation TU München 1986.

7. Kettmann, A.   Einfluß des Badnitrierens auf physikalische
   Stuhlmann, W.   Eigenschaften von Stählen mit niedrigen Kohlenstoffgehalten. Härterei-Techn. Mitt. 16 (1961) 2, S. 50-52

8. Gräbener,H.-G.   Einfluß der Stahlzusammensetzung und Behand-
   und G. Wahl   lungsparameter auf die Eigenschaften von nitrocarburierten Bauteilen. Härterei-Techn. Mitt. 44 (1989) 6, S. 331-338

9. Pakrasi, S.,   Nitrocarburieren zwischen den eutektoiden
   Jürgens, H.,   Temperaturen des Fe-N- und Fe-C-Systems
   Betzold, J.,   Härterei-Techn. Mitt. 38 (1983) 5, S. 215-219

10. Edenhofer, B.   Beitrag zum Einfluß der Stahlzusammensetzung
    Trenkler, H.   auf die Lage der Ac_1-Temperatur von Nitrierschichten. Härterei-Techn. Mitt. 35 (1980)4, S. 175-181.

11. Mittemejer,E.J.
    Collijn, P.F.   Oberflächenoxidation von Nitrierschichten. Härterei-Techn. Mitt. 40 (1985) 2, S. 77-79

12. Wahl, G.   Modifizierte Prozeßparameter beim Nitrocarburieren verbessern die Bauteileigenschaften und erhöhen die Wirtschaftlichkeit. Vortrag Härtereikolloquium Wiesbaden 1990.

Materials Science Forum Vols. 102 - 104 (1992) pp. 301-318

PLASMA NITRIDING AND PLASMA CARBURIZING OF PURE TITANIUM AND Ti6Al4V ALLOY

P. Jacquot, M. Buvron (a) and J.P. Souchard (b)

(a) Inovatique SA (HIT Group, 25, rue des Frères Lumière,
F-69680 Chassieu, France
(b) BMI (HIT Group), St. Quentin Fallavier,
F-38290 La Verpillière, France

ABSTRACT

The surface hardness, the wear and corrosion resistance of titanium and titanium alloys can be improved by thermochemical surface treatment.

Plasma heat treatment has many advantages over other surface treatment methods because of the great number of independent process parameters which enables layers with specific microstructures, hardness and properties to be produced. It is the reason why we have carried out a study to verify the role of different treatment parameters. The plasma heat treatment was performed with an industrial plasma furnace equiped with a graphite resistor to minimize the plasma power and to heat the specimens. The plasma source is given by a pulsed power supply.

The results of the investigations show the effect of different process parameters onto the metallurgical characteristics of plasma nitrided and carburized titanium alloy. It was shown that plasma surface treatment of titanium alloys, in given conditions, results in the formation of titanium nitride or titanium carbide layers.

After plasma nitriding of 6 hours of pure titanium or Ti6Al4V alloy at 900°C, the surface hardness can reach a value higher than five time the core-hardness of the base material.

INTRODUCTION

It is well known that titanium alloys have inherent advantages of light weight and low modulus. However, their tribological behaviours are characterised by a high coefficient of friction and poor wear resistance. By using plasma nitriding or ion carburizing processes, it is possible to enhance the tribological properties of titanium alloys.

Recently, the interest for improving the wear resistance of titanium alloys has been renewed by the use of the plasma nitriding technique [1] [2] [3] [4]. It seems that titanium nitride or carbide layers with high hardness and low coefficient of friction can be carried out.

In this work, we have looked for the influence of different process parameters, such as : temperature, time, pressure and electric parameters on the microstructure, microhardness, thickness of compound layer, depth of diffusion layer, grain size and roughness.

The purpose of this paper is to present the results of these investigations.

EXPERIMENTAL PROCEDURE

Specimens for treatment (30 X 20 X 20 mm) were polished and degreased with aceton at room temperature. The composition of the materials used in this study are : pure titanium (annealed, alpha alloy), alloy Ti6Al4V (annealed, alpha-beta alloy).

The plasma surface treatment was performed in a small industrial furnace (BMI - VB 50 X 50), which is shown schematically in Fig. 1. This system differs from the conventional nitriding process in two points. Firstly, the cathodic holder is made of titanium. Hence the possibility of contamination by sputtering of some foreign materials onto the parts is eliminated. Secondly, the parts are heated by an additional heating system. It consists of 12 graphite resistance elements, completely surrounding the load to give uniform heating, and connected to a 45 kW power supply. The hot zone is thermally insulated with graphite fibers insulator.

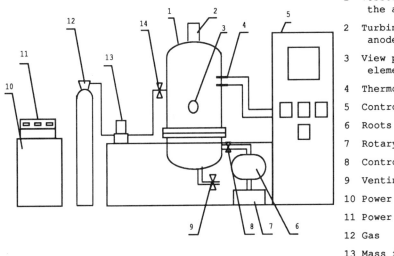

1	Vessel were used as the anode
2	Turbine were the anode
3	View port heating elements to
4	Thermocouple
5	Control panel
6	Roots pump
7	Rotary pump
8	Control valve
9	Venting valve
10	Power unit
11	Power control panel
12	Gas
13	Mass flowmeters
14	Gas control valve

Fig. 1 : Schematic representation of plasma heat treatment equipment.

The useful volume of the vacuum chamber is about 80 liters (Ø 450 x 500 mm high).

The specimens were connected to the cathodic bias of a pulsed power supply (20 kW/6-7KHz) and the heating elements to the anode.

During the plasma treatment, the voltage was set between 200 and 400 volts and current about 10 Amp. Temperature was controlled using an infrared pyrometer placed in front of a quartz view port. Nitriding and carburizing were carried out at a working pressure of 6 mbar with a constant gas flowrate of 1,4 l/mn. The gas composition was respectively either pure nitrogen or propane hydrogen mixture.

The specimens were etched during 10 mn by an argon glow discharge. After this stage, the parts were heated with the heating elements in a pure nitrogen atmosphere until the treatment temperature was reached.

The plasma nitriding of titanium alloy was performed at four different temperatures between 750°C and 900°C. Furthermore, for each temperature, the nitriding cycle time was established as following : 1, 3, 6 and 9 h.

The plasma carburizing of pure titanium was performed at 800°C, 850°C and 900°C during 6 h.

After treatment, the specimens were polished, etched using a mixture of HF and HNO3 acids and examined by optical metallographic technique. Microhardness profiles were used to determine the diffusion layer depth (depth = Hv0,05 core hardness + 100 HV).

IONIC NITRIDING OF PURE TITANIUM AND TITANIUM ALLOY

Influence of the temperature and the treatment time

The compound and diffusion layers

We have studied the influence of different temperature and treatment time on the compound layer TiN + Ti2N, achieved after nitriding of pure titanium and titanium alloy (Fig. 2).

The compound layer thickness are given in Table 1, in function of nitriding temperature and time.

Temperature	750 °C		800 °C		850 °C		900 °C	
Time (h)	Ti6Al4V	Ti	Ti6Al4V	Ti	Ti6Al4V	Ti	Ti6Al4V	Ti
1	0,5	0,5	1	1	1	1,25	1	2
3	1	1,5	1,5	1,5	2	2	2	2
6	1,5	2	2	2	2	2	2	2
9	2	2	3	3	4	4	6	6

Table 1 : Ionic nitriding of pure titanium and Ti6Al4V alloy.
Influence of time and temperature on compound layer
thickness (in μm ± 0,5)

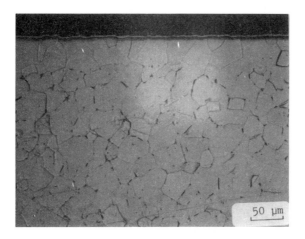

Fig. 2 : Optical micrograph of cross section of titanium plasma
nitrided in pure nitrogen for 9 h at 800 °C

We can see that the thickness of the compound layer, composed of
the nitrides TiN and Ti2N, increases with nitriding temperature and
time. If the nitriding time is less than 9 h, the thickness of the
compound layer is found to be inferior or equal to 2 μm for both
materials.

Between 800 and 900°C, and for a treatment time of 9 h, the
compound thickness increased rapidly up to 3 - 6 μm. The variation
with temperature and time of diffusion layer on pure titanium and
Ti6Al4V alloy plasma nitrided are given in Fig. 3 and 4. The
thickness of the diffusion layer is mesured with microhardness
profiles using the core hardness + 100 HV.

The curves indicates that :

- the growth of both layers is controlled essentially by a
diffusion mechanism,

- the maximal thickness of the compound layer (6 μm) is obtained
for a treatment at 900°C during 9 hours (Fig. 5),

- concerning the thickness of compound layer, there is no significant difference between pure titanium and Ti6Al4V alloy.

Sometimes, the compound layer of Ti6Al4V is a little bit thinner
than that a pure titanium.

- for the same nitriding conditions, the depth of the diffusion
layer is always thicker on pure titanium than on Ti6Al4V. Perhaps,
in the presence of alloy elements, such as aluminium and vanadium,
the Ti6Al4V nitrided surface acts as a diffusion barrier (TiN,
Ti2N,Ti2AlN) this retarding diffusion layer formation [5].

- the depth of the diffusion layer increases slowly with treatment
time, but quickly with nitriding temperature.

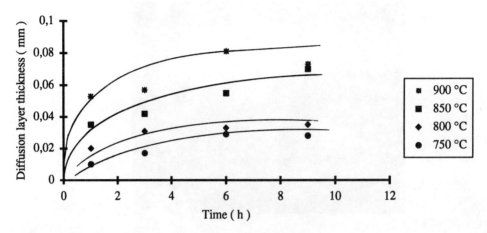

Fig. 3 : Ionic nitriding of pure titanium.
        Variation with temperature and time of the diffusion
        layer thickness.

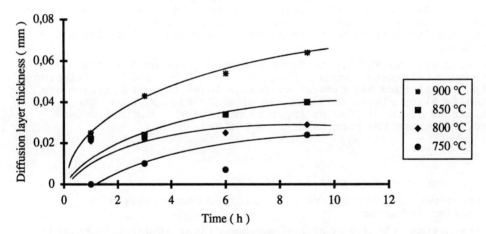

Fig. 4 : Ionic nitriding of Ti6Al4V alloy.
        Variation with temperature and time of the diffusion
        layer thickness.

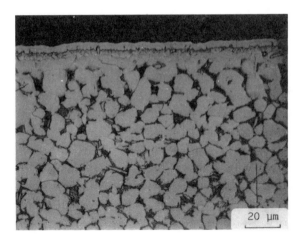

Fig. 5 : Optical micrograph of cross section of Ti6Al4V alloy
          plasma nitrided in pure nitrogen for 9 h at 900 °C.

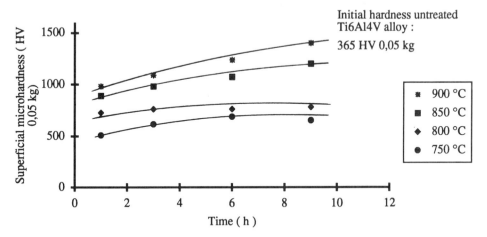

Fig. 6 : Ionic nitriding of Ti6Al4V alloy.
          Superficial microhardness (HV 0,05 kg ± 5 %) as a
          function of temperature and time.

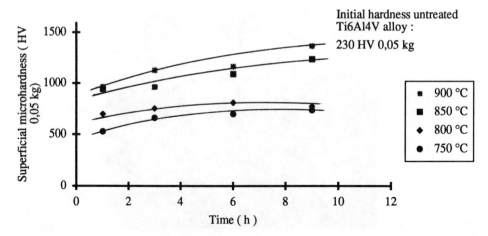

Fig. 7 : Ionic nitriding of pure titanium.
Superficial microhardness (HV 0,05 kg ± 5 %) as a
function of temperature and time.

The superficial microhardness

Microhardness of the nitrided surfaces of specimens were measured using a Vickers indenter with a 50 g load.

The influence of temperature and nitriding time on the superficial hardness is shown on figures 6 and 7.

The superficial hardness increases continuously with time and particularly with the temperature. Similar observations were made by METIN and INAL [6]. These results show that the superficial hardness level depends on the TiN + Ti2N layer thickness and the hardness of the diffusion layer.

The grain size

We have investigated the effect of temperature and time on grain size of Ti and Ti6Al4V samples. The average size of the grains increases with temperature and time of the treatment.

The grain size variations are not very important for Ti6Al4V alloy, except for the treatment at 850°C/900°C during 9 hours.

In the case of pure titanium, long treatment time at 900° C results in an important and irreversible grain growth.

The surface roughness

The surface roughness is deteriorated when the temperature and time increase (Fig. 8).

For a nitriding treatment at 900° C during 9 hours, the surface roughness is 450 % up on the initial value (0,08 to 0,35 µm).

This can be explained by the bombardment effect occuring during the ionic etching.

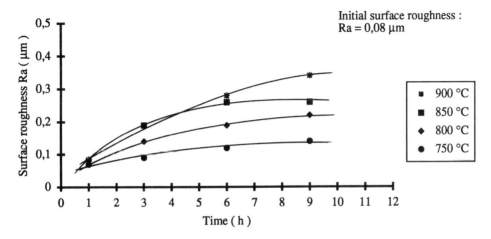

Fig. 8 : Surface roughness as a function of temperature and time - material : Ti6Al4V alloy

INFLUENCE OF THE PRESSURE

Different parameters have been tested to determine the influence of operating pressure.

Firstly, we have examined the compound thickness after plasma nitriding at 850° C during 6 hours for 3 pressure levels : 2, 4 and 6 mbar.

The results presented in Table 2 show that the thicker compound layer is obtained when the pressure decreases at about 2 mbar.

It seems that the sputtering effect occuring with the glow discharge can perhaps explained these results.

	Compound thickness ($\pm 0,5$ µm)	
Substrate Pressure	Titanium	Ti6Al4V
2 mbar	3 µm	3 µm
4 mbar	2 µm	2 µm
6 mbar	2 µm	2 µm

Table 2 : Ionic nitriding of pure titanium and Ti6Al4V alloy at
850 °C during 6 h.
Influence of operating pressure on the compound
thickness.

Secondly we have examined the influence of the pressure on the
diffusion layer depth and the superficial microhardness. No
significant difference was noticed. Finally we have measured the
surface roughness for the 3 different pressure levels. The surface
roughness increases when the pressure increases from 2 mbar to 6
mbar (see Table 3).

Pressure (mbar)	Surface roughness Ra (µm)
2	0,22
4	0,25
6	0,26

Initial roughness = 0,08 µm

Table 3 : Ionic nitriding of pure titanium at 850 °C during 6 h.
Influence of the pressure on the surface roughness.

INFLUENCE OF ELECTRICAL PARAMETERS

As previously stated, all treatments were carried out using a pulsed plasma (6-7 KHz) obtained by switching the discharge power supply on and off at microsecond intervals.

For different plasma nitriding cycles, we have studied the influence of the electrical impulse duration, corresponding to the period fraction, called : cyclic ratio or Rc (Fig. 9).

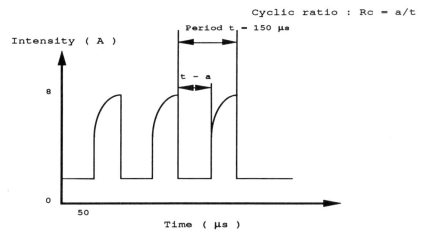

Fig. 9 : Shape of the current on an oscilloscope

Experiments were carried out to determine the influence of different cyclic ratio "Rc" (1/5, 1/4, 1/3) on the compound and diffusion layers.

It appears that these both layers thickness decrease very slowly when "Rc" decreases from value 1/3 to 1/5. This result is not very significant.

On the other hand, a surprising influence of the cyclic ratio on the superficial microhardness of the ion nitrided titanium and Ti6Al4V alloy was observed. For instance, with increasing "Rc" from 1/5 to 1/3, the superficial microhardness of the ion nitrided pure titanium increases from about 900 to more 1100 HV 0,005.

The influence of "Rc" on the surface roughness was investigated. The results are shown in Table 4.

It appears that the surface finish is deteriorated when the cyclic ratio increases. These results are probably related to the ion bombardment effect, that is more intense when the cyclic ratio or the electric impulse duration increases.

Cyclic ratio Rc	Surface roughness Ra (μm)
1/5	0,08
1/4	0,16
1/3	0,26

Initial roughness = 0,08 μm

Table 4 : Ionic nitriding of pure titanium and Ti6Al4V alloy at 850 °C during 6 h, pressure : 6 mbar, intensity : 8 A. Influence of the cyclic ratio on the surface roughness.

IONIC CARBURIZING OF PURE TITANIUM

Ionic carburizing treatment was performed in a pulsed plasma of a propane and hydrogen gas mixture.

In order to show the effects of gas carbon content on the microstructure, thickness and microhardness profile of the carburized layers, the plasma heat treatment was carried out in gas mixtures containing 3, 10 and 20 at % C in hydrogen.

After etching with an hydrogen glow discharge, the parts are heated in the same gas with the additional heating system. The specimens are all treated at 900°C for 6 hours and the gas pressure was adjusted to 6 mbar.

Microscopic examination has revealed that the compound layer produced with different carbon content in the atmosphere had an uniform and homogeneous microstructure.

After treatment, in the gas mixture with 10 and 20 at % carbon, the specimens are formed :

- immediatly on the surface, by a dense and amorphous carbon film (Fig. 10),

- by a very thin and almost pore-free TiC (Titanium Carbide) compound layer,

- and by a diffusion layer.

The thickness of these different layers is shown in Table 5.

Fig. 10 : Optical micrograph of cross section of pure titanium
plasma carburized in 20 at % C (C_3H_8-H_2) for 6 h at
900 °C.

at % C in C_3H_8-H_2	Thickness (microns)	
	TiC	a - C
3	5	0
10	1,5	6
20	1	11

Table 5 : Thickness of the TiC and amorphous carbon layers % C in
the C_3H_8-H_2　gas mixture obtained after carburizing at
900 °C during 6 h.

The use of 3 at % carbon in the atmosphere formed a TiC compound
layer of 5 μm thick and suppressed the formation of the amorphous
carbon film (Fig. 11).

Fig. 12 shows the hardness profiles of ion carburized pure titanium
with 3 different gas carbon contents. We can observe that the case
depth level is higher for decreasing carbon content in the gas
atmosphere.

The results we have obtained show that when the gas atmosphere contains more that 3 at % C, the rate of carbon mass transfer to the sample surface exceeds the rate of diffusion into the sample. It is the reason why when the treatment gas is too rich in carbon an amorphous carbon film is formed on the surface and hindered the carbon migration into the diffusion layer.

Fig. 11 : Optical micrograph of cross section of pure titanium plasma carburized in 3 at % C (C3H8-H2) for 6 h at 900 °C.

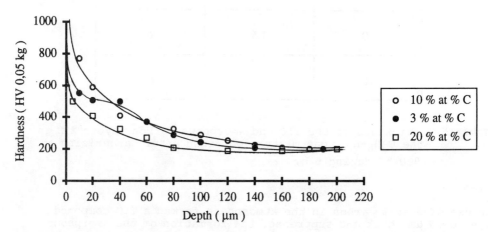

Fig. 12 : Ionic carburizing of pure titanium at 900 °C during 6h. Microhardness profiles evolution versus % C in the C$_3$H$_8$-H$_2$ gas mixture.

CONCLUSION

The effect of temperature and nitriding time on different physical parameters of plasma nitrided titanium and titanium alloy Ti6Al4V has been investigated.

The results obtained can be summarized as follows : a rise in temperature and treatment time results in an increase of :

- the compound layer thickness

- the diffusion layer depth

- the superficial microhardness

- une surface roughness

- the grain size.

A low operating pressure (2 mbar) seems to be a suitable factor to obtain the maximal compound layer thickness and a smooth surface finish.

The influence of the electrical impulse duration used in pulsed plasma nitriding was investigated. It was found that the best surface finish is obtained for the lower cyclic ratio value and on contrary, the surface microhardness is maximum for a highest cyclic ratio value.

Plasma carburizing of pure titanium in a gas containing 3 at % C results in the formation of titanium carbide in the compound layer and a carbon diffusion layer in the titanium matrix.

At higher gas mixture carbon contents, compound layer formation is slowed down by the formation of an amorphous carbon thin film.

REFERENCES

(1) E. Rolinski, Surface Eng., 1986, Vol. 2., N° 1., 35-42.

(2) K.T. Rie, Th. Lampe, S. Eisenberg, Surface Eng., 1985, Vol. 1., N° 3, 198-202

(3) A. Raveh, G. Kimmel, U. Carmi, A. Inspektor, A. Grill, R. Avni, Surface and Coatings Technology, 36, 1988, 183-190.

(4) J.M. Molarius and al, Härterei, Techn. Mitt., 41, 1986, 6; 391-397.

(5) Th. Lampe, S. Eisenberg, Intern. Seminar on Plasma Heat Treatment, Senlis (France), 21-23 sept 1987, PYC Edition.

(6) E. Metin, O.T. Inal, Proceedings Intern. Conf. on Ion Nitriding, ASM, Cleveland (USA), 15-17 sept 1989.

V. INDUCTION HEAT TREATMENT

Materials Science Forum Vols. 102 - 104 (1992) pp. 319-334
Copyright Trans Tech Publications, Switzerland

THE INFLUENCE OF TEMPERING AND SURFACE CONDITIONS ON THE FATIGUE BEHAVIOUR OF SURFACE INDUCTION HARDENED PARTS

P.K. Braisch

Institut für Werkstoffkunde, TH Darmstadt, FRG

ABSTRACT

Taking into account tempering and surface conditions a model of the fatigue strength mechanism of surface induction hardened parts has been deduced. The validity of the presented model is illustrated by experimental data. Rotating bending fatigue tests have been performed with unnotched and notched surface induction hardened specimens of different types of steel, tempered at different temperatures and with different surface roughnesses. Both model and tests lead to clear criteria for the assessment of optimal tempering conditions from a fatigue resistance viewpoint and for surface roughness from a economical viewpoint. It has been shown that the residual stresses are the governing factor of the fatigue behaviour.

1. INTRODUCTION

Typical for the cross-section of surface stregthened parts is the nonlinear distribution of the local fatigue resistance in function of the surface distance. The goal of the different methods of surface strenghtening is an optimal adaptation of this distribution on the generally as well nonlinear distribution of the load stress. It has been demonstrated in the early sixties[1] that surface induction hardening(IH) is a powerfull tool to achieve this goal.

As shown in figure 1, the local fatigue resistance at a certain surface distance can be interpreted as the concurrence of two components, which both can be determined experimentally: the local intrinsic fatigue resistance R_i^* and the contribution of the local residual stresses σ_{rs}. We call the resulting action of these two components as local effective fatigue resistance. With the goal to generalize all conclusions, we introduce the related surface distance $r(r = 0$ on the surface, $r = 1$ in the center).

It results from figure 1(right side) that from a fatigue behaviour viewpoint it is imperative but sufficient to know the amounts of the effective fatigue resistance on the surface

(1) $R_{fs}^{*} = R_{fs} - m_s \cdot \sigma_{rs,s}$

and on the core border

(2) $R_{fc}^{*} = R_{fc} - m_c \cdot \sigma_{rs,c}.$

As conventional, the contribution of the residual stresses is taken into account by the factors of influence m_s resp. m_c, which are determined as well by experimental methods and range in the order of magnitude from 0.3 to 0.5. As suggested by figure 1, the tensile residual stresses on the core border of shallow hardened parts are detrimental for the fatigue limit of the specimen, if a tensile load stress governs the considered point. The equivalent is true for the compressive residual stresses on the surface, if we suppose here a compressive load estate.

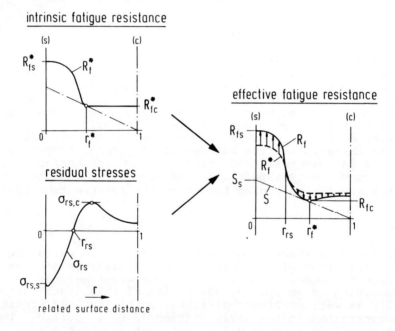

Fig. 1. Concurrence of local intrinsic fatigue resis-
        tance and residual stresses set against the
        distribution of load stresses

We call

(3) $\lambda_f^{*} = R_{fs}^{*} / R_{fc}^{*}$

respectively

(4) $\lambda_f = R_{fs} / R_{fc}$

strengthening ratio and consider for a first approach both as equal.

In a more detailed manner we have to take into account a surface degradation, for instance by oxydation, decarburation, roughness a.o. and correspondingly a reduction of the local intrinsic fatigue resistance up to a certain depth r_s. As well we have to take into account an often observed diminution of the compressive residual stresses near the surface. As suggested by figure 2, such a degradation ΔR_{fs} of the local effective fatigue resistance on the surface has no influence on the fatigue behaviour of the surface strenghtened part as a whole as long as a critical amount

(5) $\Delta R_{fs,cr} = R_{fs} - S_s$

is not achieved(S_s - nominal surface load stress).

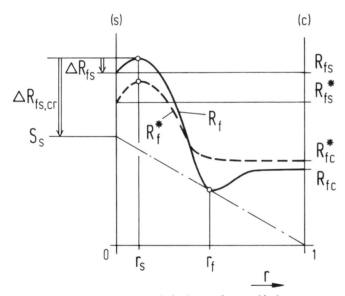

related surface distance

Fig. 2. Diminution of the local effective fatigue resistance near the surface due to roughness, decarburation or similar influences

Figure 3 shows the comparison of the distributions of the local effective resistance before and after tempering. I4 is important to realise that due to the reduction of the tensile stresses on the core border by tempering, the effective fatigue resistance on this point is increased and as a result of this improvement a considerable gain of fatigue strength of the surface hardened part as a whole is available. It is evident that the gain ΔS_s is as higher as shallower the hardened depth. On the other side, it is evident that the gain which can be achieved depends on the tolerated amount of surface degradation ΔR_{fs}.

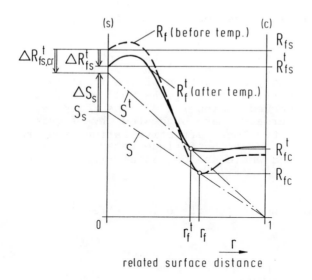

ΔS_s - gain available by tempering

Fig. 3. Augmentation of the endurance limit available by
tempering due to the reduction of tensile residual stresses on the core border

2. MODEL OF THE FATIGUE BEHAVIOUR MECHANISM

To simplify matters, we represent for the following reflections, as
done in figure 4, the distribution of the effective local fatigue
resistance as a curve with two steps and define the hardness depth
h_f as the surface distance up to the core border. A second
simplification relates to the distribution of the local load
stresses, which is represented for the first approach as linear.

From the left side of figure 4 it becomes then evident that
increasing the hardened depth, a gain ΔS_s of endurance limit is
available for the part as a whole. But this only up to a critical
hardness depth $h_{f,cr}$, as represented on the right side of figure 4,
when the probability of the localization of the crack initiation
site is equal for both the core border and the surface.

Corresponding to this fact of matter we have to distinguish between
subcritical and critical or overcritical strengthened parts. This
distinction has a fundamental importance for the understanding of
the different aspects of the fatigue behaviour of induction
hardened parts.

Conected to the related surface distance we introduce the related
strengthened depth

(6) $r_f = h_f / (d_a / 2)$

where h_f is the absolute strenghtened depth and d_a the diameter of the considered cross-section.

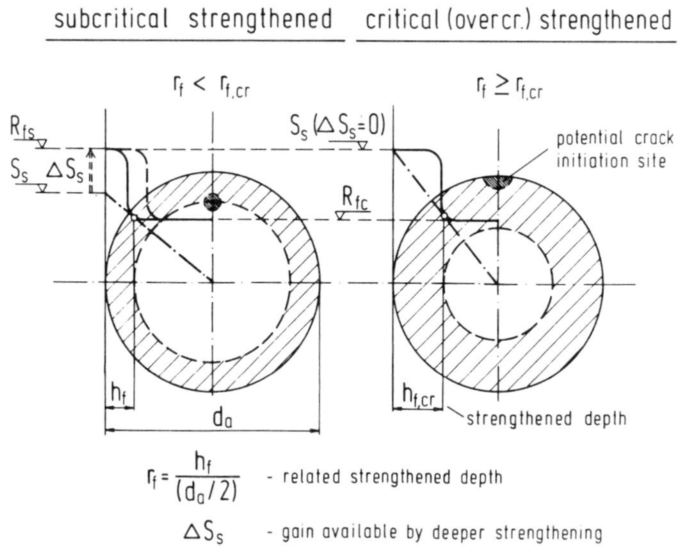

Fig. 4. Augmentation of the endurance limit by deeper strengthening

As it can easily be deduced from figure 4 and equation (6), we obtain for subcritical strengthened parts the relationship

(7) $\chi^o = S_s / R_{fc} = 1/(1 - r_f)$.

Equation (7) represents the increasing rate of the endurance limit of surface strengthened unnotched parts, χ^o, in comparison with thus of the nonstrengthened parts, taken as reference case: $\chi^o(r_f = 0) = 1$. The increasing rate is a function of the related strengthened depth and follows an hyperbolic rule.

If we turn now to notched parts and replace for simplification the curved load stress distribution in the notched area by two straight lines, we find/2, 3/ similar to equation (7) the approximation relationship

(8) $\chi^{n,o} = S_s \cdot K_t / R_{fc} = 1/(1 - r_f \cdot K_t))$,

where K_t is the stress concentration factor.

Equation (8) represents the model of the fatigue behaviour mechanism for notched(n) and unnotched(o) surface strengthened parts. The graphs of the two cases of equation (8) are shown in figure 5. The two hyperbola represent the respective subcritical strengthened ranges. It is important to realize that for a certain range of related strengthened depth the endurance limit for notched parts is higher than that of unnoched parts.

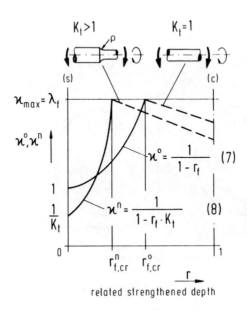

Fig. 5.   Model of the mechanism of fatigue beha-
          viour of surface induction hardened parts

As mentioned, the maximal endurance limit is achieved for the critical related strengthened depth, namely when $S_s \cdot K_t = R_{fs}$ (see figure 4). Combining this condition with eq. (4) and eq. (8) yields to

$$(9) \qquad X_{max}(r_f = r_{f,cr}) = R_{fs}/R_{fc} = \lambda_f$$

and corresponding to

$$(10) \qquad r_{f,cr}^{n,0} = (1/K_t) \cdot (1 - 1/\lambda_f),$$

the critical related strengthened depth for the notched and un-notched case.

It is important to see that the strengthening ratio is taken equal for both cases. This as a consequence of the statistical proved results, which indicate a different action of the residual stresses on the local effective fatigue resistance at the surface of a notched and an unnotched area/4, pag. 249/.

In other words, due to the fact that

$$(11) \qquad R_{fs}^n \cong K_t \cdot R_{fs}^0$$

we can set

(12) $\lambda_f^n = \lambda_f^0$,

where (^{n}) and (^{0}) indicate, as used before, the notched and the unnotched case.

We have to keep in mind that the potential crack initiation site is by definition for subcritical strengthened parts on the core border and for critical or overcritical parts on the surface. The latter is the reason for that strengthening beyond the critical related strengthened depth does not lead to a further increasing but to a reduction of the endurance limit, due to an inherent detrimental evolution of the residual stresses and/or the surface estate.

PERIMENTAL

Rotating bending fatigue tests have been carried out on notched and unnotched IH-specimens of 4 types of steels(details are shown in table 1 respectively in figure 6). In the following we represent only the main results of these tests. The results of accompanying investigations, concerning residual stress distributions and the change of theirs in function of the load level or the tempering temperature were represented elsewhere/4 to 6/.

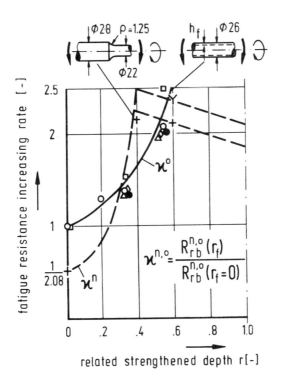

Fig. 6. Fatigue resistance increasing rate of untem-
        pered surface induction hardened specimens
        (for symbols see table 1)

The related strengthened depth r_f was varied in 4 steps between 0 and 0.65, the tempering temperature in 5 steps between RT and 300 °C and the roughness in 3 steps between 3 and 100 μm.

Table 1: Characteristics of tested specimens

K_t		1		2.08		R^o_{rb} /MPa/	λ_f
Rz/μm/		3	100	3	100	($r_f = 0$)	/-/
42 CrMo 4 V	○	●,ø	+	–		437	2.15
Ck 45 N	□	–	×	–		263	(~ 3.4)
Ck 45 V	◇	–	Y	–		370	(~ 2.4)
49 MnVS 3 BY	△	–	–	–		357	(~ 2.5)

- nominal chem. composition see DIN 17200; V - hardened and tempered; N - normalized; BY - controlled cooling from forging temperature(microalloyed steel); Rz - roughness parameter DIN 4768. Additionally characteristics see /7/.

Column 6 of table 1 shows the rotating bending fatigue resistances of the unnoched and nonstrengthened specimens, $R^o_{rb}(r_f = 0)$. Column 7 contains the corresponding strengthening ratios λ_f before tempering. From the latter only the amount for 42 CrMo 4 V is based on special experimental investigations/8/, for the other steels the amounts are estimated.

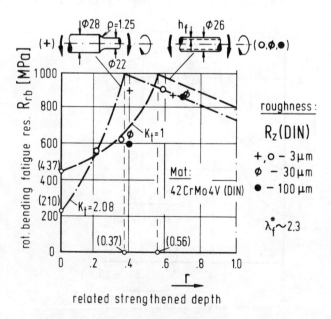

Fig. 7. Rotating bending fatigue resistance of un-
        tempered IH 42 CrMo 4(DIN) steel specimens
        of different roughnesses

Figure 6 shows the plots of the increasing rate of the endurance limit for the nontempered specimens, determined on the base of statistically analyzed S-N-curves in function of the related strengthened depths. It can be observed that the plots fit in an acceptable manner with the prediction based on equation (8).

Figure 7 extracts from figure 6 the results as absolute amounts for the steel 42 CrMo 4 V.

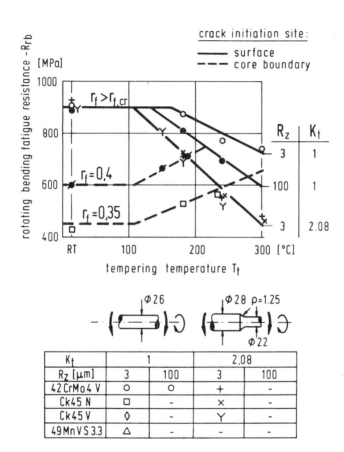

Fig. 8. General diagram of fatigue limits of IH-specimens in function of related strenthened depth,
type of steel, shape, roughness and tempering
temperature

Finally figure 8 shows in a summarizing diagram the fatigue limits for different steels, related strengthened depths, specimen shapes and roughnesses in function of the tempering temperature. It reveals that tempering affects the fatigue behaviour of subcritical and overcritical hardened specimens in a contrary way but both beginning from the same start temperature from about 120 to 150 °C.

Beginning from these temperatures starts the relieve of the residual stresses by tempering, and that in a pronounced higher proportion to the temperature than the relieve of th% hardness/4/. Accordingly the change of the fatigue limit must be attributed essencially to the variation of the residual stresses: The reduction of the compressive residual stresses on the surface of the critical strengthened specimens(i.e. surface crack initiation site) leads to a diminution and opposite, the reduction of tensile residual stresses on the core border of the subcritical strengthened specimens(i.e. core border crack initiation site) leads to an augmentation of the fatigue life.

In the case of critical strengthened specimens the relieve of the surface compressive residual stresses affects notched specimens principally in the same manner but with higher intensity as unnotched specimens with different roughnesses. We call this behaviour as a continous transition from a micro- to a macro-notched estate, which is obviously governed by the second term of the effective surface fatigue resistance in equation (2), the contribution of the local residual stresses. The first term, the local intrinsic fatigue resistance, seems to play a secondary role.

The influence of the initial metallurgical structure(i.e. before IH) on the fatigue behaviour of subcritical strengthened parts reveals from a comparison in figure 8 of the specimens with a nearly equal related strengthened depth but different structures, i.e. the quenched and tempered 42 CrMo 4 V steel(r_f = 0.4) and the normalised Ck 45 N steel(r_f = 0.35). Connected to this note the influence of the strengthening ratio λ_f in equation (10) on the critical related strengthened depth. The latter is for 42CrMo 4 V ca. 0.56 and for Ck 45 N ca. 0.7(see table 1).

It is worth to note that due to the limited hardenability(for the disponibel batch a maximum of 7 mm strengthened depth was achievable) the notched specimens of Ck 45 N steel does not achieved the critical strengthened depth for the dia. 22 mm, which is ca. 0.7x11 = 7.7 mm for the unnotched and ca. 7.7/2.08 = 3.7 mm for the notched area(see equation 10)). Due to this fact the untempered variant of these specimens does not fractured in the notched but in the unnotched specimen area/5/.

Finally we have to retain from figure 8(see trace for K_t = 2.08) that for critical resp. slight overcritical strenghtened specimens the chemical composition or/and the initial structure of the steel does not influence significantly the fatigue behaviour.

4. CONCLUSIONS AND DISCUSSION

A general survey of the results leads to the following conclusions:

4.1 The presented model of the fatigue mechanism permits an useful prediction of the fatigué behaviour of surface induction hardened(IH-) parts, taking into account the influence of the shape, the surface and the tempering estate of the specimen, all of them in function of the hardened depth.

4.2 The main factor of influence which governs the whole fatigue behaviour of IH-parts is the residual stress estate. Due to his influence

> critical and slight overcritical nontempered IH-parts
> show no difference in the absolute amount of the ro-
> tating bending fatigue limit in function of surface
> estate, type of steel or shape of the part.

A differentiation of the fatigue behaviour emerged only when the compressive residual stresses on the surface of critical IH-part are reduced by tempering(or overloading) beyond a critical limit. This differentiation can be interpreted as a

> continuous transition behaviour, first from an un-
> notched to a micro-notched and then to a macro-
> notched estate.

4.3 The intrinsic fatigue resistance of the martensitic hardened surface layer shows in the investigated range no significant influence of the chemical composition or the initial structure of the steel.

4.4 At subcritical IH-parts tempering leads, as was to be expected, to an increasing of the endurance limit, due to a substantial reduction of the tensile residual stresses on the core border. In function of the actual related strengthened depth and the surface estate(micro- or macro-notch),

> beyond a certain critical limit of the tempering tem-
> perature, the fatigue behaviour of a subcritical IH-
> part changes into the behaviour of a critical IH-part.

The latter due to the fact that on the surface of the part both the local intrinsic fatigue resistance and the contribution of the compressive residual stresses are reduced by tempering in such an extent that now the probability of crack initiation is equal for the surface and the core bord%r.

> As long as a part ranged in the subcritical streng-
> thened domain, the endurance limit is determined by the
> two terms of the effective fatigue resistance on the
> core border: the local intrinsic fatigue resistance and
> the amount of the local residual stresses and follows,
> concerning the strengthened depth, the hyperbolic rule,
> given by the model of the fatigue behaviour mechanism.

The method of practical and combined use of the two diagrams of fatigue behaviour, i.e. figure 6 and figure 8, is illustrated sche-matically in figure 9.

First we have to calculate the related strengthened depth r_{f1} and enter the corresponding diagram of the fatigue mechanism(e.g. for $r_{f1} < r_{f,cr}$, point 1 in figure 9, left side), which can be ploted in absolute, in mixt or in related quantities. Then we enter the summarizing diagram, in our example following the points 2, 3 and 4, where we find the endurance limit corresponding to a certain tempering temperature. As it can be deduced from the figure 9(right side), the endurance limit can be increased considerably(point d), if the tempering temperature is raised, but this only up to the corresponding critical amount $T_{t,cr}(r_f)$.

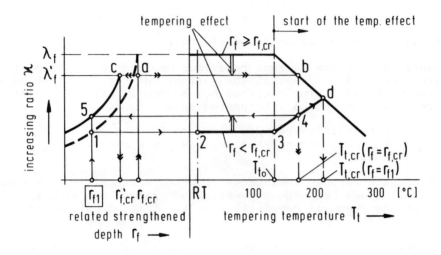

Fig. 9. Principle of the combined use of the fatigue behaviour diagrams of surface induction hardened parts

The way is similar for critical strengthened parts. In such a case we follow the points $r_{f,cr}$, a and b. If desired, we can construct on the left side of the figure with the help of the points 5 and c, the diagram for tempered specimens.

5. APPLICATIONS

To illustrate how we can use the presented results in engineering practice, we take the example of a rear axle of a vehicle. We assume pure alternating torsion load. As shown schematical in figure 10, we have to distinguish between notched(I, II and IV) and unnotched(III) cross-sections. For all sections we assume an absolute hardness depth of h_f = 5 mm. That is on the limit of the hardenability of the choiced unalloyed steel Ck 45 N. For the supposed range of diameters from 27 to 30 mm, the corresponding related strengthened depth is then about r_f = 5/13,5 = 0.37.

For the unnotched section III we assume two alternative cases: an undamaged surface estate(e.g. Rz = 3 μm) and a surface with a very high roughness(e.g. Rz = 100 μm). As suggested schematically by figure 10(bottom, left side), even the assumed high degradation of the surface estate does not injure the fatigue behaviour of the part. Due to the fact that the considered cross-section is subcritical strengthened(r_f < $r_{f,cr}$ = 0.7, see eq.(10) and table 1), the endurance limit is determined by the effective fatigue resistance on the core border. Given this restriction in the cross-section III and assuming a ratio of 1.6 of the fatigue limits of rotating bending to alternating torsion, we find from figure 8 that

the obtainable endurance limit for the part as a whole is for the
untempered estate 290 MPa and for the tempered estate(180 °C/1h)
350 MPa. It is important to see that an higher quality of the
surface estate than one corresponding to Rz = 100 µm gives no
advantage.

Considering now the notched sections I, II and IV, for which we
assume generally K_t = 2,2, we are in a different situation. As it
results anew from equation (10) and table 1, all these sections are
overcritical strengthened(r_f > $r_{f,cr}$ = 0.32). That means the
potential crack initiation site is located for these sections on
the surface. Tempering(e.g. at 180 °C) yields correspondingly to a
reduction of the endurance limit from 575 MPa to 460 MPa(see figure
10, bottom right side).

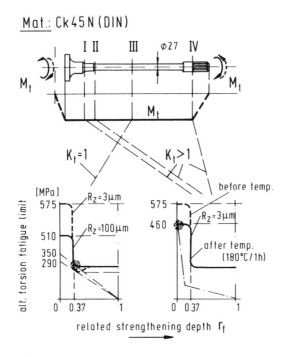

Fig. 10. Assessment of the fatigue strength of a rear
         axle considering a surface roughness(DIN)Rz =
         100 µm and tempering at 180 °C/1 h

As a conclusion of the analyzed example we have to retain that the
most endangered cross-section of the part is situated in the unnotched shaft area. If an higher endurance than the 350 MPa limited
by the unnotched shaft area is required, we have to change the
initial metallurgical structure from a normalized to hardened and
tempered estate or the type of steel, going may be to a low alloyed
steel with higher hardenability than the 5 mm achievable with the

Ck 45 N steel.

The importance of an optimized tempering is illustrated in figure 11. We assume for simplification(see note figure 11) the case of pure alternating bending load of a front spindle axle with the most endangered cross-section in point I. We suppose further similar conditions as in the preceding example(Ck 45 N steel, d_a = 30 mm, h_f = 4 mm, r_f = 0.26 $< r_{f.cr}$ = 0.7).

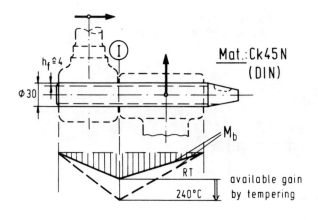

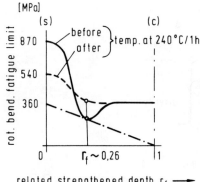

related strengthened depth r_f ⟶
(cross-section I)

Fig. 11.

As3essment of alternating bending fatigue strength of a front spindle axle, considering tempering at at 240 °C/1 h(about the influence of mean stresses see /9/)

As in the preceding example, the potential crack initiation site is located on the core border. Tempering at 240 °C, the critical tempering temperature for the assumed case, increased the endurance limit of the component as a whole from 360 MPa to 540 MPa. It is obvious that combined with the increased endurance we have achieved a substantial augmentation of toughness, an important and desired property for the considered component.

SUMMARY

It has been shown that the fatigue behaviour of surface induction
hardened parts can be predicted with the help of a model of the fatigue mechanism and experimental determined data concerning the intrinsic fatigue properties of the material and the residual stress
estate of the considered part. The employed intrinsic fatigue
properties come from investigations conducted under the strict
conditions of a short time austenitization.

It appeared that the residual stress estate is the governing factor
of influence.

The presented model of the fatigue mechanism distinguishes subcritical and overcritical surface strengthened parts.

Subcritical parts are characterized by the fact that the potential
fatigue crack initiation site is located per definition on the core
border. The endurance of such parts can be substantially increased
by tempering. A degradation of the surface estate up to considerable amounts does not injure the fatigue behaviour. Both facts can
be used for important cost saving.

At critical(or overcritical) strengthened parts the potential crack
initiation site is located on the surface. No significant influence
of the chemical compositon of the steel can be observed. In the
nontempered estate roughness and notches have no influence on the
fatigue behaviour. In the tempered estate emerged a continuous
transition of the diminution of endurance limit from the micro-
notched(roughness) to the macro-notched estate(e.g. shoulders and
grooves).

It results the presumption that other types of surface degradation
can be treated in the same manner as shown for the roughness.

REFERENCES

/1/ Wuerfel, R.K.: Characteristics of Induction Hardened Axles.
Metal Progress, December 1963, p. 93/95.

/2/ Braisch, P.K.: Über die Wirkung einer Randschichtverfestigung
auf die Schwingfestigkeit von Proben und Bauteilen, dargestellt am
Beispiel der induktiven Randschichthärtung. Dr.-Ing.-Diss. 1981, TH
Darmstadt, FRG.

/3/ Braisch, P.K.: The principal factors influencing the fatigue
behaviour of surface induction hardened machinery parts. Int. Conf.
on Fatigue of Engineering Materials and structures. Sheffield(GB),
15-16 Sept.1986. IMechE conference publication 1986-9, p. 489.

/4/ Braisch, P.K., Kloos, K.H.: Influence de l'etat de surface et
de la temperature de revenue sur la tenue en fatigue des pieces
trempees par induction. Conf. Proc. of: ATTT 90 - Internationaux de
France du Traitment Thermique, Le Mans(F), 19-21 Sept. 1990.

/5/ Kloos, K.H., Braisch, P.K.: Schwingfestigkeit induktiv randschichtgehärteter bauteilähnlicher Proben. Forschungsbericht Inst.
für Werkstoffkunde, TH Darmstadt, 1988, Darmstadt(FRG).

/6/ Kloos, K.H., Braisch, P.K.: Schwingfestigkeit induktiv randschichtgehärteter bauteilähnlicher Proben - Anlaß- und Rauheits-
wirkung. Forschungsbericht Inst. für Werkstoffkunde, TH Darmstadt,
1990, Darmstadt(FRG).

/7/ Braisch, P.K.: Zur Gestalt der Wöhlerlinie umlaufbiegebelaste-
ter, induktiv randschichtgehärteter bauteilähnlicher Proben. Conf.
Proc. of: Induktives Randschichthärten. AWT-Tagung 23. bis 25 März
1988 Darmstadt. Publ. of AWT, editors Kloos, K.H./Grosch, J.

/8/ Braisch, P.K., Simon, A., Beck, G. and Kloos, K.H.: Effect of a
Short Time Austenitization on Mechanical Properties of Steels. In:
Proc. of the 7th Intern. Conf. on the Strength of Metals and
Alloys, Montreal, Canada, 12-16 August 1985, Vol. 2, p.983/88.

/9/ Doege, E., Golze, N.: Einfluß der Oberflächennachbehandlung auf
die Schwingfestigkeit von Stahl unter Berücksichtigung vereinfach-
ter Wärmebehandlungen. Inst. für Umformtechnik und Umformaschinen,
Univ. Hannover, Forschungsbericht 5225(1987), FRG.

Materials Science Forum Vols. 102 - 104 (1992) pp. 335-344
Copyright Trans Tech Publications, Switzerland

INDUCTION HEAT TREATMENT OF CASE HARDENING STEELS

P. Archambault, M. Pierronnet, F. Moreaux (a) and Y. Pourprix (b)

(a) LSG2M, CNRS URA 159, Ecole des Mines, Parc de Saurupt,
F-54042 Nancy Cédex, France
(b) PEUGEOT SA, Centre Technique de Belchamp
F-25420 Voujeauxourt, France

ABSTRACT

The superficial induction heat treatment of mechanical parts has become more accurate with the emergence of dual frequency induction heating (contour hardening). However the development of this technique is limited by several factors (cost, time control...).
Obtaining high resistance properties can be achieved by coupling carburizing and surface induction heat treatments.
This paper deals with the influence of the heating and cooling kinetics on the structural and mechanical behaviour of the superficial zone of a case hardened cylindrical steel sample.
A large panel of heat treaments is possible thanks to a real time-controlled system developped in the LSG2M (induction heating coupled with water spray cooling). High cooling rates (200 °C/s) are obtainable and such linear coolings can be interrupted around the Ms temperature for an isothermal holding before the martensitic transformation.
We show several examples of different superficial heat treatments and their consequences on hardness penetration and on residual stress profiles in the heat affected zone.

I - INTRODUCTION

The superficial induction heating is more and more used in heat treating because of energy and rough material savings.

It allows decreasing the duration of heat treatment and so increasing the productivity. It gives also better conditions for automating the production and to make the heat treatments directly in the production line.

For some years the contour induction hardening technique (dual-pulse induction method) has been used[1]. This technique allows heating of a very thin surface layer and seems very interesting for the hardening ot teeth gears. However, only the heating is optimized, the piece being cooled by water aspersion without any cooling control.

We present here the first results obtained by induction heat treatment of case-hardened steels.

By using a real time controlled spray cooling system, one can realize very different cooling laws (linear and with an isothermal holding).

The main aim of the present study is the optimisation of the quench cooling in order to obtain a specific residual stress profile in the heat treated superficial zone leading to a better fatigue behaviour.

II - EXPERIMENTAL DEVICE

The installation consists in two parts : heating and cooling.

II.1 - Induction heating

- the generator (CELES) is an aperiodic one; its maximal power is 25 kW. The frequency varies between 100 and 500 kHz.

- the inductor coil is made of a copper tube of 4 mm ouside diameter.

- the sample is cylindrical : diameter 16mm, height 30 or 48 mm. It is fixed to a tube which allows the way for thermocouples. A thrustor permits a rapid transfer of the sample from the heating zone (up zone) to the cooling zone (bottom zone). The sample is continuously rotating (300 r/mn) to increase the thermal homogeneity all along the process (heating and cooling). The analog signals delivered by the thermocouples are transmitted to the computer through a rotating contactor.

II.2 - Spray cooling

- the nozzles are two phase-jets (GIESLER). The water enters the nozzle axially and is projected by air centrifugation. The jet homogeneity and the droplets velocity depend upon the air and water pressures and upon the distance between the nozzles and the sample [2].

- the disposition of the nozzles is optimized in order to increase the homogeneity of the jets and to obtain a sufficient cooling power in the pressure range of the two fluids. The optimal arrangement consists of two superposed rings of three nozzles each; the angle between the nozzles is 120° (Figure 1). The vertical distance between the two rings depends on the sample height (30 or 48 mm).

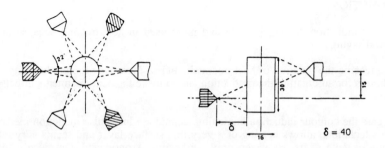

Figure 1 - Arrangement of the nozzles

The longitudinal homogeneity of the cooling is controlled by hardness measurements along one generatrix of the cylinder at 0.2 mm under the surface. In the case of a 200 °C/s cooling from 900 °C to room temperature, the hardness profile is the same than in the case of immersion quenching in water at 20 °C.

- electrovalves and proportional servovalves supply the system in air and water.

- the cooling control is performed by a sheathed thermocouple (K type - Ø 1 mm) localized at 1 mm under the surface in the mid-plane of the sample. For a better response in temperature we also use an infrared pyrometer.

II. 3 - Samples

We used cylindrical samples : Ø = 16 mm - H = 30 or 48 mm.

The experiments are led on two kind of steels :

- XC 48 normalized, the initial structure is ferrite pearlite. It is used to characterize the heating and cooling sequences.
- 20 MC5 carburized under low pressure and gaz quenched (% carbon on surface : 0.76, thickness of carburization : 0.7 mm).

III - COMPUTER CONTROLLED HEATING AND COOLING.

III.1- General remarks

In automatic mode, all the procedures (regulation, command and acquisition) are assumed by a microcomputer APPLE Macintosh II.

Because of the great number of calculations to be executed in real time, the minimal sampling time is 30 ms for heating and 50 ms for cooling.

Two other functioning modes are possible : open loop and continuous control. This gives a great flexibility to realize all kind of heat treatments .

In the open loop mode, the user enters the variations of the analog signal versus time for the control of the generator and/or the servovalves. The computer then outputs these analog signals and acquires the temperature signals synchronously (real time clock).

In the continuous control mode, the user enters a set curve (temperature versus time) and the regulation parameters. At every sampling step, the computer compares the measured temperature to the desired one. The difference between these two temperatures allows the calculation of the command signals for the HF generator (heating phase) and the air and water sevovalves (cooling phase).

III.2- PID control

The regulation process is based on several PID algorithms in order to get a more accurate response. The mean difficulty is the optimisation of the values of the PID coefficients. In the case of very hight heating or cooling kinetics, trials are necessary to evaluate these values as precisely as possible.

IV- EXPERIMENTAL RESULTS

IV.1- Controlled heating ·

Superficial heat treating of steels needs hight heating rates : rates of 200 °C/s are classical. In this case, to obtain a good regulation we have introduced two PID algorithms under and over the Curie temperature.

To identify the heating, the temperature is measured at different points along a radius in the center-plane of the cylindrical sample. The sample is cooled by immersion quenching in water at 20 °C. The variation of the surface temperature is followed thanks to a thermocouple made of thin wires directly welded on the surface itself. The temperature evolutions appears on Figure 2 for a 200 °C/s heating. The regulation uses the thermocouple located at 1 mm under the surface.

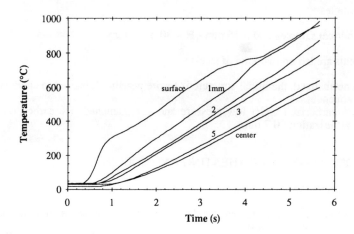

Figure 2 - Temperature variations during heating at different points of a sample radius
(XC 48 - Ø=16 mm, H=48 mm)

Figure 3 shows the radial corresponding temperature profiles at different heating times.

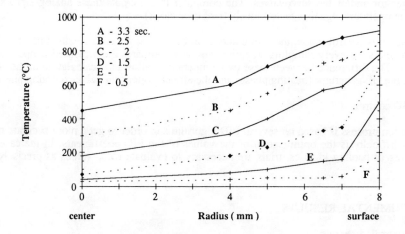

Figure 3 - Temperature profiles at different heating times
(XC 48 - Ø=16 mm, H=30 mm)

Between room temperature and 700 °C the eddy currents are localized in the contour of the sample so the temperature of the surface increases very quickly. Between 700 an 800 °C, the relative permeability of the steel decreases : it results in a reduction of the temperature gradient between the zones where the temperature exceeds the Curie point and the others. So one can observe that the temperature gradient in the surface layers becomes very low until the end of the heating.

From the analysis of these two figures, we also see that the centre temperature remains always beneath to 400 °C when the surface is at 950 °C. We are in the conditions of a superficial heating, the core of the sample is not transformed. In fact the thickness of the transformed zone is here between 3 and 4 mm. When the surface temperature is reached the sample must be cooled immediately because of the temperature gradients and so of the thermal conduction process which can induce a very fast increase of the core temperature and, by the way, its transformation.

IV.2- Controlled cooling

The real time controlled water spray cooling process allows the realization of cooling laws which are not possible through classical cooling processes with a very good reproducibility.

Two kinds of coolings are performed in the present study :
  - linear cooling : constant cooling rate between 200 and 20 °C/s.
  - linear cooling followed by an isothermal holding then linear cooling again. The holding temperature is near the martensitic transformation temperature Ms.

In the later case, and as the heating and cooling zones are separated, the isothermal holding is not perfectly controlled because of the low heat amount remaining in the sample which only allows short duration 'holdings' (less than 10 s). Of course this control is more difficult when the temperature holding is lowered. In that case one has not only to adjust the PID coefficients but also the drawing of the setpoint curve. Several trials are then necessary to obtain a good result.

An example of a controlled spray cooling (150 °C/s) is presented on Figure 4. The variations of the command signals for air and water indicate a correct regulation in the whole temperature range.

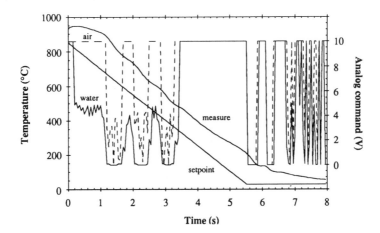

Figure 4 - Computer controlled spray cooling at 150 °C/s

V. INDUCTION HEAT TREATING OF CARBURISED 20 MC5 STEEL.

V.1- Definition of the heat treatments

- heating duration : 3.3 s. Surface temperature reached : 930 °C
- controlled coolings : linear with/without isothermal holding

V.2- Metallurgical analysis

The hardness penetration is evaluated by the measurement of the hardness radial profile in the heat affected zone (Hv 0.3 kg), after cut-off grinding of the sample in the center-plane and polishing. The results are presented on Figure 5.

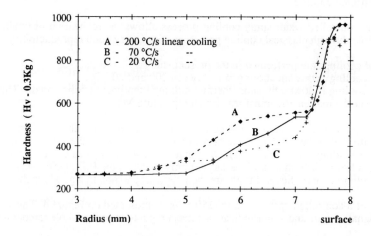

Figure 5a - Hardness profiles in a carburized 20 MC5 sample for different cooling rates
(same superficial induction heating)

The hardness decreases abruptly at the interface between the carburized zone and the 20 MC5 matrix in relation with the carbon profile in the sample. Higher is the cooling rate, higher is the hardness in the carburised zone. At 200 °C/s, the hardness level in the underlayer is characteristic of a relatively deep martensitic transformation (Figure 5a).

For the coolings with isothermal holding at 330 °C (Figure 5b) we can make the same analysis, the radial hardness profiles present the same shape. The treatments where the cooling rate before the holding is lower than the critical quenching rate of the 20 MC5 steel (120 - 140 °C/s) lead to a martensite - bainite struture in the underlayer [3]. In the case of curve A, the decrease in hardness observed at the surface is here an artefact which has been explained elsewhere [3].

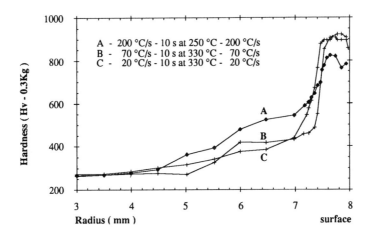

Figure 5b - Hardness radial profiles in a carburized 20 MC5 sample
after different coolings with isothermal holding
(same superficial induction heating)

V.3- Determination of the residual stress profiles

The residual stresses are measured by X-rays diffraction analysis, in the center-plane of the cy-lindre. To determine the stresses profile in the heated affected zone the measurements are perfor-med after successive chemical machinings of the sample. However, the results have to be corrected to take into account the evolution of the stress field due to the removing of metal after every machi-ning.

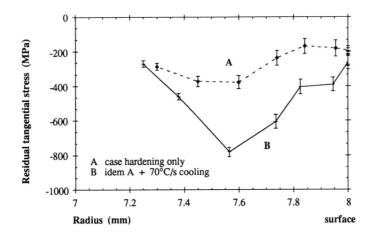

Figure 6a - Residual stress profiles in the surface layer of a 20MC5 cylinder

Two diffraction conditions are used :

    - cobalt radiation Kα (case of figure 6b) - diffraction plane 310 (martensite)
    - chromium radiation Kα (case of figure 6a) - diffraction plane 211 (martensite).

The figures 6a and 6b gives the results for a carburized sample (curve A) and for the same heat treated sample: induction heating and cooling at 70 and 200 °C/s respectively (curve B).

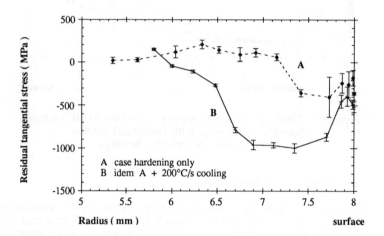

Figure 6b - Radial residual stress profiles (20MC5 cylinder)

For the case hardened condition, the amount of compressive stress does not exceed 400 MPa for the two samples.

For the other treatments, the compressive stress level can reach 1000MPa. One can see that the depth in compression is larger :

    - compared to the carburized sample,
    - when increasing the cooling rate, which is coherent with the hardness results.

Unfortunately, we have not yet the results of the stress profile determination in the case of a linear cooling interrupted by an isothermal holding. The advantage of such a treatment is to not superpose a thermal gradient to the Ms variations due to the carbon profile in the carburized zone. We have seen in that case that the hardness penetration is similar (figure5) but one can suppose that the stress profile is certainly different and by the way, the fatigue behaviour. Work is continuing to characterize this effect.

VI- CONCLUSIONS

We showed the first results of the superficial induction heat treatment of case hardened cylinders.

The real time control of the heating and cooling procedures gives reproducible results. Very different cooling laws can be realized which allows acting on the order of the transformations in the carburized zone and in the underlayer.

The determination of the stress profiles in the heat affected zone showed an increase of the level of the compressive stress and also of the width of the compressed zone with respect to the carburized samples.

Complementary stress measurements associated with fatigue behaviour experiments are then necessary to evaluate the performance of the coolings with isothermal holding.

All these results will be compared with the contour hardening technique.

REFERENCES

1) Oakley, G.A.: Heat Treatment of Metals, 1990, 4, 99.

2) Didier, G., Archambault, P., Moreaux, F. : Rev. Gén. Therm., 1983, 256, 333.

3) Pierronnet, M., Archambault, P., Moreaux, F., Clément, B., Beauget, M.: Traitement Thermique, 1988, 222, 35.

Materials Science Forum Vols. 102 - 104 (1992) pp. 345-364
Copyright Trans Tech Publications, Switzerland

SELECTIVE SURFACE TREATMENT OF GEARS BY INDUCTION PROFILE HARDENING

G.D. Pfaffmann

Tocco, Inc., Madison Heights, MI, USA

ABSTRACT

The presentation will outline the technique and capability of an innovative process for hardening of gears. This process has been proven and is in production in the U.S.A. The technique uses a multi-frequency/multi-cycle surface gradient profile hardening process that provides increased hardness and strength at the pitch line with optimum strength gradient at the root fillet, without excessive tip temperature or tooth form brittleness. Profile hardening is a new development which can be highly automated and merges three distinctive technologies; programmed preheat (audio frequency - low frequency), high intensity final heat (radio frequency - high frequency) and incremental hardening. Also included can be an induction tempering operation to assure the proper level of hardness and toughness. The technique can be a stand-alone system or totally integrated in a manufacturing cell or a in-line automated system. The system provides the following advantages in comparison with existing furnace carburizing or other selective treatments:

- Consistent reduced dimensional change.
- Improved metallurgy.
- Higher strength and improved operating durability.
- Higher quality.
- Lower manufacturing costs.
- Reduced in process inventory.

Present day demands for improved manufacturing processes has motivated both equipment manufacturers and gear manufacturers to search for new innovations in the manufacturing process heat treatment and materials used in gears. This has resulted in a new innovation in the heat treatment technique for gears which is now capable of producing a hardened tooth form whose operating capability is substantially better than that previously obtained, either by carburizing and/or previous induction hardening techniques. This paper addresses the advantages of this technique along with an explanation of the technical analysis that supports the objectives.

BACKGROUND HISTORY

The technique described herein uses a multiple frequencies along with multiple cycles to product the improved techniques. The use of more than one frequency for producing a contoured hardened pattern on the gear tooth form is not new technology. Dual frequency hardening for treating gears was developed in the late 40's and early 50's and used primarily on heavy duty gears used in final drive trains in the tractor industry. Further refinement and enhancement of the concept of dual frequency hardening gears were subsequently largely curtailed because of the economic attractiveness of the low fuel cost of the continuous gas carburizing.

During the mid 60's and early 70's, the technique of root by root selective hardening of large gears was successfully implemented. In addition, the application of selective induction hardening the internal teeth on planetary ring gears for automatic transmissions was developed and implemented. This allowed the conversion from alloy steels to pearlitic cast irons for this application. In most of these cases, however, the hardening was only applied to the active profile face (pitch line). Root strength was not a requirement, since the torque was applied through a number of planet opinions.

Development in current manufacturing technology is rapidly moving to more effective and cost efficient techniques in manufacturing systems. Gears are a high priority in this quest for advances in processing methodology because of the high cost associated machining, heat treatment, inventory, and metrology control. To accomplish this task will require an in-line integrated flexible manufacturing processes incorporating automated handling and improved quality through inter-active process control. The objective is to replace the present inventory intensive carburizing system with an in-line surface heat treat process incorporating "batches of one".

NEW DEVELOPMENTS

In industries quests for improved manufacturing technology and more flexible manufacturing systems, gear manufacturers

recognize a number of the liabilities in the existing
carburizing techniques. These are listed as follows:

- Labor intensive - batch process
- Large "in process" inventory
- Expensive metal cutting operations
- High prime material cost
- Expensive and low productivity grind, honing, or
  hard gear finishing
- Removal of surface beneficial residual stresses on
  finishing operations
- Detrimental hi-temperature transformation products
- Inconsistent dimensional change

These challenges motivated TOCCO, a induction heating
manufacturer, to re-evaluate the "state of the art" of the
existing concept and developed a program to improve the
technology. This effort led to the successful development and
implementation of the TOCCO Profile Hardening (TPH) technique
(TPH) utilizing a Multi-Frequency/Multi-Cycle (MFMC) method to
produce a surface hardening gradient profile that truly
complements the gear operation requirements.

GEAR PERFORMANCE REQUIREMENTS

In understanding gear performance requirements as related to
material characteristics and strengths of material, my
intention is to evaluate three (3) functional areas. These are
as follows:

A.   Tooth Bending Fatigue Strength (tooth breakage)
B.   Pitch Line Surface Degradation (Pitting)
C.   Pitch Line Subsurface Failure (spalling)

As one can understand, these requirements can vary in various
gear designs and each has to be evaluated depending on
operating requirements and gear design characteristics.

Tooth Bending Fatigue Strength - Tooth bending

Tooth bending is caused by a load applied along "the line of
action" which generates a bending force on the tooth form. The
mode of failure tends to be the classic root fillet crack
propagation, as noted in **Figure 1 (11)**.

It is important to note that the applied stress can be very
high at the surface and it also falls off relatively rapidly as
shown on **Figure 2 (2)**. This point is particularly significant
when considering type of heat treatment for improved
performance. These include: higher harness for increased
strength, improved microstructure and generation of relatively
high level of beneficial residual compressive stress. Breen
provides a graphic representation of the stress/strength
gradient relationship as shown in **Figure 3 (3)**.

**CLASSIC BENDING FATIGUE
TOOTH FRACTURE**

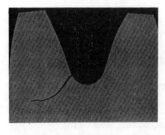

FIG # 1

BENDING STRESS GRADIENT IN GEAR TEETH

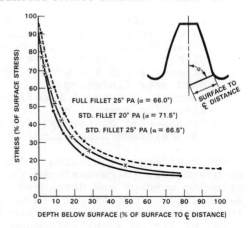

FIG # 2

**GEAR TOOTH ROOT STRENGTH
AND APPLIED STRESS GRADIENT**

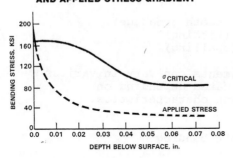

FIG # 3

**PITTING FATIGUE
INTERSECTION OF APPLIED STRESS &
ALLOWABLE STRENGTH AT PITCH LINE**

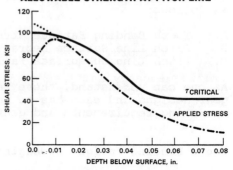

FIG # 4

Surface Degradation - Pitting

This condition is also well described by Breen in the **Figure 4**
(4). The intersection of applied stress and allowable heat
treated material strength extremely close to surface.

**PITCH LINE STRENGTH GRADIENT
VERSUS APPLIED SHEAR STRESS GRADIENT**

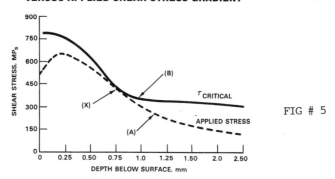

FIG # 5

SHOWS INTERSECTION OF STRESS AND ALLOWABLE STRESS NEAR CASE-CORE
INTERFACE. STRENGTH (τCRIT.) IS CONVERTED FROM HARDNESS GRADIENT.

When evaluating surface performance requirements, at the pitch
line, one must consider the surface contact stress that occurs
when the gear surfaces are in mesh as noted by author Breen,
this system is complicated by the traction caused by sliding
between teeth and therefore it is greatly influenced by
lubrication and surface finish. When traction is low (as in
roller bearings) or at the gear pitch line, the curve in **Figure
4** reverses and is noted that the curve slopes downward to
surface. According to Breen, the maximum stress occurs at or
about .008" - .010" (0.2 to 0.25mm) below surface, depending on
profile geometry and load(s). Also note, when sliding is
present and co-efficient of friction is significant, the curve
of applied stress continues to rise, exceeding the heat treated
material strength. The surface mechanical action at the active
face in the pitch line area is a combination of rolling and
sliding action.

The actual combination of conditions occurring at Pitch Line
Active Face depends on gear tooth profile modification, contact
ratio, pressure angle. Importantly, the lubrication and
surface finish conditions also are a major factor.

Subsurface Failure - Spalling

As explained by Breen and as noted in **Figure 5 (6)** there is an
additional mechanism that can occur at the active profile face
as a result of the fact that the applied stress level falls off
relatively slowly such that the applied stress can approach and
exceed the critical fatigue strength of the selected material
and its heat treatment. In carburizing gears, this can occur
at the case core interface. This additional mechanism occurs
due to the octagonal shear stress generation and its attendant
subsurface failure mode and is normally termed spalling.

Summary of Review of Design Considerations

In viewing the above operating conditions and attempting to provide the optimum arrangement of surface treatment technique, the strength/depth requirements are not uniform. What is actually required is a gradient heat treatment technique which provides the varying depth of strength improvement in accordance with performance requirements.

A high strength material capability at the surface in the area of root fillet radius of metallurgically sound microstructure. Essentially this requires material of high hardness/high yield strength along with a clean clear martensitic structure with no case or near surface discontinuities either physically or metallurgically. In addition, it is preferable to compliment this result with a reasonably high level of residual compressive stresses to improve endurance and/or fatigue life. However, as noted above, the depth requirement for this surface treatment is relatively nominal since you will note, per **Figure 2**, the applied stress falls off quite rapidly.

The situation at the pitch line, active profile area, is considerably different. In this case, you need a combination of high hardness at the surface to handle the surface contact stress and wear plus a substantial adequate strength sub-surface to prevent sub-surface failures (spalling).

Therefore, in contrast, surface treatment techniques that produce a uniform contour heat treated patterns around the tooth geometry profile are not necessarily the optimum. The optimum arrangement is a "profile gradient" wherein the depth of surface treatment compliments the operating requirement of the gear form. This requires substantially greater depth at the active pitch line profile area than that which is required for the root. This is the proposed concept of the TPH technique described herein.

Technique of Application

Selective surface treatment of gears or surface hardening by induction is not a new concept. Very early in induction heating technology development it was recognized that uniform depth of hardening the convoluted surface of a gear is difficult to obtain using a single frequency approach. Early developments could do a fairly acceptable job on particularly coarser tooth diametral pitches using a combination of two (2) frequencies; a low frequency (or even a furnace) for preheat and a much higher frequency for subsequent post final heating. Mass preheat was used to elevate the entire gear to some intermediate temperature (400-800°F) depending on the diametral pitch which compensates only for a portion of the subsequent final heat (RF) thermal inertia of the root heating plus some reduction in the rate of energy flow out of the root because of the lower temperature differential between the root and the core.

With the normal dual frequency method, attempting to force more

energy into the root area for good austenitization
transformation, you run the risk of over heating the tips with
the RF frequency. In addition, any compromises to this thermal
program results in slower austenitization rates.

Realizing that the above concept has a number of substantial
limitations, an innovative new concept was evolved using a
multi-cycle/multi-frequency approach. This approach involves
setting up a specific, dynamic thermal profile to produce the
optimum gradient hardened profile plus adding the advantages of
accelerated quench-cooling by the beneficial core quenching.
This cold core thermal extraction technique contributes to more
rapid core quenching action which results in improved
metallurgical structure plus a higher level of beneficial,
residual, compressive stresses. In addition to this
characteristic, there is also an attendant benefit of high
speed austenitization rates which enhance material properties
and improve hardness.

Multi-Frequency technique is now possible with flexible solid
state power supplies that allow a broad range of medium
frequency power supply frequencies to accommodate the various
gear diametral pitch (module) and involute profile
modifications. The initial objective is to generate
substantial energy into the root preferentially to compensate
for the thermal inertia of the root associated mass geometry in
this area. This is accomplished by use of a specific frequency
and power profile to set up the proper thermal dynamics. Also,
the temperature of the tooth form is partially elevated to
increase its electrical resistivity, as explained below. An
additional requirement is to allow a minimal heat input into
the core of material situated below the root annular mass. A
graphic description for this preheat thermal profile is termed
"hot root, cold core, cool tip", see **Figure 6**.

Subsequently, at the proper point in this thermal profile
generation, the high intensity RF power is applied and as a
consequence, the energy follows the profile of the tooth form.
Note the current will not short circuit the tooth form because
of the higher electric current reference depth resulting from
the slightly elevated tooth temperature and its attendant
higher electrical resistivity. In addition, the depth of
current flow in the root is also increased since the root is at
higher temperature and its depth of current flow is also
greater. The root will also exceed the critical temperature
quicker because of the hot root condition since once the
material becomes non-magnetic the current flow reference depth
step jumps virtually by a factor of 10.

The system has considerable flexibility for various gear
designs, diametral pitches module, profile modification,
contact geometry modifications, etc. The process has a wide
range of frequencies that can be used for preheat along with
various power profiles plus the high intensity RF portion has
substantially wide range of upper limits (Phase II) due to the
utilization of sequential energy application (progress
hardening).

Induction Stress Relief/Temper

Also an additional (optional) operational improvement can be incorporated in the in-line heat treat system to perform the low temperature stress relieve/temper operation. This has been proven on the first major TPH production installation.

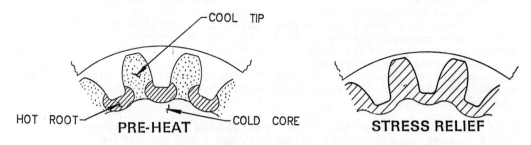

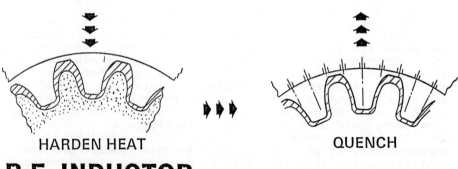

FIG # 6

TPH MCMF PROCESS RESULTS

PHASE I - As stated previously, the process uses a multi-cycle programmed energy input utilizing two (2) or more frequencies. The original concept utilized the traditional approach for processing external diameter gear surfaces and internal ring gear (spur and helical) where the energy is applied to the part, first in a low frequency coil with the gear in a static position axially within the coil and then moved to a second coil for radio frequency application. This technique was very successful in application both to straight spur gears and helical gears. This is termed our Phase I approach as illustrated in **Figure 7**.

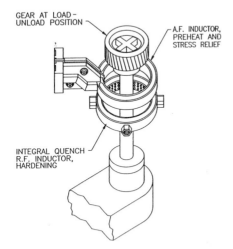

GEAR AT LOAD –
UNLOAD POSITION

A.F. INDUCTOR,
PREHEAT AND
STRESS RELIEF

INTEGRAL QUENCH
R.F. INDUCTOR,
HARDENING

FIG # 7

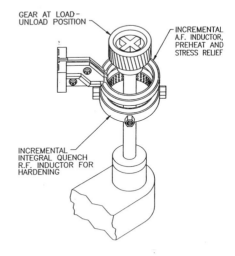

GEAR AT LOAD –
UNLOAD POSITION

INCREMENTAL
A.F. INDUCTOR,
PREHEAT AND
STRESS RELIEF

INCREMENTAL
INTEGRAL QUENCH
R.F. INDUCTOR FOR
HARDENING

FIG # 8

PHASE II - In a future economic analysis of this technique, it soon became apparent that one of the major liabilities was the requirement of a large kilowatt rating RF power supplies in the order of 600 - 1000 kW would be required for the high intensity application of the final R.F. heat. This large power supply, incidentally, was only used for a fraction of a second in many cases and cost per kilowatt is very high.

Phase II of this concept was developed wherein the energy was programmed into the part incrementally (progressively) such that the part was moved progressively through the heating coil or coils into the quench in a progressive manner such that the energy was applied incrementally along the axial surface of the gear teeth. This technique is shown in **Figures 8 and 9**. This technique allowed the use of smaller RF power supplies and as shown in **Table A:**

PHASE II

INCREMENTAL HEATING

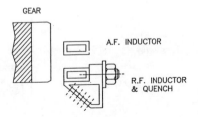

GEAR

A.F. INDUCTOR

R.F. INDUCTOR & QUENCH

FIG # 9

TABLE A

PHASE II INCREMENTAL VERSUS PHASE 1 STATIC PROFILE GEAR HARDENING

SAMPLE GEAR DEVELOPMENT DATA

GEAR DIMENSIONS:

Major Diameter	-	12,676"
Root Diameter	-	12.00"
Face Width	-	2.800"
Surface Area	-	111 IN²
Material	-	4150 Quench & Drawn

PROCESS REQUIREMENTS:

Surface Hardness	-	58-62 Rc After 300°F Temper
Case Depth	-	.030" Minimum in Root
Contour Pattern	-	No More Than 2/3 of Tooth Height

PHASE I
POWER REQUIREMENTS:

Low Frequency	-	1176 kW
High Frequency	-	784 kW

PHASE II
POWER REQUIREMENTS:

Low Frequency	-	480 kW
High Frequency	-	170 kW

still achieved the higher power density requirements, in fact, it provided an additional advantage in that there was less total energy in the part at any one time which improved core/mass quenching action as well as better thermal stability of the part. This configuration can be utilized for either OD gears, ID gears or other combinations.

PHASE III - In our work with automatic transmission designs, it soon became apparent that another challenge was in the offering with respect to the Phase I and Phase II of the concept. This involved transmission design improvements wherein several functional parts were either combined into one (1) integral part and/or part designs wherein additional integral attachments of part design geometry prevent the use of a dual coaxial, a coil assembly of passing over an OD gear or passing through a ring gear. For instance, in the case of a planetary gear assembly of a pot design.

To solve this challenge, TOCCO then developed a technique for using the MCMF approach utilizing a single coil with a unique system for applying the separate frequencies into the coil in a programmed manner per **Figure 10**.

PHASE III

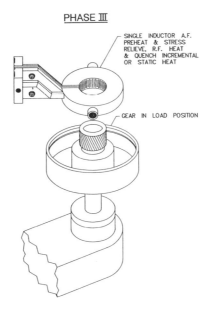

SINGLE INDUCTOR A.F.
PREHEAT & STRESS
RELIEVE, R.F. HEAT
& QUENCH INCREMENTAL
OR STATIC HEAT

GEAR IN LOAD POSITION

FIG # 10

Therefore, with our Phase III development we have broad flexibility by using static or incremental heating, 1 coil - 2 coils - 3 coils, or any combinations thereof to optimize the results for handling a wide range of gear and multiple function gear components.

MFMC PROFILE HARDENING RESULTS

The process has been successfully applied on a substantial number of gear designs and configurations covering an extensive list of prototype developments and several production installations. These include two (2) automated in-line facilities that have been purchased by a major U.S. automatic transmission manufacturer, each installation processing two (2) different gear configurations. One of these installations is presently in successful operation for hardening a highly stressed O.D. sun gear and the ID teeth on a planetary gear ring gear.

This technique can be applied to a wide range of materials of both plain carbon and alloy types as well as cast irons and even powdered metal materials. Examples of several applications are outlined below.

SAE 4150 Pinion Gear- 6.5 DP

The hardness traverses on this heavily loaded pinion gear of 6.5 diametral pitch, is as shown on **Figure 11**. You will note in these hardness traverses the relatively high surface hardness generated by the incremental Phase II technique that was used on this application. We believe that this hardness to be a result of a complete austenitization of the micro structure along with the excellent core quenching action of this particular gear geometry (integral opinion on a shaft).

PROFILE GEAR HARDENING
ROCKWELL SURVEY

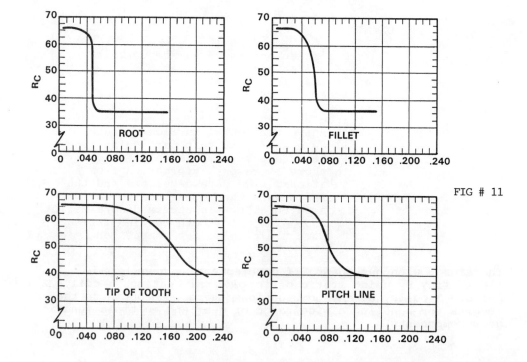

FIG # 11

Also note the relatively flat plateau in the hardness traverse (for total heat affected depth) which provides excellent hardness/strength at the surface as well as a substantial depth of this high hardness. This gradient concept provides an important additional depth of hardness at the pitch line to meet the requirements of subsurface stress generation as outlined in previous **Figure 5.**

As a matter of comparison of a carburized and carbo-nitreded case hardness traverse is shown on **Figure 12 & 13 (9).**

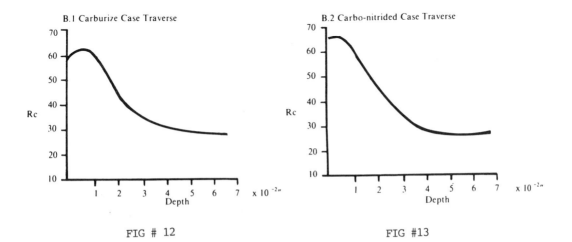

FIG # 12   FIG #13

COMPARISON OF DATA

Surface residual stress measurements were made on this particular gear by use of "strain relief" technique and this strain was measured in the circumferential direction to evaluate residual compressive stress in the operational mode of the gear. A comparison of the residual stress measurements, including those obtained on this gear design, are illustrated on **Figure 14 (7)** which shows the actual surface residual stresses to be 156 KSI as measured on this particular gear configuration. Note, the data presented is only for general comparisons only since these represent information from a number of gear configurations. The data developed and the residual stress measurements were then converted to fatigue strength calculations based on allowable stress diagrammatic techniques. The comparison of single tooth bending fatigue strength is shown on **Figure 15 (8)** also for comparison.

RATIONAL FOR PERFORMANCE IMPROVEMENT

Projecting the benefits of the gradient profile hardness hardening technique as demonstrated in **Figure 11** in comparison to carburized case depths, we have projected graphically the

COMPARISON OF ROOT COMPRESSIVE
STRESS LEVEL FOR VARIOUS HARDENING TECHNIQUES

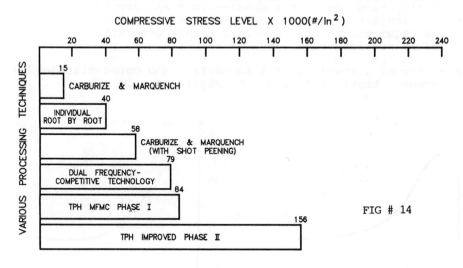

FIG # 14

COURTESY OF R.C. ASSOCIATES

COMPARISON CHART
SINGLE TOOTH BENDING FATIGUE STRENGTH

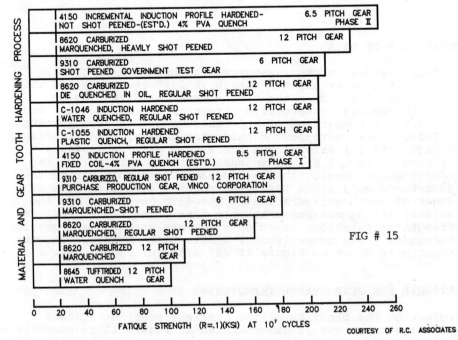

FIG # 15

COURTESY OF R.C. ASSOCIATES

improvement on **Figures 16 and 17. Figure 16** projects the
relationship between the gradient profile hardness strength
improvement at the root fillet in comparison with conventional
heat treatment, we suggest the technique, as described above,
provides substantially higher hardness/strength which thereby
substantially improves tooth bending load capability. The
Figure does not necessarily include the additional benefit of
residual compressive stress.

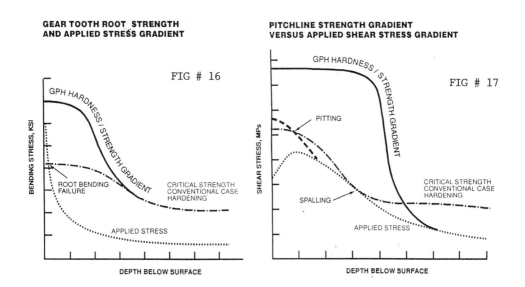

Figure 17 shows a comparison between conventional heat
treatment and gradient profile hardness, strength gradient
improvement, and the pitch line area. We suggest that the
substantial higher hardness with complete martensitic
transformation plus no surface transformation products along
with the substantial hardness continuing to the substantial
depth provides operational improvements from both a pitting and
spalling failure mode situations.

At this time, we do not have quantitative specific data to
verify this analysis. Although there is considerable data
available on contact fatigue and bending fatigue on carburized
gears there is none known at this time using induction heat new
processes. We have plans underway to run comparative studies
on contact fatigue and tooth bending to support this analysis.

Implementation in Manufacturing Systems

As noted in our introduction, "state of the art" manufacturing
systems are being directed toward in-line, flexible, flow
through manufacturing systems. The optimization of these
concepts require elimination of batch process techniques and
"in process inventory". The TPH-MFMC process inherently

incorporates virtually all the necessary features that make such a system practical. These include: batches on one, in-line manufacturing integration, high speed processing rates, real time individual part process monitoring, plus new advances in process quality assurance. These include newly introduced advances in inter-reactive and even inter-proactive process parameter adaptive control systems. See **Figures 18 and 19** for in-line concepts for automation.

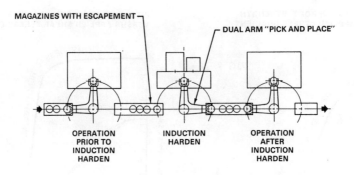

"PICK AND PLACE" AUTOMATION   FIG # 18

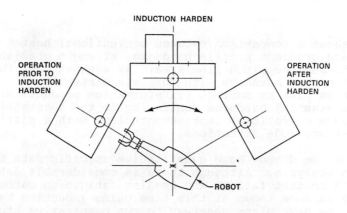

ROBOT AUTOMATION

FIG # 19

SAMPLE PHOTOS

The following photos illustrate various TPH patterns obtained
on designs as listed:

      Figure 20 -   Shaft Pinion 4150 material 6.5 DP
                     2.625" P.D., 5.0" F.W.

      Figure 21 -   Drive Gear 1050 material 5.0 DP 9.50"
                     P.D., 3.5" F.W.

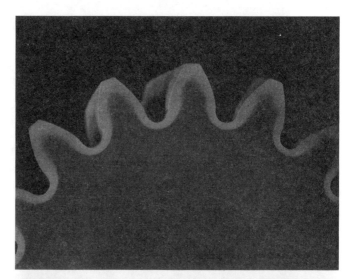

FIG # 20

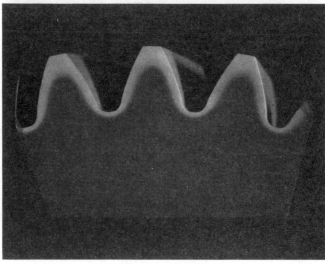

FIG # 21

Figure 22 -   Sun Gear Pearlitic Cast Iron 18.6 DP
                1.560" P.D., 1.0" F.W.

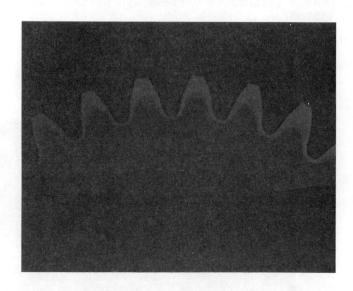

Figure 23 -   In-line Automated Transmission Internal
                Ring Gear Hardening Machine

Figure 24 -   Helical Engine Gear

Sectioned in multiple steps through teeth
face width to show high degree of profile
axial uniformity.

Advantages of TPH - MFMC Process

The profile hardening process described provides the following
advantages:

1.   Consistent part size change and reduced distortion.
2.   Optimum strength gradient at the root fillet.
3.   High beneficial residual compressive stress in the
     critical tooth root fillet for improved tooth bending
     fatigue strength.
4.   High hardness and strength at greater depth in pitch line
     area than conventional gear hardening processes.
5.   No excessive heating and/or brittleness at tooth tip.
6.   Improved metallurgical characteristics and elimination of
     undesirable heat treat transformation products.
7.   Allows for improved quality control methods, interactive
     process control and batches of one.
8.   Low installation costs.
9.   A major reduction in "in process" inventory.
10. Usually no finishing operation is required after
     hardening.
11. Low material costs (medium carbon steel or even cast iron
     as opposed to alloyed materials).
12. In-line, fully automated processing capability,
     incorporating process induction temper.
13. Energy efficient process with reduction in total energy
     requirements.

REFERENCES

1.   Dale H. Breen   Gear Research Institute SAE Paper No.
                841083
     **Fundamentals Of Gear Stress/Strength Relationships-
     Materials**
     Published in Gear Design Manual AE-15
     Society Automotive Engineers Warrendale, PA 1990

2.   R. A. Cellitti Manufacturing Technology Consultant -
     R. C. Associates

3.   IBID, 1, P44

4.   IBID, 1, P45

5.   IBID, 1, P45

6.   IBID, 1, P45 - 46

7.   IBID, 2

8.   IBID (2)

9.   L. Skip Jones, Lindberg Heat Treating Company SME Paper
     CM81-998, Reprinted with Permission of the
     Society of Manufacturing Engineers. Copyright 1981
     Published in Gear Design Manual AE-15 Society of
     Automotive Engineers, Warrendale, PA 1990
     **Selection of Materials and Compatible Heat
     Treatments for Gearing**

10. U.S. Patents 4,675,488; 4,749,834, 4,785,147; 4,855,551,
     4,757,170; 4,855, 556, 4,894,501

11. European Patent Application EPO 0,384,964A1

Materials Science Forum Vols. 102 - 104 (1992) pp. 365-372
Copyright Trans Tech Publications, Switzerland

INDUCTION HARDENING OF CAST IRON CYLINDER HEAD VALVE SEATS

Charanjit Charlie Lal

Consolidated Diesel Co., Whitakers, NC, USA

ABSTRACT

This presentation at the ASM Conference, dealt with the equipment, procedure, and statistical quality control during induction hardening of cast iron cylinder head valve seats for diesel engines. Various aspects of selecting coil design, frequency, power, duration of heating, and coil proximity were discussed. The detailed procedure of sectioning valve seats and the application of optical microscope and micro-hardness testing machine to evaluate microstructure, hardness, and case depth were described. The methods of dealing with the production problems involving shallow case depth and overcase conditions were discussed. Experiences with the eddy current equipment and magnetic particle inspection machines in sorting out defective parts and disposition of discrepant material involving surface cracks, non-conforming microstructure, and unacceptable case depths were shared. The photomicrographs of the induction hardened case showing excessive retained austenite, free ferrite, microcracks, and incipient melting conditions were exhibited. The application of statistical process control based on moving range and average charts for case depth data of intake and exhaust valve seats was critically evaluated. The adverse effect of improper valve seat hardening in the performance of cylinder heads was demonstrated in the light of some field failures.
- - - - - - - - -

In manufacturing of diesel engines with power ratings of 80 to 300 hp, the cylinder head valve seats can be induction hardened to impart increased wear resistance and enhanced fatigue strength properties. The added advantage of electromagnetic induction hardening of intake and exhaust valve seats is reduced manufacturing cost compared to the installation of highly alloyed valve seat inserts. A desired heating pattern and case depth can be achieved by varying the coil design, operating frequency, alternating current-power input, heating time, and the proximity between the inductor and the valve seat surface. The flux concentrators such as Ferrocon or Fluxtrol are generally utilized in inductor floating assembly to improve the case depth uniformity around the circumference despite the mass variation around the valve seats. Due to the large mass of the cylinder head, the hardening takes place by a self-quenching mechanism. In order to achieve case depths of 1.5 to 2.0 mm for a 45 degree valve seat about 30 mm in diameter, a typical process using TOCCO equipment consists of--frequency: 350 KHz, power: 10 Kw, heating time: 8 seconds, and inductor back off: 0.125 mm.

The following induction hardened valve seat inspection procedure is used at the Consolidated Diesel Company:
* Inspection frequency: one head from each day's production.

* Inspect all intake and all exhaust valve seats for hardness, microstructure, case depth, and any surface defects which could adversely affect performance.
* Section the head 10 mm behind the combustion face to remove valve seats; mark each seat with the cylinder number to maintain identity.
* Conduct magnetic particle inspection to check for any cracks or other surface defects resulting from over heating of the valve seats.
* Make metallurgical mounts from each cylinder, so that the 0, 90, and 270 degree circumferential positions from each valve seat could be examined.
* Grind and polish the samples with the final step using 0.05 micron alumina slurry.
* Etch the samples with a 3% nital solution for microstructural examination.
* Measure the total case depths at 50X using metallurgical microscope.
* Check the percentages of martensite, ferrite, and retained austenite phases for all valve seats.
* Measure the overlap of the hardened pattern at the 0 position on both the combustion chamber and valve guide sides.
* Measure surface hardness of all valve seats within 125 microns depth, from the surface, using the Knoop microhardness tester with a 300 gram load; the hardness indentations must be in the matrix area away from the graphite flakes.
* Check the hardness at a minimum specified case depth distance of 0.89 mm.
* Record the obtained date on the inspection report form and write applicable comments.
* Enter the date into the "Datamyte" computer files; examine all average and moving range charts for statistical process control; if an out-of-control condition is observed, notify the quality resource to initiate proper corrective actions.
* The corrective action steps could include: (a) power or time adjustments, (b) inductor change, or (c) maintenance corrections.

An ideal microstructure of an induction hardened valve seat consists of ASTM Type "A" graphite flakes randomly dispersed in a martensitic matrix with hardness of 57 to 62 HRC (converted from microhardness). Lack of heating can result in ghost pearlitic partially transformed microstructure with poor wear resistance. Excessive heat and alloy segregation at the cell boundaries can produce varying percentages of retained austenite in the micro-structure. Low heat and higher ferrite contents in the base material can cause free ferrite in the microstructure. Free ferrite amounts exceeding 3% in the martensitic matrix would adversly affect the wearing properties of the valve seat. Faster rates of heating during induction hardening can cause micro-cracking at the tips of the graphite flakes. The hardened pattern and various microstructures are exhibited in Figures 1-8. The incorporation of an energy monitor in the induction hardening equipment would reduce the variability of the case depth and the amount of undesirable phases in the microstructure.

Figure 1: Cross-section of the valve seat showing
induction hardened pattern.   15X

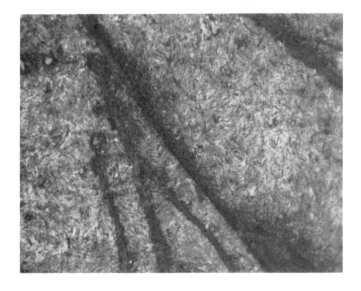

Figure 2. Photomicrograph showing martensitic matrix
of the induction-hardened valve seat.   400X

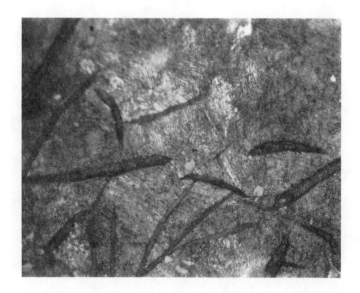

Figure 3: Photomicrograph showing martensitic plus ghost pearlitic matrix due to insufficient heating. 200X

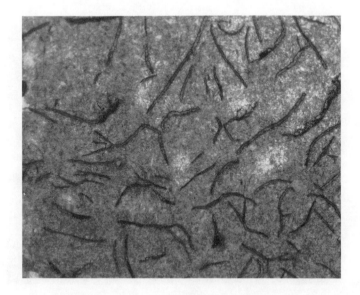

Figure 4: Photomicrograph showing about 5% retained austenite in a martensitic matrix.   200X

Figure 5: Photomicrograph showing about 15% retained
austenite in a martensitic matrix. 200X

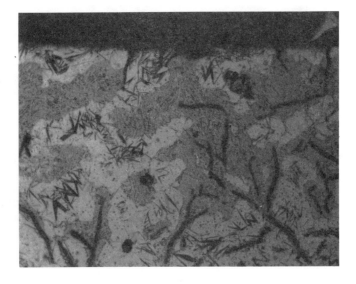

Figure 6: Photomicrograph showing about 40% retained
austenite in a martensitic matrix and incipient melting
resulting from excessive heating. 200X

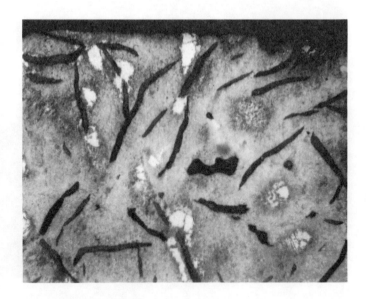

Figure 7: Photomicrograph showing free ferrite in a martensitic matrix resulting from excessive ferrite in the microstructure prior to induction hardening. 200X

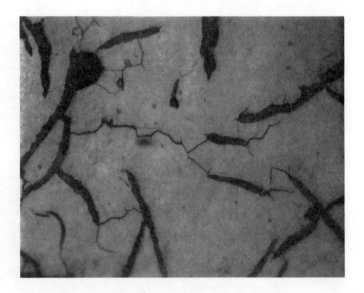

Figure 8: Photomicrograph showing micro-cracks initiating at the tips of the graphite flakes. 400X

Grain Growth in Polycrystalline Materials

Eds. G. Abbruzzese & P. Brozzo

Proc. 1st Intl. Conf. , Rome, Italy, 1991
ISBN 0-87849-640-8
1992, 930 pp, 2-vol. set
SFr 340,00 (ca. US$ 220.00)

The volumes present 19 invited lectures and 97 research papers (all reviewed) covering experimental and theoretical investigations on grain growth phenomena and their observation in various materials: metals and alloys, ceramics, sintered materials, thin films, etc.; normal and abnormal grain growth including twinning, texture, particle and other drag effects as well as analysis of topological aspects and grain size and grain orientation correlations, experimental and theoretical investigation of grain boundary structure, mobility and interaction with particles and impurity atoms. Experimental methods (e.g. ECP, HVEM, X-ray diffraction and topography, etch pitting, etc.) applicable to measurements of grain size, orientation of individual grains, etc. and cobination of physical and topological parameters are also covered.

Some Topics from Invited Lectures:

Statistical Theory and its Topological Foundation, *K. Lücke*
Synchrotron Study of Grain Growth, *J. Gastaldi*
Modelling Grain Growth, *D. Weaire & J.A. Glazier*
Bubble Model of Grain Growth, *W.W. Mullins*
Random Walk in Grain Growth, *N.P. Louat*
Texture Effects on Grain Growth, *G. Abbruzzese*
Microstrucral Path Concepts, *R.A. Vandermeer*
Grain Boundary Character Distribution and Analysis, *T. Watanabe*
Grain Boundary Migration, *D.A. Smith*
Grain Growth in Thin Films, *C.V. Thompson*
Electron Back-scatter Diffraction, *V. Randle*
Grain Boundary Energy and Mobility, *G. Gotstein*
Phase Transitions on Migrating Boundaries, *L.S. Shvindlerman*
Grain Growth in Multiphase Alloys, *S. Ankem*
Effects of Particles and Solutes, *B. Ralph*
Soap Froth and Potts Models, *S. Ling*

Trans Tech Publications Ltd

Segantinistr. 216 - CH-8049 Zürich - Switzerland
Fax: (++41) 1 342 05 29

Materials Science Forum Vols. 102 - 104 (1992) pp. 373-382
Copyright Trans Tech Publications, Switzerland

TFX
AN INDUCTION HEATING PROCESS FOR THE ULTRA RAPID HEAT TREATMENT OF METAL STRIP

R.C. Gibson, W.B.R. Moore (a) and R.A. Walker (b)

(a) Electricity Research and Development Centre, Capenhurst, UK
(b) Davy McKee (Poole) Ltd. UK

ABSTRACT

Transverse flux induction heating, a compact and efficient electrical technique with excellent controllability and response, has been developed for ultra rapid heating of metal strip at very high power intensities. The principles of this mode of induction heating are described, together with typical applications from a few kilowatts to several megawatts. As well as being used for annealing and other conventional heat treatments, TFX induction heating takes advantage of heating and quenching times down to less than a second, to make available a wide range of new metallurgical processes. Other applications include strip coating, galvanising, and other surface treatments. Economic and process advantages are discussed, quoting examples from commercial plant experience.

Introduction.

The processing of metal strip is a vast industry, providing material for such diverse fields of application as steel automobile panels and aluminium beverage cans. At some time in its production history this strip usually has to be heated at least once for one reason or another, such as for annealing or for curing of surface coatings.

For some processes the coils can be batch heated in an oven without being unwound, although heating the virtually solid mass is a lengthy business which precludes the possibility of very rapid heating or cooling. Alternatively, the strip can be threaded through a process line which runs more or less continuously, the line including a suitable type of radiant or convective strand-heating furnace. These lines can run at speeds up to several hundred metres per minute and, to obtain adequate heat transfer a conventional furnace must be very long, giving an overall line length sometimes extending up to 400m.

TFX is a registered name for transverse flux induction heating equipment made by Davy McKee (Poole) Ltd under licence from the Electricity Association Services Ltd.

Although large lines may offer some economies of scale they have obvious disadvantages arising from their sheer size, namely high capital cost, very large buildings, long start-up time, vulnerability to breakdown (a strip break may take hours to repair) and particularly the inability to adapt quickly to process changes. This is becoming more significant with the advent of 'just-in-time' manufacturing techniques which depend on the plant to produce material rapidly to an individual customer's specification, so that short-term orders can be fulfilled quickly without the need to stock or pre-batch materials.

A radical change in approach, from the very large inflexible line to one capable of instant adaptability, implies a need for compactness and the use of process stages with short-cycle capability, coupled possibly with changes in the strip material itself to make it responsive to ultra-rapid processing. Improvements have been made in a number of treatment areas, for example in strip-cleaning methods, but the greatest saving is to be made in the furnace section which occupies the bulk of a conventional annealing line.

If a suitable intense heating method were used, coupled with a quench technique of matching performance, it would be possible to reduce the overall process sequence from 4-5 min down to a few seconds. The significance of this in relation to particular materials is reviewed later but generally the aim is to improve product quality and yield, to make it possible to exploit novel material developments and, of course, to reduce greatly the physical bulk and cost of the process equipment and so obtain production flexibility. Rapid turnround also results in a dramatic reduction in the inventory of material held in process or queuing for treatment.

Process lines operate at substantial tonnage throughput, a fairly modest aluminium annealing line handles 20 tonnes/hour and needs a power input to the strip of over 2MW. It is evident that, to transfer power at megawatt levels into thin strip, within a transit time of a few seconds, demands a heating technique of remarkably high power density. In fact the only viable industrial method with the necessary capability is induction heating, which is unmatched for intensity by any other practical means and has many other virtues such as cleanliness, energy efficiency, absence of heat outside the workpiece itself and ability to function in vacuo or in a protective gas atmosphere if required.

Effective induction heating of thin strip at high powers presents special problems, and a particular mode of induction known as transverse flux has been adapted especially to handle this particular application.

Principle of transverse flux induction heating.

It is possible to heat thin strip in a conventional wrap round coil. However this requires the use of high frequencies, possibly extending into the radio band, particularly when non-magnetic material is to be heated. Frequency convertors of sufficient power are not readily available. The efficiency of heating in thin strip is inherently low, particularly for materials of low electrical resistivity, as the longitudinal magnetic flux which threads the coil axially only 'sees' a slender cross-section of the strip, most of the flux passing parallel to the strip through the gap required for thermal insulation and to accommodate strip movement.

The transverse flux technique makes it possible to heat thin strip efficiently at low frequency. This is done by directing the magnetic flux at the face of the strip so that a broad area of strip is exposed to its influence. It uses an inductor, in effect a large electromagnet split into two halves (figure 1) which are placed on each side of the strip. Heavy currents are induced in the plane of the strip, generating self-heating at great intensity, of the order $1-1.5MW/m^2$. Because the induced currents do not have to eddy within the confines of the thickness of the strip, frequency-related skin effects are largely irrelevant and the load impedance is such that efficient power transfer is obtained at frequencies in the low to medium band up to about 3000 Hz. Efficiencies better than 75% are easily obtained. There is no problem in procuring medium-frequency convertors at this power level, as they are in regular use for billet heating and melting furnaces.

The transverse flux principle does not demand a very tight air gap. The two halves of the inductor can operate with a reasonable gap between them of about 100mm daylight, giving comfortable room for the strip to pass through freely without marking. Another advantage of transverse flux induction heating is that power can be instantly switched on and off. Hence the line can be stopped at any time with no damage to the strip.

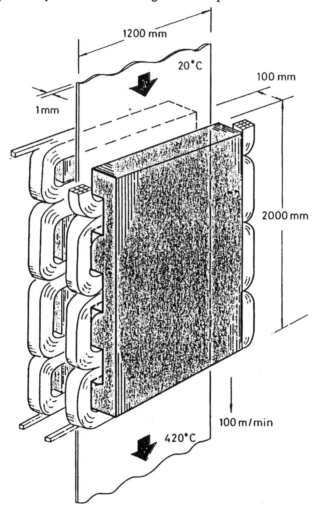

Figure 1. 3MW TFX inductor for aluminium strip heating.

Transverse flux induction heating practice.

Although the principles of transverse flux induction heating have been known for some time [1], early practical applications have been few and short lived. In the 1970's [2], work started at the Electricity Research and Development Centre to produce an industrially practical process. This led to a 1MW 6tonnes/hr development line for aluminium strip [3] in an Alcan plant at Kitts Green, Birmingham, UK.

The success of this work [4] resulted in a licence being granted to Davy McKee (Poole) Ltd, UK who now market transverse flux induction heating equipment under the registered name TFX.

In 1984 a 1.8MW aluminium heat treatment line was supplied to NLM in Japan [5], followed by a 2.8MW similar plant at SIDAL in Belgium [6]. Both are now in production. The plant at NLM has recently been extended to increase the product range.

Tests at the Davy McKee pilot line on the heat treatment of copper alloys have shown that the rapid response of TFX offers the unique facility to programme process temperature against line speed and so regulate grain size in the annealed material. A TFX inductor system and power unit has been shipped to Sweden for further development of this technique at pre-production level [7]. Other specialist applications with plant size down to a few kW are also being investigated. A new pilot line at the Electricity Research and Development Centre will treat coils up to 1m width and 2tonnes weight, allowing potential users to assess the merits of TFX for a proposed application.

A clear description of TFX inductors, plant layout, associated equipment, and their control is given by Ireson of Davy McKee [8].

Thermal treatment of aluminium alloy strip using TFX.

A major advantage that a continuous annealing process has over batch treatment is the ability to cool the strip very rapidly. TFX strip annealers, because of their very short heating length of a metre or so, can utilise immediate water quench in the same vertical pass without strip distortion. With quench rates of the order of 1000°C/sec it is possible to obtain very fine grain structure. Figure 2 shows the reduction in grain size from 53μm in batch annealed 3103 aluminium alloy to 11μm in TFX annealed strip. This eliminates the unacceptable 'orange peel' effect during forming [4].

Brawers [6] reports the use of TFX for full annealing, and partial annealing to tempers between half-hard and full-hard, with equal properties across the width and the length of the strip. Stabilising treatments as well as solution treatments of alloys in the 6000 and 7000 series (AlMgSi and AlMgZn types) are performed. Oilstaining and magnesium discoloration belong to the past. He quotes that the strip at the exit of the drying section needs no proper levelling most of the time, showing that strip distortion is absolutely minimal, with unlevelled strip being recoiled without problem. He also says that although the TFX was expected to be noisy, the fitting of an efficient acoustic enclosure ensured that 'even at full power the line can be considered to be silent'.

Another advantage of TFX is that the processing of a coil takes less than half an hour instead of the 2 or 3 days heating and cooling required in a batch process. With a high value product such as aluminium, the capital savings due to reduced stock inventory can help defray the additional cost of a continuous process.

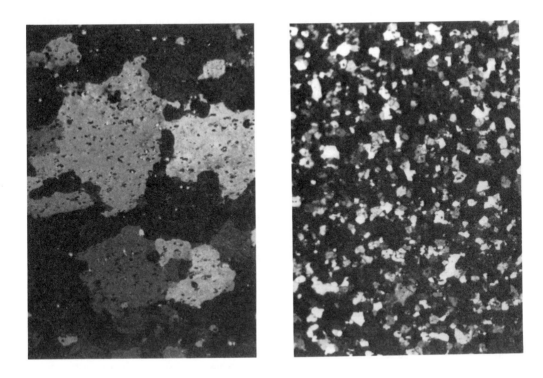

Batch annealed TFX annealed

⊢ ⊣ 50μm

Figure 2. Comparison between grain sizes of batch annealed and TFX annealed 3103 aluminium alloy strip.

Continuous annealing of steel strip.

Before 1950 virtually all the cold rolled steel was annealed in batch furnaces, a process that can take up to a week. During the early '50s continuous annealing was introduced for tin plate production. This process was limited to the production of a relatively high strength, low ductility material, beneficial for three piece tin-can production.

In the early 70's Japanese steel makers introduced continuous annealing techniques for a wider range of products, resulting again in greatly reduced processing times. By incorporating several previously independent stages, pickling, cold rolling, annealing and temper rolling, modern mills [9] are processing hot rolled coils in an hour, rather than the previous norm of twelve days. The advantages are both economic and metallurgical, with massive reductions in capital tied up in work in progress, faster response times, reduction in labour costs and the production of a wider range of metallurgical specifications [10]. The properties of the final strip are the outcome of many complex and interactive metallurgical variables. Commercial continuous annealing plants produce a wider range of properties and offer more opportunities for reliably and consistently controlling properties than conventional batch annealing techniques [10,11]. By the end of 1990 a total of 45 continuous steel annealing plants were reported to be in operation or under construction, throughout the world, all of which were gas or oil fired.

The application of TFX to ultra-rapid annealing of steels is currently under keen development. It seems likely that the best advantage will be obtained when TFX is used in conjunction with formulation changes introduced in the original steel making. The rewards of success will be far reaching in terms of lower product cost, improved properties, flexible production scheduling and reduced maintenance. TFX plant with throughput capability equivalent to that of a conventional annealing line can be built at substantially lower capital cost and housed in low-level building occupying only 30% of the volume. It will also be possible to build viable plants of lower throughput, to suit the needs of the smaller steel industries or to fulfil niche markets for specialised products.

The various continuous annealing plants already in operation use a range of operating conditions to achieve the required final properties. Process speeds vary from 100 to 300 metres/minute. In thinking of possible applications for TFX for steel strip annealing, it is useful to consider the basic process and its major variations in terms of the three stages, heating the steel to temperature, holding the steel at high temperature, and overaging and cooling to ambient.

Heating the Steel to Temperature.

In gas or oil fired furnaces, strip is heated to temperatures between 500 and 850°C. in 1 to 3 minutes. TFX can reduce this time to a second or so. As one of the objectives of the continuous annealing philosophy is to reduce line size, there is scope to capitalise on the rapid heating capability of TFX. Indeed in his assessment of future developments, Mould [11] looks forward to increased heating rates. The use of TFX may also have consequential metallurgical effects worthy of investigation. These could result from the rate of heating and/or the method of heating.

Compared with 'conventional' batch annealing, the current range of continuous annealing techniques use much more rapid heating rates and shorter times at temperature. It is claimed that this enables the annealing temperature to be increased with a consequent increase in the rate of recrystallisation and grain growth with less tendency to intragranular carbide coarsening. By reducing the time to reach temperature, an electromagnetic induction heating technique would enable even higher annealing temperatures to be adopted with beneficial increase in grain size, and no penalties in terms of increased carbide sizes or increase in the plant dimensions.

The nature of the heating process in TFX may also have interesting metallurgical effects. Recrystallisation, grain growth and carbide precipitation and growth are all influenced by the concentration, movement and generation of vacancies and dislocations. Both are discontinuities in the atomic arrangement of the metal. As such they will cause a disturbance at atomic level in the paths of the induced currents and will represent discontinuities in the mechanical forces generated by the applied magnetic field. Thus the use of induction heating may directly impinge on those metallurgical features that dominate important high temperature processes within the steel, or indeed any other metal. The Electricity Research and Development Centre are investigating this effect to see if this leads to new ways to exploit the unique nature of induction heating.

Holding the Steel at Temperature.

In gas or oil fired furnaces, steel strip is held at temperature for 1 to 5 minutes in a holding portion of the furnace, which may be direct fired or heated by electrical radiant tubes. The effect of induction heating at an atomic level on the behaviour of metallurgical features as discussed above could be equally relevant in this portion of the process.

Alternatively, the capability of greatly increasing the heating rate may enable a higher maximum temperature to be used with sufficient grain growth at a significantly reduced, perhaps zero holding time.

Overaging of steel.

The initial cooling rate, even at the slowest rate used in gas or oil fired continuous annealing plant, can be sufficiently rapid to retain significant carbon in solution in the ferrite. This may subsequently cause quench and/or strain aging of the strip. The consequence of either aging process is an increase in strength and a reduction in formability. This may or may not be advantageous, depending on the subsequent manufacturing processes and applications.

In general an overaging heat-treatment is applied to reduce subsequent aging in manufacture or service. This requires a heating cycle of between 1 and 5 minutes at temperatures between 200 and 450°C. For a particular steel formulation and product requirement, the faster the primary quench rate the lower the overaging temperature and/or the shorter the overaging treatment. For certain applications, particularly high strength tin plate, an overaging heat-treatment is not used.

If an overaging treatment is used the steel strip will need to be reheated to temperature, if it had been quenched to or near ambient. Alternatively, if the primary quench had been to the overaging temperature it will simply need to be held at that temperature for the required time. Those lines that use an overaging treatment after quenching to or near ambient currently use fossil fuel to reheat the strip, taking of the order of a minute to reach temperature. TFX could achieve temperature within seconds enabling process times and plant dimensions to be reduced.

The comments in the above sections concerning the possible metallurgical effect of the induction heating process at the atomic level, and the dominant metallurgical effects, apply to this process. The Electricity Research and Development Centre is investigating the interaction of induction heating and steel metallurgy for this section of the plant.

Furthermore, because induction heating offers the opportunity to employ much more rapid heating rates than presently available, it may be possible to avoid the need to over-age. Use of TFX would enable the temperature of the steel strip to be raised above Ac_1 in about a second. Above this temperature the solubility of carbon in ferrite decreases. Thus it is possible to postulate that one could heat strip sufficiently rapidly to a temperature at which there is no tendency for carbon, in the form of carbide particles, to dissolve in the ferrite but at which recrystallisation of the ferrite still will occur.

If the strip is cooled sufficiently rapidly through the temperature zone in which the solubility of carbon in ferrite increases, there will be no or little excess carbon in the annealed ferrite. In this case there will be little tendency for aging to occur, nor need to apply an over-aging heat treatment. If this is proves to be the case there would be a significant reduction in capital cost, size and operating costs. It would also open up new markets.

Extra deep drawing quality steel (EDDQ) is not readily available because of it's susceptibility to aging at ambient temperatures in relatively short times. Thus the market tends to be limited to major users such as car manufacturers. These can still face problems with stock control. For example if there is a hold-up in production, material kept in stock may begin to age and have to be returned to the steel mills for reprocessing. Smaller manufacturers or re-workers such as electro-platers and coaters simply cannot use EDDQ material from current production. A compact TFX unit offers both small and large manufacturers the opportunity to produce EDDQ material at the point of use.

Metallurgical Factors.

It has been known since the 1960's, from work at the Steel Company of Wales [12], that cold rolled rimmed steel could be recrystallised in less than a second if heated sufficiently rapidly at rates around 1600°C/sec. More recently Shi et al [13] have shown that a cold rolled decarburised steel, 0.005 wt% carbon, could be recrystallised in less than a second with sufficiently high heating rates. This is despite the low concentration of carbide particles which can provide nucleating sites for recrystallisation. Until comparatively achieving such rapid heating rates was not practical on an industrial scale.

With the availability of TFX, heating rates of 1000°C/sec are already available, and higher rates could provided if beneficial. The evidence [12,13] is that recrystallisation could be achieved much faster than is currently the practice. This has been confirmed by laboratory trials using both resistance heating and a prototype TFX [14], which produced satisfactory annealing in seconds rather than the minutes of continuous annealing plants or the days of batch furnaces. Semiatin [14] reported some loss of tensile properties with 0.05% carbon aluminium killed steel which he ascribed to lack of time at temperature - fine grain size and rapid cooling and overaging - fine intra-granular carbide precipitation. However the heating rates were restricted to 500°C/sec and his cooling rates to about 250°C/sec. Work currently being undertaken by the authors with substantially faster heating and cooling rates with similar steels shows no evidence of a yield point and enhanced elongation.

Conclusion.

TFX systems are in commercial production for the thermal processing of aluminium alloys and have been shown to produce material of superior quality, with remarkable production adaptability and efficiency. This method of heating has been further developed for copper based alloys and other non-ferrous materials, and is under investigation for the steel industry. TFX has also been demonstrated in the rapid curing of paint coatings, particularly on aluminium substrates for which conventional induction is unsuitable.

The financial rewards for even minor improvements to continuous steel annealing plants is huge. Consequently there is considerable research effort into the process being carried out world wide. The rapid heating capabilities of TFX technology appear to offer specific advantages to the initial heating and, to a lesser extent, the over aging heating phases of the production route of steel strip.

Generally, the emergence of TFX as an advanced high-power heating method is in line with the worldwide trend towards ultra-rapid process cycles in a wide range of industries. It offers the prospect of very compact installations of high energy efficiency, operating to flexible schedules in favourable working conditions, and producing materials of superior quality at lower cost.

The use of this ultra rapid heating method opens up immense possibilities for the metallurgist to expand his metal heat treatment horizons.

Acknowledgements.

The authors wish to thank the managements of the Electricity Research and Development Centre, Capenhurst and Davy McKee (Poole) Ltd, for the permission to publish this paper, and gratefully acknowledge the assistance of their colleagues.

References.

1)   Baker, R.M.: 'Transverse flux induction heating', Trans AIEE, 1950, vol 69, part 11.

2)   Jackson, W.B.: 'Transverse flux induction heating of flat metal products', 7th Int Conf Electroheat, Warsaw, 1972.

3)   Gibson, R.C. et al.: 'Transverse flux induction heating of aluminium alloy strip', Heat Treatment '81, Birmingham 1981.

4)   Walker, D.J. et al: 'Metallurgy of Rapid Heat treatment of aluminium strip by TFX', Inst of Metals, London, March 1986.

5)   Yamagishi, T., Kitajima, Y., Nagahama, K., Ishii, H., Ikeda, H.: 'Induction heating CAL for rolled aluminium sheet', Japan Light Metal Association, Tokyo, April 1988.

6)   Brawers, T. et al: 'Experience with a 2.8MW TFX transverse flux continuous thermal treatment line for aluminium strip', METEC, Dusseldorf, May 1989.

7)   Ireson, R.C.J.: 'Developments in transverse flux (TFX) induction heating of metal strip', Metallurgia, Feb 1991, vol 58, No 2, p68.

8)   Ireson, R.C.J.: 'Induction Heating with Transverse Flux in strip metal process lines', IEE Power Engineering Journal, March 1989.

9)   Anon: Steel Times Int. July 1990 p22.

10) Takechi, H. et al.: Steel Times Int. July 1990 p12-19.

11) Mould, P.R.: Proceedings of a Symposium on the Metallurgy of Continuous-Annealed Sheet Steel 1982. P3-33. Edited by Bramfitt, B.L. and Mangonon, P.L., AIME.

12) Andrew, T.O., Atkinson M.: B.I.S.R.A./I.S.I. Physical Metallurgy Meeting, London 23 Nov 1966.

13) Shi, H. et al.: Recrystallisation '90 edited by Chandra A. Published by the Minerals, Metals and Materials Society.

14) Semiatin, S.L. et al.: Adv. Materials & Processes, Vol 4 1987 p43 - 50.

Materials Science Forum Vols. 102 - 104 (1992) pp. 383-392
Copyright Trans Tech Publications, Switzerland

PRODUCTION AND CONCENTRATION OF MAGNETIC FLUX FOR EDDY CURRENT HEATING APPLICATIONS

R.S. Ruffini and R.J. Madeira

Fluxtrol Manuvacturing, Inc.

ABSTRACT

Any material object, part or "charge" that is able to conduct electrical currents can be heated due to the resistive power dissipation of reaction eddy currents that flow due to externally produced time varying magnetic flux passing through the body or surface of the object or part.

For a particular application the amount of time varying magnetic flux needed to do the heating job is fixed, and therefore the amount of power that is needed to be transferred via the time varying magnetic field to the part is also fixed. The excess amount of power produced by the electrical generator driving the heating "coil", excess beyond the actual heating power transferred to the part, is then a measure of the inefficiently of the overall induction heating process. The objective of any induction heating system design is to minimize this excess generated power and therefore minimize the operating cost of the heating process. The design of the induction heating system is then reduced to design of the heating "coil" itself and matching its resultant current demand characteristics. If the reluctance of this path can be reduced then, for the same amount of total resultant magnetic flux, by the Magnetic Ohms Law, the drive coil ampere turns would mean a reduction in drive current; the number of coil turns would remain constant. Less drive current means less drive coil resistive heating (I R loss) and thus less excess drive generator power; which in turn means a more efficient induction heating system.

One of the best and easiest ways to reduce the reluctance of the air path for the magnetic flux in an induction heating setup is to simply shorten the path by positioning the drive coil in close proximity to the part.

A statement which is true 100% of the time is this: "the air path" reluctance in any induction heating setup can always be reduced through the use of soft magnetic flux concentrator materials include silicon-steel laminations, ferrite and iron powder composites.

Case histories will be presented from various industries showing power usage, production data, and metallurgical results of using flux contractors.

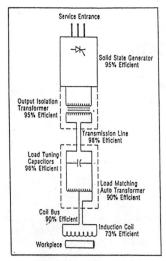

Efficiency of components in induction heating system.

FIGURE 1

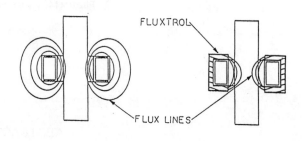

FIGURE 2

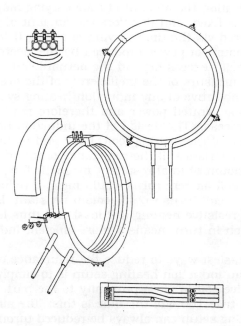

FIGURE 3

Any material object, part or "charge" that is able to conduct
electrical currents can be heated due to the resistive power
dissipation of reaction eddy currents. Eddy currents flow due to
externally produced time varying magnetic flux passing through the
body or surface of the part or "charge". This process is commonly
referred to as "Induction Heating". Though the scientific basis
of Induction Heating is quite simple (eddy current flow due to
time varying magnetic flux) the economic production of the needed
time varying magnetic flux is anything but simple. Each new
"charge" that one works with presents a different challenge wholly
dependent of where, on or in, the particular part one wishes the
induction "heat" to be concentrated.

INDUCTION HEATING EFFICIENCY

Magnetic flux is produced by driving electrical current through a
"coil" placed near or surrounding the target charge. We used
quotes around the word coil in the previous sentence since in some
cases this "coil" is but a simple loop or even just a straight
piece of conductor bar. In any event, the purpose of the "coil"
is the same - - production of magnetic flux. For a particular
application the amount of time varying magnetic flux needed to do
the heating job is fixed, and the amount of power that is needed
to be transferred via the time varying magnetic field to the part
is also fixed. The amount of power required to heat a part can be
calculated. Therefore the excess amount of power produced is then
a measure of the inefficiency of the overall Induction Heating
process. Figure 1 depicts the magnitudes of these inefficiencies
and illustrates that it is entirely possible that the complete
induction system is only 50% efficient at the coil. The objective
of any induction heating system design is to maximize the ratio of
energy input to transferable power output and minimize the
operating cost of the heating process.

FACTORS INFLUENCING HEATING EFFICIENCY

The effective depth of penetration of eddy currents in a target
charge, from the surface inward, is proportional to the inverse
square root of the frequency of the time varying alternating flux
produced by the heating "coil". Thus, dependent on the desired
depth of heat for the charge the operating frequency for a given
application is load dictated. The design of the induction heating
system is then reduced to design of the heating "coil" itself and
matching its resultant current demand to a more or less standard
set of drive generator output characteristics. The design of the
heating "coil" involves both science and art. The science portion
stems from needed adherence to the so called Magnetic Ohms Law and
the art portion stems from the experience of the coil builder
(i.e. what has worked before in similar situations). The Magnetic
(form of) Ohms Law states that the total operating ampere turns of
a "coil", the coil current in amps times the total number of
effective turns, is proportional to the total flux produced by and

passing around the coil. The proportionality constant in this law/equation is called the total magnetic resistance, or reluctance, of the complete magnetic circuit. Magnetic resistance or reluctance of the flux path in a magnetic circuit is proportional to the physical length of the path and inversely proportional to the cross sectional area of the path and to the magnetic permeability of the path medium.

The magnetic permeability of a material of medium is a measure of the ease with which magnetic flux flows through that material or medium. The complete magnetic circuit of and induction heating system consists of the flux path around the flux producing coil, the flux path between the coil and the part to be heated, and the flux path through the part itself. Consider a one-turn coil which coaxially surrounds a steel rod charge. The flux produced by this coil would flow in paths denoted by the lines shown in the cross sectional view in Figure 2. The total reluctance of these paths is the sum of the reluctance of the air path and the path through the steel charge. Whether or not the charge is a magnetic material (one which has a very low magnetic resistance or reluctance to the flow of flux) the reluctance of the magnetic path through the part is functionally beyond control. Since flux must be pushed through the part, the reluctance of the path through the part is a fixed feature of the given load. However, the reluctance of the air path of the flux is not necessarily a fixed portion of the problem. If the reluctance of the path can be reduced then, by the Magnetic Ohms Law, the drive coil total ampere turns is correspondingly reduced for the same amount of total resultant magnetic flux. In general, a reduction in drive coil ampere turns would mean a reduction in drive current; the number of coil turns would remain constant. Less drive current means less drive coil resistive heating (I^2R loss) and thus less excess drive generator power, which in turn means a more efficient induction heating system.

MEANS OF INCREASING HEATING EFFICIENCY

One of the best and easiest ways to reduce the reluctance of the air path for the magnetic flux in an induction heating setup is to simply shorten the path by positioning the drive coil in close proximity to the charge. This technique reduces the path lengths between the drive coil and the charge but does not help reduce the path lengths for the flux which must go around the coil (Figure 2). It can be stated that the magnitude of the reluctance pertaining to the complete magnetic circuit, due to the effective cross sectional path, depends entirely on the specific application and the specific details of the coil design and the coil proximity to the charge. The air path reluctance in any induction heating setup can always be reduced by the use of a high permeability magnetic material flux concentrator. The magnitude of which is dictated by the resultant flux path of a particular coil design. Examples of magnetic flux concentrator materials include silicon-steel laminations, ferrites and iron powder composites. These materials can be placed "around" the drive coil in what

would be the air path portions of the total flux path.   Since the
flux concentrator has a low reluctance, lower than that of the air
path which the flux concentrator material replaces,   the total
reluctance of the magnetic circuit is reduced.

CONTROLLED HEATING IN SELECTED AREAS

Different specific applications may call for different flux
concentrator   materials.   Another benefit of the use of flux
concentrator material is the ability to focus flux generation   and
therefore heat, in selected areas within a charge (Figure   2).
This can be attained by a number of means, specialized coil design
being perhaps the most obvious; however, movement of easily placed
segments of flux concentrator material is perhaps the   simplest.
Flux concentrators are also used to retard heating.   Selective
placement of concentrator material about the drive coil can
effectively shield areas of a charge surface from heavy eddy
current production by steering the flux away from the surface area
one wishes to protect (Figure 3).

CONSIDERATIONS IN SELECTING CONCENTRATOR

The choice of a concentrator material is dependent on several
factors: magnetic permeability (the higher the better),   maximum
sustainable level of flux density (saturation level),   loss under
magnetic excitation, machinability, ease of installation and
removal, ability to withstand induction heating temperatures and
environments, and last but not least, cost.
When choosing a concentrator material for a particular application
the   selection factors previously mentioned should   all   be
considered carefully.   But major consideration should be given to
the magnetic properties of the material.   These properties can be
determined from the manufacturer's data sheet or measured with a
suitable test set.
A particularly informative presentation of magnetic data for a
material is the magnetization curve or B/H curve (Figure 4).   It
shows the response of the material, measured in units of magnetic
flux density (flux per unit of cross sectional area), to an
applied magnetic stimulus measured in units of magnetic field
strength.   The better a magnetic material is, the more resultant
flux density it will produce for a given level of magnetic drive.
The slope of the magnetization curve depicts the   magnetic
permeability of the material. Since the material can only respond
up to its saturation limit in resultant flux density   the
permeability decreases as the applied field is increased.   Good
magnetic materials have both a high slope or permeability and a
high saturation flux density. One other magnetic property is also
discernible from the magnetization curve: the relative hysteresis
loss of the material. Magnetization of a material is accomplished
by stimulated alignment (due to the applied magnetic field) of
microscopic magnetic "domains" within the material.   The material
saturates when all of these domains are aligned, but once these

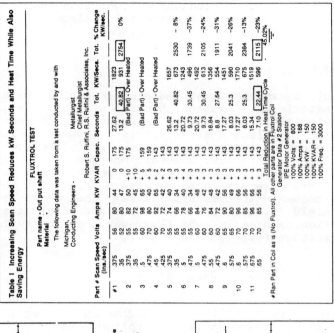

Table I Increasing Scan Speed Reduces kW Seconds and Heat Time While Also Saving Energy

FLUXTROL TEST

Part name - Out put shaft
Material :

The following data was taken from a test conducted by and with

Michigan,
Conducting Engineers -

Metallurgist
Chief Metallurgist
Robert S. Ruffini, R.S. Ruffini & Associates, Inc.

Part #	Scan Speed (ins./sec)	Volts	Amps	KW	KVAR	Capac.	Seconds	Tot.	KW/Secs.	Tot.	% Change KW/sec.
#1	.375	56	80	44	0	175	27.62		1823		0%
	.35	52	80	47	0	175	13.2	40.82	931	2754	
2	.375	55	82	50	+10	175	(Bad Part) - Over Heated				
	.35	60	72	45	+10	175					
3	.5	60	98	65	+5	159	(Bad Part) - Over Heated				
	.475	70	80	40	+5	159					
4	.45	70	92	65	+2	143	(Bad Part) - Over Heated				
	.425	70	72	42	+2	143					
5	.375	70	74	40	+4	143	26.62		1657		-8%
	.35	55	78	34	+4	143	13.2	40.82	673	2530	
6	.475	55	66	34	+3	143	20.72		1243		-37%
	.5	60	84	42	+3	143	9.73	30.45	496	1739	
7	.475	60	84	42	+3	143	20.72		1492		-24%
	.55	60	76	42	+3	143	9.73	30.45	613	2105	
8	.475	60	84	42	+3	143	18.84		1356		-31%
	.6	65	72	42	+3	143	8.8	27.64	554	1911	
9	.575	65	90	56	+3	143	17.27		1451		-26%
	.6	65	80	49	+3	143	8.03	25.3	590	2041	
10	.575	70	96	56	+3	143	17.27		1710		-13%
	.675	70	96	66	+3	143	8.03	25.3	675	2384	
11	.675	70	85	56	+3	143	15.34		1519		-23%
	.65	70	85	56	+3	143	7.10	22.44	596	2115	

Ran Part in Coil as is (No Fluxtrol). All other parts are in Fluxtrol Coil

Total Reduction in Heat Time Cycle = 45.02%

Generator Data #2 Station
IPE Motor Generator
100% Volts = 800
100% Amps = 188
100% KW = 150
100% KVAR = 150
100% Freq. = 3000

FIGURE 6

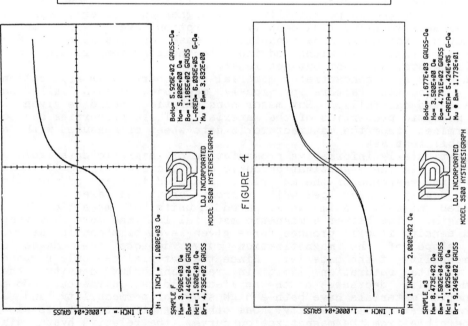

Figure 4

LDJ INCORPORATED
MODEL 3500 HYSTERESIGRAPH

H: 1 INCH = 1.000E+03 Oe
B: 1 INCH = 1.000E+04 GAUSS

SAMPLE F
Hm = 3.990E+03 Oe
Bm = 1.449E+04 GAUSS
Hc1 = 1.500E+01 Oe
Br = 4.735E+02 GAUSS

Bo:Ho = 5.926E+02 GAUSS-Oe
Ho = 5.000E+00 Oe
Bo = 1.185E+02 GAUSS
L-RRER = 6.035E+05 G-Oe
Mu @ Bm = 3.632E+00

Figure 5

LDJ INCORPORATED
MODEL 3500 HYSTERESIGRAPH

H: 1 INCH = 2.000E+02 Oe
B: 1 INCH = 1.000E+04 GAUSS

SAMPLE #3
Hm = 8.473E+02 Oe
Bm = 1.502E+04 GAUSS
Hc1 = 1.025E+01 Oe
Br = 9.249E+02 GAUSS

Bo:Ho = 1.677E+03 GAUSS-Oe
Ho = 3.580E+00 Oe
Bo = 4.791E+02 GAUSS
L-RRER = 5.424E+05 G-Oe
Mu @ Bm = 1.773E+01

domains are aligned it takes energy to reverse the process and align them in the opposite direction. This reversal energy is referred to as the hysteresis loss and is indicated by the width of the opening in the closed magnetization curve. A typical magnetization curve, represents one cycle of magnetization in one direction, a reversal of the applied field up to the maximum level in the original direction, and the other reversal back to the same direction and magnitude of the original field (Figure 4). The wider the opening in the magnetization curve the more hysteresis loss per cycle of flux reversal. Figure 5 shows less efficient material as compared to the previous B/H curve shown in (Figure 4).

TEST RESULTS IN USING FLUX CONCENTRATOR

Significant efficiency increases can be achieved by utilizing a flux concentrator and result in heat times reduced by as much as 54 percent. These results were obtained using Fluxtrol magnetic flux concentrator applied to a single turn static heating radio frequency coil. The output drive generator power was not reduced in this case.
A typical test sequence as depicted in Figure 6, shows that increasing the input power of the generator actually reduced the total kilowatt-seconds needed to arrive at the desired case depth and Rc. Total Kilowatt-seconds were reduced by increasing the scan speed. Please pay particular attention to the heat time that went from 40.82 seconds to 22.44 seconds for an overall cycle reduction of 45.02 percent while using 23 percent less energy.
Flux concentrators are by no means new but resistance to this technology is common. A resistance to change is normal, however, it can only be detrimental to the growth and competitiveness of our industry. This science requires an open mind and willingness on the part of both the supplier and user to objectively experiment and test concentrator technology. In most cases the utilization of flux concentrators is a "no loose" proposition for the induction processor. If your company has not established a serious approach to qualifying flux concentrators in your induction process you are loosing a real technological edge.

VI. LASER-BEAM TREATMENT

Materials Science Forum Vols. 102 - 104 (1992) pp. 393-400
Copyright Trans Tech Publications, Switzerland

REDUCTION OF THE TENSILE STRESS STATE IN LASER TREATED MATERIALS

J.Th.M. De Hosson, J. Noordhuis and B.A. van Brussel

Dept. of Applied Physics, Materials Science Centre, University of Groningen, Nijenborgh, 18,
NL-9747 AG Groningen, The Netherlands

ABSTRACT

Despite the advantages of laser processing for the production of wear resistant materials, surface melting results in tensile stresses. These tensile stresses may lead to serve cracking of the material and to deleterious effects on the wear behaviour. Our basic idea is to convert the high tensile stresses in the laser melted surface into a compressive state either by implantation or by shot-peening afterwards. Neon implantation into laser melted RCC steel turns out to reduce the wear rate substantially. Shot-peening showed several advantages over implantation: the influenced layer is thicker (100 μm compared to 0.1 μm), and the technique is more effective since shot peening results in a compressive residual stress state whereas implantation only reduces the tensile stress.

INTRODUCTION

Many successful wear resistant materials consist of particles of a hard phase dispersed in a more ductile matrix. Such dispersions can be prepared by powder metallurgical techniques or by solidification of an eutectic structure from a melt. However, in the former technique the coatings produced by spray processes remain separated from the substrate by a sharp interface which is always a potential source of weakness. In the latter type of technique materials are prepared from the melt and the proportion of hard phases is controlled by equilibrium thermodynamics. A different approach, as presented here, is to modify the surface layer by using a 1.5 kWatt CW CO_2 laser beam. Among the available laser applications, laser surface melting turns out to be a powerful technique for the production of wear resistant layers since it combines the advantages of local hardening, the possibility of surface alloying and the use of high quench rates. The latter may result in new metastable phases with novel wear properties. Indeed, not only the hardness, but also the ductility and internal stresses play an important role in the wear process. These phenomena are also affected by laser surface treatment.

Despite the advantages of laser processing for the production of wear resistant materials, laser surface melting results in tensile stresses. As a result high tensile stresses in the surface layer are generated which may lead to severe cracking of the material. Tensile stresses in the order of several hundreds MPa are possible, which detrimental influence the wear behaviour: The basic idea is to convert these tensile stresses produced by laser melting into compressive ones either by ion implantation or by shot-peening afterwards.

When pressurized bubbles of implanted ions are nucleated in the surface layer, one can imagine that the corresponding compressive stress field could annihilate the tensile stress field of the laser melted material. Furthermore, the surface layer might also be strengthened during wear, due to the interaction between moving dislocations and the bubbles. On the other hand shot peening the laser melted material induces substantial compression that may reduce the tensile stress state as well.

The material under investigation is RCC-steel (2.05 wt% C, 11.05 wt% Cr, 0.62 wt% W and bal. Fe). It has been chosen, because it shows a constant hardness profile after laser melting. Furthermore this material does not have a martensitic phase after laser melting, and compression due to the martensite is not expected.

EXPERIMENT

In this study a transverse flow Spectra Physics 820 CO_2 laser system has been used. The parameters of the treatment are: a measured power on the surface of 1350 W, a focal length of 127 mm, a focus point at 15 mm above the surface, resulting in a melted track of 2 mm. Tracks are made adjacent to each other at a distance of 1 mm apart under a protective argon atmosphere. After laser melting the micro structure consists of a dendritic structure of retained austenite, which is surrounded by M_3C carbides. The measured dislocation density is 10^{13}-10^{14}/m^2. A high vacancy concentration is to be expected due to the high quenching rate during laser melting. As a matter of course, the ion implantation process produces point defects as well. The surface roughness created by laser melting is smoothed by grinding with SiC paper and finally polished using diamond paste.

Implantations are carried out using an Extrion 200 kV implanter with doses of $3\ 10^{16}$, $1\ 10^{17}$ ions/cm^2 at an energy of 50 keV per Ne$^+$ - ion. Stereo-transmission electron microscopy for depth profile analysis is done using a JEM 200 CX operating at 200 keV.

As our second approach shot peening has been carried out on laser melted RCC using glass beads with a diameter of 700 µm which were shot during 5 min at the metal surface using air at a pressure of 3 bar. Regular filtering of the shot material was done to remove broken glass beats.

Wear performance is tested on a conventional pin-on-disk wear tester. A ruby crystal ball with a diameter of 5 mm is pressed upon a rotating sample. The ball does not show any significant damage during the experiments. Before each test the ball is rotated or replaced to have a well defined starting condition. A constant speed of 5.0 cm/sec or 0.5 cm/sec and a constant load of 2.3 N have been chosen. According to Hertzian stress analysis it can be shown that in the present situation no plastic yielding due to the ball pressure occurs. The normalized force F of this configuration is about 0.03-0.06. The effect of humidity is reduced by applying absolute ethanol. In addition the experiments are carried out under a dry nitrogen atmosphere. The profile of the wear track is determined with an interference microscope. Measurements with a standard profilometer failed because of the small resolution in depth.

OBSERVATIONS AND ANALYSES

Noble gas implantations

In a laser melted-Ne implanted layer Ne bubbles with a radius up to 34 nm are observed by TEM. Coalesced bubbles are found with the same appearance as found by Johnson, Mazey and Evans [1]. To determine the implantation depth the bulk side thinning method has a disadvantage since there is no guarantee that the front surface near the polished hole will not be affected by the polishing solution. Therefore thin foils have been implanted as well. Figure 1 shows a TEM-picture of an implanted foil exhibiting the depth distribution of the bubbles throughout an austenitic cell, preferentially nucleated at dislocations. The bubble density is 5.7×10^{20} bubbles/m^3. The mean bubble radius is 20 nm and the mean volume is $4.5 \ 10^4 \ nm^3$.

The wear performance is depicted in figure 2. The wear rate decreases with increasing Ne dose. At a dose of $3 \ 10^{16} \ Ne^+$ ions/cm^2 the wear rate increases after 6000 turns. The depths of the wear track is then about 160 nm, which is two times the calculated implantation depth.
Hereafter the wear rate increases strongly, which is due to the interfacing layer causing both compressive and tensile stresses to be present in the wear track. Such a transition is also found with the highest dose implantation.

Extended measurements done at lower loads show, after 15000 turns, when the depth of the wear track is about the same as compared to the lowest dose implantation, an increase in wear rate. The wear rate increases gradually to a value of the not implanted laser melted steel.

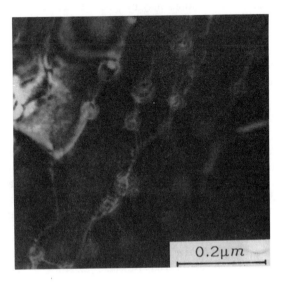

Fig. 1   TEM picture of neon bubbles in an austenite cell. Implanted dose is $3.10^{17} \ Ne^+$ ions/cm^2.

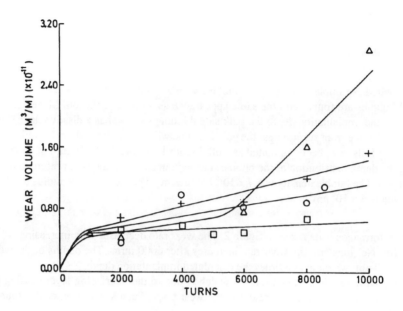

Fig. 2   Measured wear volume of neon implanted laser melted steel. + not implanted, Δ 3.10^{16} ions/cm², o 1.10^{17} ions/cm², $\square$ 3.10^{17} ions/cm².

Clearly the stress fields of the Ne bubbles interact with the stress field of moving dislocations during wear and plastic deformation. Dislocations may by-pass the bubbles either by pure glide (Orowan process) and cross slip or by shearing the bubbles. The stress field around a bubble determines whether a dislocation will reach the bubble and penetrate or will by-pass by bowing around the bubble. Usually in aged-hardened alloys the contribution of modulus hardening to the shearing force, required to cut a particle by glide dislocations, is small. The modulus hardening arises from the differences between the elastic moduli of matrix and particle. Since the elastic energy of a dislocation is a function of the shear modulus, a change in energy and hence a force will be associated with a dislocation interacting with a particle whose shear modulus differs from that of the matrix.

The extreme case is the interaction with a void or bubble, in which the elastic energy is reduced to zero and the system can be described with image forces. So, the modulus hardening in this case of Ne bubbles is a crucial contribution to the hardening since the dislocations experience in the neighbourhood of the Ne bubbles always an attractive force component towards the bubbles. In addition, dislocations may bow around a bubble. The latter is experimentally observed pointing at the Orowan mechanism as well as dislocations pinned by bubbles.

More details regarding the pressure field around the bubbles, the dislocation-bubble interaction and a comparison with N⁺ - implantation is published elsewhere [2,3,4].

Shot peening

Stress measurements were carried on the laser-melted and afterwards shot peened RCC material using a well aligned Philips X-ray diffraction system (PW1830 equipped with a θ drive). The diffractometer is equipped with a fine focus copper tube operated at 50 kV, 20 mA and a graphite monochromator in the diffracted beam which filtered all radiation except Kα. Copper radiation was chosen because of the small penetration depth. (Penetration depth at 90% intensity loss, Cu on steel: 5 µm, Cr on steel 10 µm.) This is especially important for stress measurements of implanted steels (the implantation depth of 100 keV Ne^+ ions in steel is approx. 0.1 µm), but also in cases where stress gradients perpendicular to the surface are expected.

Layer removal was carried out by chemical polishing (85 v.% Acetic acid, 5 v.% H_2O, 10 v.% perchloric acid). The removed layer thickness was determined by weighing the sample before and after polishing. The advantage of this method of layer removal over mechanical methods is that no new stress components are introduced. A disadvantage, however, is that the determination of the removed layer thickness becomes less accurate since after several polishing passes, some preferential etching/polishing takes place.

The stress σ_ϕ is measured using the $\sin^2\psi$ method [5]:

$$\varepsilon_{\psi\phi} = \frac{d_{\psi\phi} - d_0}{d_0} = \frac{(1-\nu)}{E}\sigma_\phi \sin^2\psi,$$

where E represents the modulus, ν Poissons ratio, d_0 the strain free [hkl] plane distance, $d_{\psi\phi}$ the strained [hkl] plane distance, ε the strain and σ_ϕ the stress to be determined. The stresses normal to the surface are assumed to be zero. E=220 GPa and ν = 0.3 were used for the stress analysis.

Line profiles of the {420} diffraction peak (2θ≈143°) were measured at six equidistant $\sin^2\psi$ values from 0 to 0.5. Both positive and negative values of the ψ angle were used for the measurements but only peak positions of the positive ψ angles were used for the stress analysis. The use of negative ψ angles can be informative as the penetration depth varies with ψ angle giving the smallest perpendicular penetration for the negative ψ angles. In the underlying case the deviation of the negative branch in the ε versus $\sin^2\psi$ plot is small which indicates a homogeneous stress situation within the 5 µm layer.

The stress in the shot peened RCC sample turned out to be 1200 MPa (fig. 3), large enough to produce cracks in the surface ground using SiC paper. Subsequent shot-peening prevented cracking. The stress measured was -900 MPa (fig. 3). Layer removal (fig. 4) revealed an increase in compressive stress in the first few µm and at a depth of 5 to 70 µm a nearly constant value of -1200 MPa. Below 70 µm the stress increases to zero indicating the end of the zone affected by shot peening. Although the values of stress deeper then 70 µm are not correct due to the effect of layer removal on the general stress state, the values ranging from 5 to 70 µm are correct due to the lack of strong gradients. Therefore, no correction for layer removal has been performed.

The wear experiment was performed on a sample that was fully laser treated, but only half peened. In this way environmental differences are totally excluded. On the other hand, one must be aware of the possibility that the wear mechanism of the not peened half is influenced by the peened half, or vice versa.

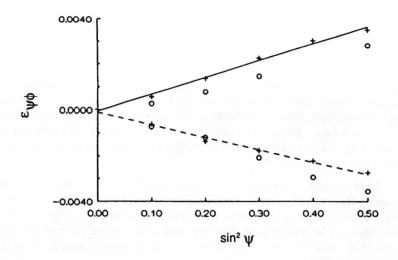

Fig. 3 $\varepsilon_{\psi\phi}$ vs. $\sin^2 \psi$ after laser treatment (solid line) and after subsequent shot peening (dashed line).

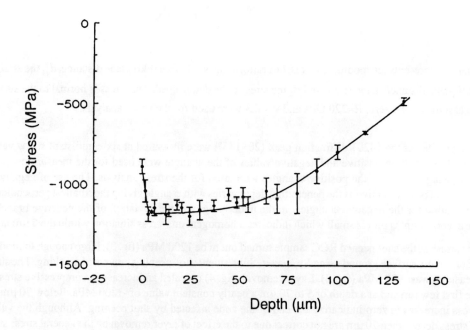

Fig. 4   Stress as a function of depth.

Additional experiments lead us to the conclusion that this is not the case. The wear volumes as a function of the number of turns, as measured with a interference microscope, are depicted in figure 5. It is clearly visible that both the steady-state wear and the running-in wear have decreased by roughly a factor two. This is caused by a higher hardness, a changed micro-structure and a different stress state.

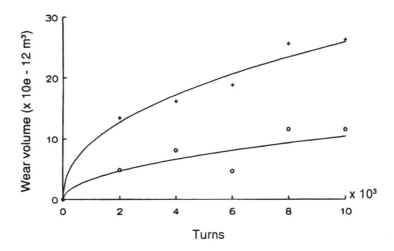

Fig. 5   Measured wear volume of laser melted RCC steel: + not peened, o peened.

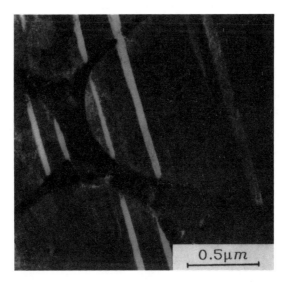

Fig. 6   Darkfield TEM micrograph, showing stress induced α-martensite in austenitic cells.

The micro-structure has been examined with TEM. Specimens are prepared using a dimpler followed by ion milling in such a way that various depths below the surface can be inspected (fig.6). At depths at least up to 15 μm, stress induced martensite is found with the same morphology as observed in a worn laser treated RCC sample [6]. At a depth of 30 μm only an increase in dislocation density was found: more than $10^{14}/m^2$ after peening compared to about $10^{13}/m^2$ before.

CONCLUSION

Neon implantation contributes to compressive strengthening by the formation of bubbles. The volume of the bubbles is high enough to compensate for the tensile stress state after laser melting and the bubbles contribute to a further hardening of the austenitic cells. Orowan looping has been observed by in situ TEM deformation experiments, but more dominant is the pinning of dislocations at bubbles by image forces, a mechanism which in aged hardened alloys is known as modulus hardening. Wear measurements after neon implantations even show a strong improvement in wear which can not only be ascribed to an increase in hardness, but is more caused by a conversion of the tensile stress state into a compressive stress state.

The benefit of laser melted steel can be enhanced even better by a simple technique like shot peening. The wear rate reduces substantially and surface cracking does not occur. Given the experimental conditions the enhancement is such that no tensile stress is left after shot peening and that the layer under compression is thick enough for further machining.

REFERENCES

[1]   P.B. Johnson, D.J. Mazey, J.H. Evans, Rad. Eff. 78, 147 (1983).

[2]   H. De Beurs, J.Th.M. De Hosson, Appl. Phys. Letters 53, 663 (1988).

[3]   H. De Beurs, J. Hovius, J.Th.M. De Hosson, Acta Metall. 36, 3123 (1988).

[4]   J. Noordhuis, J.Th.M. De Hosson, Acta Metall. 38, 2067 (1990).

[5]   M. R. James and J. B. Cohen, in: Experimental Methods in Materials Science, Vol 1, ed. H. Herman , Academic Press, New York, pp 1 (1980).

[6]   H. De Beurs and J.Th.M. De Hosson, High power lasers, Eds. A. Niku-Lari and B.L. Mordike, Pergamon Press, Oxford, 27 (1988).

Materials Science Forum Vols. 102 - 104 (1992) pp. 401-408
Copyright Trans Tech Publications, Switzerland

LASER SURFACE NITRIDING OF TITANIUM AND A TITANIUM ALLOY

V.M. Weerasinghe (a), D.R.F. West (b) and M. Czajlik (c)

(a) Polytechnic of East London, London, UK
(b) Imperial College of Science, Technology and Medicine, London, UK
(c) Technical University Kosice, Czechoslovakia

ABSTRACT

Surfaces of titanium and titanium alloy IMI829 have been nitrided by laser melting in nitrogen and in a gas mixture containing various proportions of argon and nitrogen. The nitrided layers have been characterised with respect to their hardness, microstructure, chemical composition and surface finish. Observations are also reported on the laser remelting in argon of the as-formed nitrided layers and laser melting of nitride layers preformed by physical vapour deposition and furnace heating.

INTRODUCTION: Surface modification of titanium to improve its poor wear resistance has been the subject of a number of investigations including the use of high power lasers [1-7]. Laser surface nitriding in particular is an attractive process because of its simplicity and the possibility of producing hard layers of substantial depths (>1mm) with minimum effect to the 'bulk' of the material.

Laser surface nitriding was initiated by the work of Katayama et al [3] and since then there has been a number of other investigations [e.g. 4-7]. The laser nitriding process involves using the intense optical energy of the laser to melt the surface in a nitrogen containing atmosphere. Extensive and rapid solution of nitrogen occurs in molten titanium and dendrites of TiN are produced at traverse speeds of the order 100mm/s [4]. As in all laser surface modification processes, surface layers are produced by overlapping of single melt tracks, typically by about 40%.

Previous work has shown the formation of non-stoichometric nitrides of titanium dendrites in an α- Ti matrix. Hard layers of thicknesses 0.2 - 1.0mm and having hardness values >700VHN have been produced using mainly CO_2 lasers.

Cracking in the nitrided layers has been identified as a major problem, although crack free layers were reported to be possible when the volume fraction of the hard phase TiN was kept low [4].However to achieve this, it was found to be necessary to reduce the laser interaction time [4], thereby also reducing the hard layer thickness, which is detrimental in particular to load bearing applications. In the present work , the volume fraction of the hard phase TiN was controlled independent of the laser interaction time by diluting the nitrogen with various proportions of argon. Furthermore, the present work aimed to investigate the effects of laser melting in argon of preformed nitrided layers. Nitrided layers were preformed on cp titanium (IMI125) and titanium alloy Ti - 5.5Al - 3.5Sn - 3Zr - 1Nb - 0.3Si wt.% (IMI829) by: laser melting, PVD technique and furnace heating at 1000° C for 2 hours in nitrogen.

EXPERIMENTAL PROCEDURE: Experiments were performed using a fast axial flow type 2kW continuous wave CO_2 laser. The laser processing parameters were: laser power 2kW, defocussed spot diameter 1-2mm, traverse speed 15 - 100 mm/s. Specimens were in the form of 15 x 25 x 6mm thick coupons cut from rolled plate and the surfaces were sand blasted and ultrasonically cleaned prior to laser treatment. Some specimens were also annealed at 750°C for 1 hour prior to laser treatment.
A shroud system was used to provide the work area of the specimen with the required gas atmosphere i.e. 'oxygen-free' nitrogen, nitrogen diluted with argon or argon, during laser processing. Effective shroud systems for laser processing of materials such as titanium have been developed at Imperial College. One such system, for example, uses a rotating centrifugal fan to disperse the surrounding air while the required gas is injected via multiple holes in to the central region [8].
Physical vapour deposition (PVD) of TiN was carried out using a industrial facility at Tecvac (UK) Ltd., Cambridge.
The nitrided layers were characterised with respect to their hardness, microstructure, chemical composition and surface finish. Microhardness measurements were taken mainly on transverse sections using a 50g load. Optical and scanning electron microscopy was used to study the microstructure and the chemical composition. Specimens were etched with a reagent consisting of 10ml HF and 5ml HNO_3 in 85ml of water. The surface finish and topography were evaluated using a Talysurf instrument.
Commercially pure titanium (IMI125) and titanium alloy IMI829 were used. The main part of the work was concentrated on cp titanium.

RESULTS AND DISCUSSION
General: Figure 1 shows a laser nitrided cp titanium specimen produced by melting in nitrogen. The processing parameters were: Laser power 2kW, spot diameter 2mm and traverse speed 25mm/s. The nitrided layer is 0.6mm thick and the substrate thickness is 6mm. Dendrites of TiN were observed in the nitrided layer and hardness values of 750 - 950VHN were measured depending on the local dendrite 'density'. A number of cracks were observed confined to the nitrided layer and mostly transverse to the traverse direction.

Fig 1 *A laser nitrided layer on cp titanium.*

Remelting of nitrided layers: Figures 2a,b show longitudinal sections of an as-formed layer and a remelted layer on IMI829 alloy. The as-formed layer was produced by laser melting in pure nitrogen with a traverse speed of 44mm/s. The as-formed layer was then remelted in a transverse direction in argon with a small addition of nitrogen. The remelting speed was 20mm/s. With reference to figures 2a,b the following observations are made:

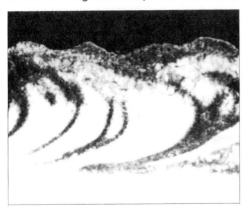 

Fig 2a *Longitudinal section of a laser nitrided layer on IMI829. 100X*

Fig 2b *Longitudinal section showing the effect of remelting 100X*

* The remelting has diluted the original as-formed layer by penetration into the substrate. By comparison of thicknesses, the dilution is about 22%.
* The surface of the remelted layer appears smooth compared to noticeably heavy rippling on the as-formed layer surface.
* The remelted layer was more homogeneous with respect to its hardness and microstructure. The microhardness of the remelted layer was 800 +/- 50 VHN as compared to a scatter of 630 - > 950 VHN measured in the as-formed layer.

The surface rippling and the inhomogeneity in the as-formed layer are attributed to surface tension driven fluid flow. Surface tension gradients are set up by temperature gradients and appear to be much enhanced by nitrogen concentration gradients on the melt pool surface. Nitrogen concentration gradients are expected to exist because the solution of the gaseous nitrogen is temperature and time dependent. The usual definition of the interaction time between the workpiece and the laser is the beam diameter divided by the displacement velocity. In fact the melt pool is elongated and extends beyond the beam diameter in the longitudinal direction and consequently the time for which nitrogen can go into solution is greater and furthermore a nitrogen concentration gradient is set up in the longitudinal direction. These gradients are 'neutralised' in a cyclic manner by surface tension driven fluid flow.

Preliminary results from electron microprobe analysis have indicated no substantial loss of alloying elements in the IMI829 due to the laser treatment. This is contrary to the almost complete loss of Al and Sn reported in [7]. Further aspects of remelted layers on IMI829 and cp titanium are shown in figures 3a,b and 4a,b. With reference to these figures the following observations are made:

* Absence of a continuous surface film of TiN in the layers which were remelted in argon. Surface 'films' of TiN (~5 micron) have been previously observed on laser nitrided layers [4]. The surface film is consistent with the existence of a concentration gradient of nitrogen in the melt pool from the surface downwards, during exposure to the gaseous nitrogen and furthermore according to the Ti - N equilibrium diagram, TiN is formed over a wide range of compositions.

* In the as-formed layers, large dendrites of TiN, with high aspect ratios, appear to grow normal to and downwards from the surface. In the remelted layers, the TiN dendrites appear to be more equi-axed.

Fig 3 *Effect of re-melting, IMI829. As-formed(a),44mm/s. Remelted(b),20mm/s. 1000X*

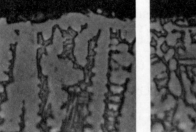 

Fig 4 *Effect of re-melting, cp titanium. As-formed(a),20mm/s. Remelted(b),16mm/s. 1000X*

The peak to valley height of the surface finish of the as-formed layers was at best 20 microns compared to 5 microns in the remelted layers. The topography of the surfaces is shown in figure 5.

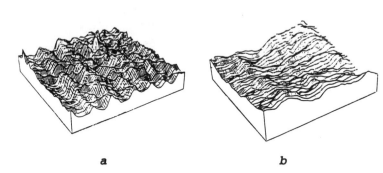

Fig 5 *Surface topography of an as-formed surface (a, 100X - vertical) and a remelted surface (b, 400X - vertical).The surface area shown is ~5 x 5mm.*

a   b

After remelting in pure argon, the surfaces appeared white whereas the as-formed surfaces were gold coloured possibly due to the presence of a surface film of near stoichometric TiN formed in equilibrium with the gaseous nitrogen.

Nitriding with dilute nitrogen: Figure 6 show section profiles of melt tracks which were produced with increasing amounts of nitrogen mixed with argon.The tracks become larger and flat bottomed as the nitrogen content is increased. This is attributed to the exothermic heat of solution and convective fluid flow.

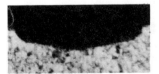

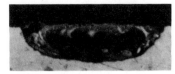

Fig 6 *Effect of progressively increasing the nitrogen content in the nitrogen/argon gas mixture on the single melt track section profile. The section with the most dilute nitrogen content is on top left. 10X*

Nitrided layers having no TiN dendrites were produced on cp titanium with very dilute nitrogen. The thickness of these layers was typically 0.3mm and the hardness was 480VHN. No cracking was observed. A continuous film (~ 1 micron) assumed to be TiN was observed on the surface.

Figure 7 shows a layer which was produced with an increased nitrogen content in the argon/nitrogen mixture and in which TiN dendrites appear to have formed near the surface. The thickness of the dendritic layer was 25 microns. No cracking was observed. The hardness of the melt zone sublayer where no dendrites were observed was 480 VHN. The microstructure of the sublayer consisted of α and/or transformed β .

50X   400X

Fig 7 *A nitrided layer produced with dilute nitrogen showing TiN dendrites near the surface.*

Pronounced surface tension ripples were observed on the top surface of the melt track when a small amount of nitrogen was added to argon, figure 8. The ripples indicate simultaneous transverse and longitudinal flow pointing towards the traverse direction. When the nitrogen content was increased, reduced rippling was observed. This may be due to a stabilising effect caused by the increased amount of TiN dendrites present in the molten titanium.

Fig 8 *Surface tension ripples on the top surface of a melt track produced with dilute nitrogen. 50X*

The technique of nitriding with various proportions of nitrogen in a nitrogen/argon mixture makes feasible the production of so called 'directionally gradient layers' i.e. a sublayer to support a harder and more brittle top layer.

Melting of preformed TiN layers: Figure 9 show a PVD TiN layer (~ 3 microns) on cp titanium, laser melted in argon and in argon with a small addition of nitrogen. The nitrogen was added to counteract a possible loss of nitrogen by way of dissociation of the ~ 3 micron thick PVD TiN layer during laser melting. The traverse speed was 58mm/s. The hardness of the layer melted in argon was 480 VHN to a depth of 120 microns. The dendritic zone in the layer melted in dilute nitrogen was 15 microns and no cracks were observed.

Similar results were obtained for laser melting of TiN layers which were preformed by furnace heating. The objective of laser melting preformed TiN layers was to obtain a more uniform hardness profile to a greater depth.

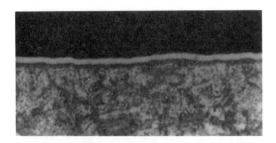

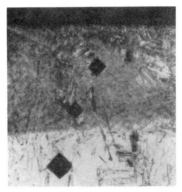

Fig 9 *Top left: PVD TiN coating on cp titanium 1000X, Top right: PVD coating laser melted in argon 200X, Left: PVD coating laser melted in dilute nitrogen 1000X.*

CONCLUSIONS

Various modifications to the process of laser nitriding of titanium have been successfully explored.

1) Remelting of nitrogen alloyed layers in argon can produce a better surface finish and more uniform microstructures.

2) Nitriding by laser surface melting in argon containing low concentrations of nitrogen provides an effective route for preventing cracking and for controlling the microstructure and the hardness.

3) Laser surface melting of furnace nitrided titanium or titanium pre-coated with a TiN layer can be used to increase the depth of hardening, with a uniform hardness distribution.

ACKNOWLEDGEMENTS: The authors wish to thank PERA International, Melton Mowbray, UK, for financial support; Tecvac (UK) Ltd. and Struers Ltd. respectively for PVD coating and metallographic specimen preparation; and IMI Titanium Ltd. for supplying cp titanium and IMI829 titanium alloy.

REFERENCES

1) Rolinski E.:Surf. Eng.,1986,Vol.2,No.1,p35
2) Bharti A.,Goel D.B.,Sivakumar R.S.:Proc ICALEO'88,ed Bruck G.,1988,p42
3) Katayama S. et al: Proc ICALEO'83,ed Metzbower E.A.,1983,p127
4) Seiersten M.: Proc First Nordic laser material processing conference, ed Thorstensen B.,1988,p66
5) Walker A.,Folkes J.,Steen W.M.,West D.R.F.: Surf. Eng.,1985,Vol.1,No.1,p23
6) Bell T.,Bergmann H.W.,Lanagan J.,Morton P.H.,Staines A.M.:Surf. Eng.,1986,Vol.2,No.2,p133
7) Ubhi H.S.,Baker T.N.,Holdway P.,Bowen A.W.:Proc Sixth world conference on titanium,France,1988.
8) Weerasinghe V.M.,Steen W.M., U.S.Pat.No.4803335,1989

Materials Science Forum Vols. 102 - 104 (1992) pp. 409-416
Copyright Trans Tech Publications, Switzerland

FATIGUE PROPERTIES AND RESIDUAL STRESSES IN LASER HARDENED STEELS

R. Lin and T. Ericsson

Dept. of Mechanical Engineering, Linköping University
S-581 83 Linköping, Sweden

ABSTRACT

This paper deals with the study of fatigue properties and residual stresses in laser hardened steels. Notched fatigue specimens were made of Swedish steels SS 2225 and SS 2244 which were low alloy steels with similar chemical composition but different carbon content. They correspond to AISI 4130 and AISI 4140 respectively. Laser hardening was performed on the gauge sections. Results from the studies showed that compressive residual stress existed in the hardened zone at notches and marked improvement in the fatigue behaviour was achieved with laser hardening. The fatigue limits were increased by 64% for SS 2225 and 69% for SS 2244. It is concluded that laser hardening can greatly improve the fatigue properties of steels, which is attributed to the hardened case with high compressive stresses induced by laser hardening.

1. INTRODUCTION

Laser hardening can be used to strengthen the surface of steels. It is a technique with unique features such as the ability for hardening selective areas or specimen with complex shape[1]. Its high cooling rate due to self-quenching also allows hardening of steels with low hardenability. As a consequence of the rapid thermal cycle created by the process, residual stresses can be introduced in the hardened component. Depending on the component and the laser hardening process, the obtained residual stress field can be very different. It has long been recognized that the fatigue behaviour of a component depends not only on the resulting metallurgical features but also on the mechanical states such as the magnitude and distribution of the residual stresses. Therefore, the present work aims at studying the effect of laser hardening on the fatigue properties of steels with special regards to residual stresses induced by laser hardening.

2. EXPERIMENTAL DETAILS

The materials used in the study were two Swedish steels, SS 2225 and SS 2244 which correspond to AISI 4130 and 4140 respectively. They were low alloy steels with similar alloying contents except for carbon. The chemical compositions are listed in Table 1.

The fatigue specimens were made from hot rolled bars which were in a quenched and tempered condition with a hardness of about 300 HV0.3. The geometry of the notched specimen is shown in

Figure 1. The theoretical stress concentration factor K_t was 1.56. The gauge section was finely ground to 0.8 µm.

Table 1	Chemical compositions of the materials in wt%						
Material	C	Si	Mn	P	S	Cr	Mo
SS 2225	0.26	0.19	0.71	0.004	0.02	1.13	0.17
SS 2244	0.45	0.28	0.65	0.013	0.02	1.02	0.16

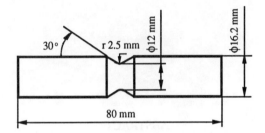

Figure 1. The geometry of the fatigue specimen

Laser hardening was performed with a Rofin Sinar HF1700 laser. A line focussing optical system was used which produced an elliptical shape of beam profile with rather uniform energy distribution. The long axis of the ellipse which paralleled to the specimen axis was 10 mm and the short axis was 2.5 mm. The power of the laser beam was 1150 ± 25 watts and the rotating speed of the specimen at the notch was 0.5 m/min. To enhance the absorption of laser energy the specimen surface was coated with soot from burning acetylene. Because of the small dimension of the specimens, the gauge section was treated by two continuous half turns. After the first half was scanned and the specimen was cooled the second half was treated with exactly the same laser parameters. They will be referred as the first and the second track respectively in the paper. Water was sprayed immediately on the specimen each time after the the laser beam had been removed.

Residual stresses caused by laser hardening were investigated with a computer controlled diffractometer. Cr-k_α radiation was used to obtain the (211) peak. The standard $\sin^2\psi$ method[2] was used with 7 ψ angles ranging from -40 to +40 degrees for the axial direction and from -45 to +45 degrees for the tangential direction. A 0.5 mm collimator was used to give an irradiated area of about 0.7 mm in diameter at $\psi=0$. To obtain the depth profile of residual stresses layers of material were removed by electropolishing. No correction was done for residual stress relaxation due to material removal.

Two groups of specimen from each material were subject to fatigue loading. One was in a quenched and tempered condition and the other was after laser hardening. Fatigue testing was performed with completely-reversed-plane bending load. That is the mean stress value was zero. The specimens were placed in the fatigue machine in such a way that the overlap zone was on the neutral line of bending. The up-and-down strategy [3] was used to locate the fatigue limit at 10^7 cycles. All the fractured specimens were examined with a scanning electron microscope.

3. RESULTS AND DISCUSSIONS

Metallurgical Studies
Figure 2 shows the longitudinal cross section of a laser hardened specimen. A rather uniformly hardened zone was obtained at the notch. The hardening depth varied somewhat from specimen to

specimen for the same material, mainly because of variations in coating thickness. Misalignment of the laser beam could also contribute to the variation. The hardening depth was 0. 53 mm for SS 2225 and 0.46 mm for SS 2224. It was defined as the depth at which the hardness dropped to 50 HV above the base hardness and was an average over 6 hardened tracks. The peak hardness obtained was about 550 HV for SS 2225 and 750 HV for SS 2244. There was no obvious hardness difference between the first and the second track, which indicated that the input heat from hardening the second track had little tempering effect on the first track.

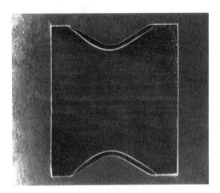

Figure 2. Longitudinal cross section of the laser hardened specimen of SS 2225. 2x.

Microstructure in the hardened zone was essentially martensitic. A bright layer in the order of ten micrometers in thickness was observed in the surface of some specimens. X-ray diffraction showed that it contained a large amount of retained austenite. The size of martensitic needles in both steels decreased with increasing depth due to faster cooling towards the core.

X-ray diffraction studies

Retained austenite measurements were carried out on the surface of some hardened specimens. It was found that the content of retained austenite varied from specimen to specimen. Even though it could be very high, up to e. g. 45%, it decreased very rapidly to less than 5 % within a couple of hundredths of millimeter. The existence of such a layer could be explained by a carbon rich surface which resulted from reactions between the surface and the coating during the laser hardening process.

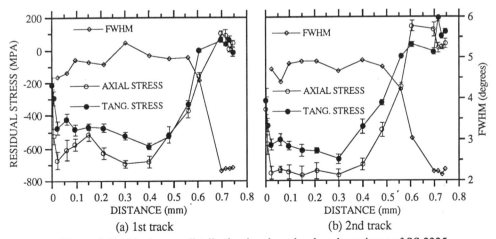

(a) 1st track   (b) 2nd track

Figure 4. Residual stress distribution in a laser hardened specimen of SS 2225.

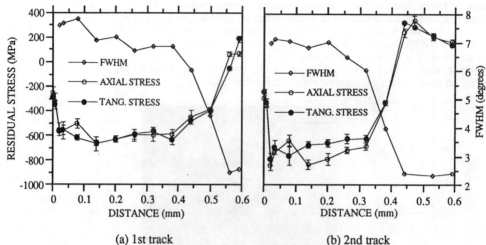

(a) 1st track   (b) 2nd track
Figure 5. Residual stress distribution in a laser hardened specimen of SS 2244.

The depth profiles of residual stress were shown in figures 4 and 5, together with the distribution of the peak breadth that was measured as the full width at the half maximum intensity. The peak breadth was about the same for the first and the second track, which further confirmed that there was little tempering effect on the first track. On the other hand it was different for the two materials. Due to higher carbon content in SS 2244, microstrains induced by martensitic transformation was larger and therefore the measured peak breadth was bigger.

Figures 4 and 5 also showed that residual stresses were compressive in the hardened zone of all tracks. Partly due to the larger amount of retained austenite in the surface, the magnitudes of compressive stress were lower in the surface. In fact, for specimens with lower surface content of retained austenite the surface stresses were higher. As contrast to the retained austenite distribution the compressive stresses reached maximum values within a few hundredths of millimeter and then remained about the same with further increasing depth. They started to change towards tensile when reaching the transition zone. Compared with the peak breadth distribution, it can be seen that the size of compressive zone roughly matches that of the hardened zone.

Although the input heat from hardening the second track had little tempering effect on the first track, it did change its stresses. As shown in both figures 4 and 5, the magnitude of stress were lower in the first track. The stresses had somewhat been relaxed on heating the second track.

There was also some difference in stress values between the axial and tangential direction. It was higher in the axial direction. The same tendency was also reported on shot-peened-notched specimens[4]. This was perhaps due to the different constraint to deformation at the notch.

Fatigue Behaviour
Results from bending fatigue testing are plotted in figure 6 as S-N curves. The stresses shown in the vertical axis are nominal values. The horizontal lines give the fatigue limits which are also listed in table 2. The finite life part of the curves was calculated by least square fitting. It can be seen that the bending fatigue behaviour of the specimens was greatly improved after laser hardening. Both low and high cycle fatigue lives were increased.

For both steels in quenched and tempered condition the bending fatigue limits were similar. After laser hardening they were increased by 64% and 69% for SS 2225 and SS 2244 respectively. The increase was almost the same for the two steels. It can be concluded that the difference in carbon content between the two steels had little effect on the fatigue limits after the laser hardening although

it did result in different peak hardness. Studies carried by the authors [5] on smooth specimens also
showed that laser hardening had the same effect on both steels.

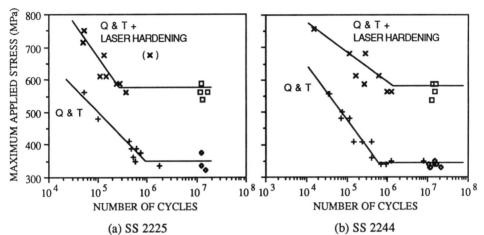

(a) SS 2225   (b) SS 2244

Figure 6. S-N curves of a) SS 2225 and b) SS 2244. Crosses refer to failure and open symbols to
runouts. The point in parentheses was not used for curve fitting.

Table 2	Bending fatigue limits of notched specimen			
Material	SS 2225		SS 2244	
Condition	Q & T	Q & T + laser hardening	Q & T	Q & T + laser hardening
Fatigue limit (MPa)	350 ± 62	575 ± 21	343 ± 10	581 ± 29

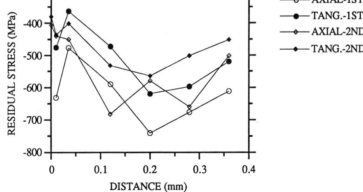

Figure 7. Residual stress distribution in the 1st and 2nd tracks of a laser hardened specimens
of SS 2225 after fatigue loading at 587 MPa and 1.25×10^7 cycles.

The effect of a residual stress on fatigue resistance is similar to that of an equivalent mean stress.
Usually a compressive mean stress will prevent any existing or nucleated cracks from propagating.

However, residual stress can play a role only when it is not completely relaxed by the fatigue loading. To check the stability of the residual stress field induced by laser hardening, residual stresses in the hardened zone of a SS 2225 specimen after fatigue loading at 587 MPa and 1.25×10^7 cycles were measured and the results were shown in figure 8. The higher surface stress was due to the lower surface austenite contents in this particular specimen. As shown by the figure, the magnitudes of compressive stresses in both tracks are comparable to each other. In comparison with figure 4, it can be seen that relaxation had occurred only in the second track and the magnitudes of the stresses after fatigue loading were still considerable.

Fractography showed that for the reference specimens, all fatigue cracks started from the notch surface as expected. For laser hardened specimens of SS 2225, fatigue origins were also confined to the surface as can be seen in figure 8(a), while fish eyes were observed in most of the laser hardened specimens of SS 2244 which broke in the vicinity of the fatigue limit, see figure 8(b). Since the magnitudes of residual compressive stresses were about the same for the two materials, it can be reasonably assumed that this difference was caused mainly by the difference in peak hardness between the two steels. For SS 2225 steel, the increase in the local fatigue strength at the notch surface due to laser hardening was not enough to wipe out the effect of the stress concentration. Therefore, fatigue failure started from the surface. On the other hand, the local fatigue strength in the notch surface of the SS 2244 specimens was higher due to a higher hardness. Cracks then initiated from inclusions at a depth between 0.56 to 0.71 mm, which was in or near the transition zone. Around this region the hardness dropped to a low value and residual stresses were usually not favourable as shown in figure 5. It can be concluded that although both steels had obtained a similar increase in the fatigue limit after laser hardening, the fact that fatigue cracks started from subsurface in SS 2244 indicates that there is still potential for further improvement in the fatigue properties of the steel. By increasing the hardening depth, cracks can be suppressed from starting in the subsurface where the local fatigue strength is lower and forced out to the surface where local fatigue strength is higher. In this way, a better fatigue resistance will be obtained.

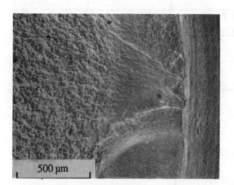 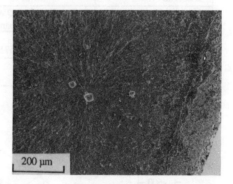

        (a) SS 2225, laser hardened   (b) SS 2244, laser hardened
Figure 8. Photos showing fatigue crack started at (a) surface, and (b) an inclusion below surface.

Observations on the fractured surfaces also showed that the main cracks were often found in the first track for both steels. By comparing the stress distribution for the two tracks in figures 4 and 5, it can be seen that higher compressive stress yields a better resistance to fatigue.

From the above discussion we can see that the marked improvement in the fatigue resistance was attributed to a hardened case with a rather stable residual stress field of high compression.

4. CONCLUSIONS

The bending fatigue properties and residual stress distribution in two laser hardened steels were investigated. The following conclusions have been drawn from the study.

1). Hardening depth of 0.53 mm for SS 2225 and 0.46 mm for SS 2244 were respectively obtained at the notch of specimen by laser hardening. The peak hardness was 550 HV for SS 2225 and 750 HV for SS 2244.

2). High compressive stresses were found in the laser hardened notch of both materials, the depth of which depended on the depth of hardening zone.

3). Stress relaxation had occurred in the second track during fatigue loading. Though the magnitudes of residual stresses were still quite high after fatigue loading.

4). The bending fatigue properties of notched specimens were greatly improved after laser hardening. Fatigue limits were increased by 64 % for SS 2225 and 69 % for SS 2244. The effect of the laser hardening process employed here was about the same for the two steels. Further increase in the fatigue resistance of steel SS 2244 is possible.

5). The marked increase in the fatigue strength was attributed to the hardened and compressed surface layer induced by laser hardening.

ACKNOWLEDGMENTS

This paper is part of a joint project of Linköping University, Atlas Copco AB, and Saab-Scania Truck Division. The financial support and technical helps from the cooperators are very much appreciated. The authors would specially like to thank Mr A. Brunnberg from Saab-Scania Truck Division for his work on the fractography.

REFERENCES

1) Mazumder, J.: J. Metals, 1983, 5, 18.

2) Noyan, I.C. and J.B. Cohen: Residual stress, Springer-Verlag, 1986.

3) Dixon, W.J. and A.M. Mood: Journal of American Statistical Association, 1948, 43, 109-126.

4) Bergström, J.: Surface Engineering, 1986, 2, 115-120.

5) Lin, R. and T. Ericsson: to be presented at NOLAMP Conference, August 21-22, 1991, Lappeenranta, Finland.

Materials Science Forum Vols. 102 - 104 (1992) pp. 417-432
Copyright Trans Tech Publications, Switzerland

COUPLED GROWTH IN EUTECTIC SYSTEMS

S.M. Copley, E.Y. Yankov, J.A. Todd and M.I. Yankova

Dept. of Metallurgical and Materials Engineering, Illinois Institute of Technology, Chicago, IL 60616, USA

ABSTRACT

Previous research on surface melting and resolidification of silver-copper alloys of near eutectic compositions by Beck, Copley and Bass using a laser beam and Boettinger, Shechtman, Schaefer and Biancaniello using an electron beam has revealed that the aligned composite plate microstructure that forms at slow growth speeds (< 2.5 cm s^{-1}) is replaced by a banded microstructure at higher speeds, while at speeds exceeding 50 cm s^{-1} it is replaced by a microsegregation-free structure. Employing transmission electron microscopy to study the microstructure of laser melted trails in a Cu-61.7 at. pct. Ag alloy, we have found that the banded microstructure is not due to a periodic cellular breakdown of the interface, as suggested previously, but consists of alternating regions of the extended metastable solid solution, gamma, and a coupled growth structure containing thin plates of gamma and the copper-rich phase, beta prime. The spacing of the coupled growth structure is approximately 100 Å, which is less than half of the minimum spacing previously observed for coupled growth from the melt in Ag-Cu. A theoretical model explaining these effects is presented based on a finite elements solution of the diffusion equation in the melt, the interface response functions for continuous growth derived by Aziz and Kaplan, the free energy - composition relationships developed by Murray, and a recently developed theory for coupled growth that includes far from equilibrium interface conditions by Yankov et al.

INTRODUCTION

The impetus for the work reviewed here came from a series of papers that treated the interphase precipitation reaction in vanadium steels as a type of coupled growth [1-5]. Recently, this work has been extended to eutectic systems [6-8]. The relevant experimental work for eutectic systems has been carried out on Ag-Cu alloys [9-11]; however, the results of our work are believed to explain observations in alloy

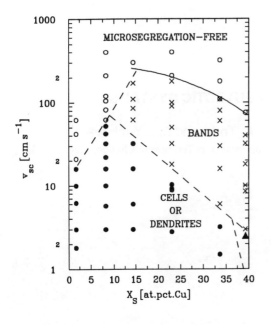

Figure 1:   Microstructures observed in electron beam melted trails of Ag-Cu alloys by Boettinger et al. [11].

systems such as Al-Cu [12,13], Al-Pd [14], Al-Zr [15] and Al-Mn [16].

To place our results in context, the results of Boettinger et al. [11] have been replotted on an atomic percent scale in figure 1. This figure indicates the microstructures they observed in electron beam melted trails as a function of beam velocity (v) and alloy composition (X). At low velocities, cellular or dendritic growth was observed over a wide range of compositions giving way to coupled lamellar growth of alloys near the eutectic composition. In dilute alloys, a transition from cellular-dendritic to microsegregation-free growth took place at a critical velocity that increased with increasing Cu content. Beyond 9 atomic percent Cu, a banded microstructure intrudes consisting of light and dark etching bands marking sequential positions of the melt-solid interface. Boettinger et al. concluded that the dark etching bands had the same composition as the light etching bands in contrast to the conclusion of an earlier investigation by Beck et al., who observed bands in laser melted trails and reported a mean increase in Cu in the dark etching bands of at least 1.1 atomic percent in a Cu- 61.7 at. pct. Ag alloy [9].

The results reviewed here will be presented in several sections: (1) Observation of the finest coupled growth; (2) Criterion for microsegregation-free growth; (3) Specification of the Aziz-Kaplan interface response functions; (4) Far-from-equilibrium eutectic growth; (5) Bands; and, (6) Interphase precipitation.

OBSERVATION OF THE FINEST COUPLED GROWTH

Yankov et al. [6], have recently reinvestigated the nature of bands formed in laser melted trails of the Cu-61.7 at. pct. Ag alloy studied by Beck et al. [10] using transmission electron microscopy. The TEM investigation of the boundary between the light and dark etching regions revealed that the darker region consisted of a lamellar, coupled growth structure, while the lighter region was featureless as shown in figure 2. The spacing of the lamellar structure was 10 nm, which is one half the finest spacing observed previously for coupled eutectic growth in the Ag-Cu system [11]. The structure was produced at a much higher beam scanning velocity than was used in the previous work (34 cm s⁻¹ versus 2.5 cm s⁻¹). Selected area diffraction patterns indicated the phases present to be the extended metastable solid solution, gamma, and the copper-rich phase, beta prime.

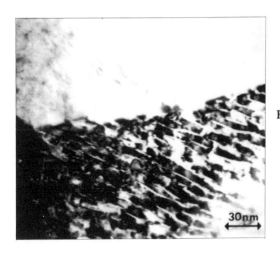

Figure 2: A TEM micrograph showing the lamellar, coupled growth region and the featureless region [6].

CONDITIONS FOR MICROSEGREGATION-FREE GROWTH

A method is presented here for predicting the beam scanning velocity required to bypass the banded structure and to solidify the alloy in a microsegregation free manner based on the continuous growth model developed by Aziz and Kaplan [17]. When this velocity has been measured, as in the case of the Ag-Cu system (figure 1), the method can be used to evaluate the speed of solute-solvent redistribution across the interface at infinite driving force, an important parameter in the Aziz-Kaplan model.

The solidification of a 2-component alloy is governed by four differential equations describing heat flow and interdiffusion in the melt and solid phases. These equations must be solved subject to initial conditions (e.g., alloy composition) and boundary conditions at the melt solid interface. The boundary conditions assure temperature continuity, continuity of heat flow, and continuity of mass flow. In additions there are two equations that specify solute concentrations at the interface as a function of interface velocity and temperature. These equations are called "interface response functions". The continuous growth model of Aziz and Kaplan gives two interface response functions for a solid phase and a liquid phase involving: solid composition, X_S, liquid composition at the interface, X_L, velocity, v and temperature, T. If two of these parameters are known, then the other two can be calculated.

The first of the Aziz-Kaplan response functions is for the partition coefficient:

$$k(v) = \frac{X_S}{X_L} = \frac{\dfrac{v}{v_D} + g_e}{\dfrac{v}{v_D} + 1 - (1 - g_e)X_L} \tag{1}$$

where v_D is the speed of solute-solvent redistribution at infinite driving force and g_e is a partitioning parameter. The parameter g_e is given by:

$$g_e(X_L, X_S, T) = \exp[-(\Delta\mu'_B - \Delta\mu'_A)/RT] \tag{2}$$

where A represents the solvent component, B represents the solute, R is the gas constant, Δ is used to indicate differences of quantities across the interface, i.e. quantities in the solid at the interface minus those at the liquid at the interface. The parameter μ' is called the "redistribution potential" and is given by the difference between the actual chemical potential and the entropy of mixing:

$$\mu'(X,T)=\mu(X,T)-RT \ln X \tag{3}$$

The second function is for the interface velocity. Its form depends on whether or not "solute drag" is taken into account. In a recent paper, [18], Kaplan et al. showed theoretically that including solute drag results in violation of Onsager's reciprocity relations [19]. Hence, our calculations here use the form "without solute drag":

$$v=v_0[1-\exp(\Delta G_{DF}/RT)] \tag{4}$$

where ΔG_{DF} is given by:

$$\Delta G_{DF}=X_S \Delta \mu_B +(1-X_S)\Delta \mu_A \tag{5}$$

The Aziz-Kaplan interface response functions can be evaluated if the parameter v_D and analytical equations for the composition dependance of the Gibbs free energy are known. The parameter v_D is equal to the interdiffusion coefficient in the liquid, D_L, divided by the width of the solid-liquid interface, L. The width of the solid-liquid interface has been shown recently to vary by over two orders of magnitude in different systems [21].

Interface response functions have been represented by plotting interface temperature *versus* interface compositions for specific interface velocities [17]. The resulting plot is similar to the phase diagram. In fact, the liquidus and solidus curves on the phase diagram correspond to the liquid and solid interface compositions for growth in the limit of zero interface velocity. In this paper, interface response functions are represented in a different form. Interface velocity is plotted *versus* the interface compositions for specific interface temperature. This form has been very useful in understanding the criterion for microsegregation-free solidification and analyzing far-from-equilibrium eutectic growth.

The Aziz-Kaplan interface response functions for the Ag-rich and Cu-rich phases of the Ag-Cu system are evaluated here using the Gibbs free energy *versus* composition functions developed by Murray [20]. Two examples are plotted in figure 3 for interface temperatures of 1054 K (the eutectic temperature) and 1030 K. The parameter v_D was set equal to 30 cm s^{-1} in plotting these functions. The Ag-Cu phase diagram based on Murray's Gibbs free energy *versus* composition equations is also plotted in figure 3 so that it can be compared to the Aziz-Kaplan interface response functions.

At low interface velocities, the difference in solute content between the liquid and solid phases at the interface is greatest. As the velocity increases, this difference decreases owing to solute trapping. With increasing velocity, the solute content of the solid at the interface first increases, passes through a maximum value, X_m, and then decreases. The velocity at which the solute content is maximum is designated v_m. It should be noted that the effect of decreasing the interface temperature is to shift the velocity *versus* interface composition curves towards the center of the diagram.

In comparing the Aziz-Kaplan interface response functions to the phase diagram it should be noted that the interface compositions at zero velocity given by the interface response functions plot correspond to intercepts on the liquidi and solidi curves at the interface temperature. Also, at high velocities, the interface compositions are approximately equal to the T_0 curve intercepts at the interface temperature. The T_0 curve is the locus of composition-temperature points where the liquid and solid phases have equal Gibbs free energy.

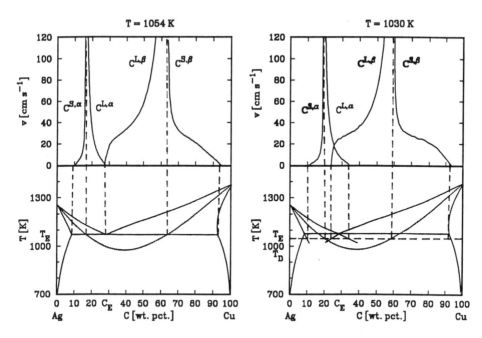

Figure 3:   A diagram showing the effect of interface temperature on the Aziz-Kaplan interface response functions and relating these functions to the phase diagram.

For a binary Ag-Cu alloy with solute content X_0 to solidify at steady state, $X_0 \leq X_m$. If this condition is not satisfied, then the solute content of the growing solid cannot equal that of the liquid from which it is formed. As a consequence, the liquid will be continuously enriched or depleted in solute and steady state solidification is impossible.

For the alloy to solidify microsegregation-free, a second condition must be satisfied; namely, $v \geq v_m$. This condition can be deduced by considering the effects of various perturbations on steady state growth.

The situation for $v \geq v_m$ is summarized in Table I. If, for example, the temperature of the interface increases slightly during steady state growth, then the velocity *versus* liquid composition curve (see figure 3) shifts to the left. Thus, liquid at the interface solidifies at a lower velocity then the isotherm velocity causing the temperature at the interface to decrease. The perturbation causes a compensating reaction to take place so that stable, plane front solidification can be maintained. A slight temperature decrease, a slight solute concentration increase and a slight solute concentration decrease all result in compensating reactions.

Perturbation	Reaction	Result	Behavior
$\delta T > 0$	$\delta v < 0$	$\delta T < 0$	STABLE
$\delta T < 0$	$\delta v > 0$	$\delta T > 0$	STABLE
$\delta C_L > 0$	$\delta C_s < 0$	$\delta C_L < 0$	STABLE
$\delta C_L < 0$	$\delta C_s > 0$	$\delta C_L > 0$	STABLE

Table I: Effect of perturbations for $v \geq v_m$.

Perturbation	Reaction	Result	Behavior
$\delta T > 0$	$\delta v < 0$	$\delta T < 0$	STABLE
$\delta T < 0$	$\delta v > 0$	$\delta T > 0$	STABLE
$\delta C_L > 0$	$\delta C_s < 0$	$\delta C_L > 0$	UNSTABLE
$\delta C_L < 0$	$\delta C_s > 0$	$\delta C_L < 0$	UNSTABLE

Table II: Effect of perturbations for $v < v_m$.

The situation for $v < v_m$ is summarized in Table II. As in the case $v \geq v_m$, increasing or decreasing the temperature slightly results in a compensating reaction. If, however, the solute concentration in the liquid increases slightly, the solute content of the growing solid decreases, causing the solute content of the liquid at the interface to increase still further beyond its steady state value. Similarly, unstable behavior results from a slight decrease in solute concentration. Thus, for $v < v_m$, the planar interface is unstable with respect to concentration perturbations.

The effect of changing v_D on the Aziz-Kaplan interface response functions is illustrated in figure 4. Although changing v_D has little effect on the shape of the interface response functions curve it does change the scale factor of the velocity axis. Decreasing v_D, decreases the velocity corresponding to a specific interface composition.

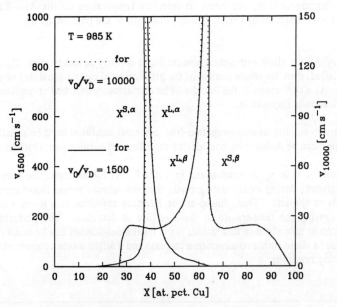

Figure 4:   A plot showing the effect of changing v_D on the Aziz-Kaplan interface response functions for an interface temperature of 985 K.

SPECIFICATION OF THE AZIZ-KAPLAN INTERFACE RESPONSE FUNCTIONS

The boundary between banded and microsegregation free solidification regions has been established experimentally by Boettinger et al. [11] and is represented by the solid line in figure 1. Based on the conditions proposed for microsegregation-free growth, the points on this boundary are identified as $\{X_m, v_m\}$. From X_m, the interface temperature is determined that shifts the velocity - solid interface composition curve sufficiently towards the center of the diagram (figure 1). The parameter v_D is determined to adjust the scale factor on the velocity axis so that the velocity corresponding to x_m equals v_m. The best fit of experimental data is attained by assuming that v_D is independent of composition but depends on temperature according to the equation $v_D = 0.33 \, T - 314.5$ expressed in cm s^{-1}. This gives a variation in v_D of between 18 and 65 cm s^{-1}, as a function of temperature and, for the first time, provides a means of determining the width, L, of the solid-liquid interface. As the liquid diffusion coefficient and its temperature dependence have been relatively well established [22], it is possible to calculate L as the ratio D_L/v_D. It can be seen from figure 5, that L varies between 13 and 78 Å as a function of temperature in the range considered.

Alternatively, if the width of the solid-liquid interface has been determined experimentally, then it is possible to predict the minimum scanning velocity of the beam at which steady state, microsegregation-free solidification will occur, as a function of alloy composition.

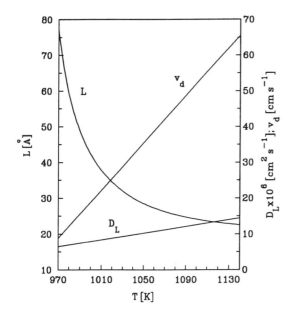

Figure 5: Variation of L, v_D, and D_L with the temperature of the interface.

FAR-FROM-EQUILIBRIUM EUTECTIC GROWTH

Consider the coupled growth of α and β plates with respective widths a and b into a liquid of solute concentration C_0. The characteristic spacing of the eutectic structure is denoted as λ. The mean concentrations in the solid α and β phases are respectively $C^{S,\alpha}$ and $C^{S,\beta}$, while the mean concentrations in the liquid just in front of the interface are $C^{L,\alpha}$ and $C^{L,\beta}$. It is assumed that no diffusion takes place in the solid phases. If we express the ratio of b/a as θ, then we can write the following relationship for the mean concentration of the growing solid, $C^{S,M}$:

$$C^{S,M} = \frac{1}{1+\theta} C^{S,\alpha} + \frac{\theta}{1+\theta} C^{S,\beta} \qquad (6)$$

Similarly, for the mean concentration of the liquid in front of the interface we can write:

$$C^{L,M} = \frac{1}{1+\theta} C^{L,\alpha} + \frac{\theta}{1+\theta} C^{L,\beta} \tag{7}$$

As we are looking for steady state solutions, $C^{S,M}$ must be equal to C_0. Substituting C_0 for $C^{S,M}$ in equation 6, we obtain the following expression for θ:

$$\theta = \frac{C_0 - C^{S,\alpha}}{C^{S,\beta} - C_0} \tag{8}$$

In the near-equilibrium mode of growth the concentrations in the liquid in front of the α and β phases can be approximated by the eutectic concentration C_E, so the mean liquid concentration is defined. However, in the far-from-equilibrium regime, the deviations of $C^{L,\alpha}$ and $C^{L,\beta}$ from C_E become quite large, so the mean liquid concentration is no longer the eutectic concentration. In this case we can solve for $C^{L,M}$ by inserting the expression for θ into equation 7 to get:

$$C^{L,M} = \frac{C^{L,\alpha}(C^{S,\beta} - C_0) + C^{L,\beta}(C_0 - C^{S,\alpha})}{C^{S,\beta} - C^{S,\alpha}} \tag{9}$$

The mean liquid concentration can be determined as long as the liquid and solid concentrations of both growing phases can be identified, applying the interface response functions specified for the Ag-Cu system.

The characteristic spacing of the coupled structure, λ, can now be calculated by taking into account the solute transport at the interface. To simplify the calculation, an approximate treatment of mass transport is employed in which solute transport parallel to the interface takes place by diffusion over an average distance $\lambda/2$ in a liquid layer of thickness, δ. The mass balance in this case yields the following expression:

$$\frac{4D_L(C^{L,\alpha} - C^{L,\beta})\delta}{\lambda v} = (C^{L,\alpha} - C^{S,\alpha} - C^{L,M} + C_0) \frac{\lambda}{1+\theta} \tag{10}$$

where D_L is the temperature dependent liquid diffusion coefficient and we have assumed unit depth.

In our model, it is assumed that at low velocities the width of the diffusion layer is constant while at high velocities it is inversely proportional to the growth rate. The arctan function has the correct asymptotic behavior, and thus δ is described by the equation:

$$\delta = \frac{2\delta_0}{\pi} \arctan\left(\frac{\kappa\pi D_L}{2\delta_0} \frac{1}{v}\right) \tag{11}$$

where δ_0 is a constant and κ is a dimensionless constant. For Ag-Cu at the eutectic concentration, the best agreement with the experimental data results from setting $\delta_0 = 425$ nm and $\kappa = 1$.

Substituting δ from equation 10 into equation 11 and rearranging gives the following expression for the characteristic spacing of the coupled structure, λ:

$$\lambda = \left\{ \frac{8 D_L \delta_0}{\pi v} (1+\theta) \left(\frac{C^{L,\alpha} - C^{L,\beta}}{C^{L,\alpha} - C^{S,\alpha} - C^{L,M} + C_0} \right) \arctan \left(\frac{\kappa \pi D_L}{2 \delta_0} \frac{1}{v} \right) \right\}^{\frac{1}{2}}$$
(12)

The undercooling specified in this analysis, ΔT_D, is that required to produce a sufficient concentration difference, $C^{L,\alpha} - C^{L,\beta}$, to drive the diffusion process transferring solute from the α to the β phase. The total undercooling, ΔT_T, has a second component required for the formation of the interphase boundaries, ΔT_B.

The contribution to the bulk free energy due to the formation of interfaces is given by:

$$\Delta G_B = \frac{2 \sigma_{\alpha\beta} W}{\lambda \rho} [\text{J mol}^{-1}]$$
(13)

where W is the molar weight, ρ is the density of the alloy, and $\sigma_{\alpha\beta}$ is the α/β interface energy. Knowing this free energy change we can determine the interface undercooling, ΔT_B, corresponding to a particular λ by comparing ΔG_B at C_0 to the free energy diagrams for the system at different temperatures and finding the matching one. The free energy versus composition functions for this system have been developed by Murray [20].

In this way we can calculate for a number of values of the diffusional undercooling, ΔT_D, and velocities of the interface, the corresponding lamellar width, λ, and the total undercooling, $\Delta T_T = \Delta T_D + \Delta T_B$. From these data, the value of λ that gives the smallest total undercooling ΔT_T at a specific interface velocity is determined. This will be the same λ which gives the largest velocity at a specific total undercooling.

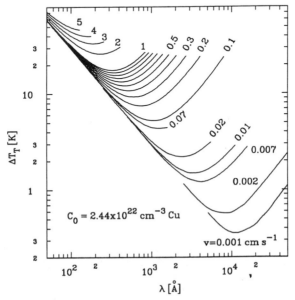

Figure 6:   Total undercooling, ΔT_T, *vs* lamellar spacing for different growth rates, assuming that the spacing for a specific velocity will be that which grows at a minimum total undercooling.

In figure 6, the total undercooling for solidification is plotted against the lamellar spacing for different growth rates and a near eutectic concentration. It can be seen that the minimum lamellar spacing decreases monotonically with velocity, while ΔT_T is increasing. In surface melting and solidification of Ag-Cu specimens using an electron beam, Boettinger et al. [11] found experimentally that the maximum velocity for coupled growth in this system was 2.5 cm s^{-1} with a spacing of 20 nm. The present model obtains the same velocity for a spacing of 20 nm but gives no indication why this growth rate should be a maximum for the Ag-Cu system. Boettinger et al. [11] calculate that for this velocity, 2.5 cm s^{-1}, there is an undercooling of 254 K, which they suggest causes a decrease in the diffusion coefficient leading to a termination of the steady state eutectic growth. Our model predicts a total undercooling of 27 K at this velocity, which is an order of magnitude less than the one calculated by Boettinger et al. [11].

The observation of a coupled growth structure in Ag-Cu with a spacing of 10 nm (figure 2), although not formed at steady state, suggests the feasibility of attaining lower spacings than 20 nm.

In figure 7 the spacings which are realized at the minimum undercoolings are plotted against growth rate and are compared to the existing experimental data for Ag-Cu from Cline and Stein [23], Cline and Lee [24] and Boettinger et al [11]. It can be seen that the model curve fits the results better than the classical relation $\lambda^2 v = $ constant, especially in the region of $v = 0.1$ cm s^{-1} where a change of slope can be observed.

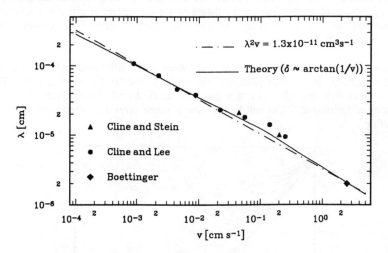

Figure 7:   Lamellar spacings at minimum total undercooling *vs* growth rate. The comparison of our model with the experimental data is shown.

BANDED STRUCTURE

The banded structure can be explained as a two step growth process. First, we have growth of the extended metastable solid solution and second, the growth of the coupled structure. The single phase growth is modeled using a finite difference solution of the diffusion equation in front of the moving solid-liquid interface in order to calculate the values of the solute concentration in the liquid in front of the interface, $C^{L,\gamma}$, at all points along successive time rows for a calculated temperature profile. After each step, the interface response functions specified for the Ag-Cu system are applied in order to calculate the velocity of the interface, v, and the concentration of the growing solid, $C^{S,\gamma}$. In this way, the changes of the concentrations and the velocity can be determined as the interface moves ahead.

Modeling the growth of the coupled structure is based on: (a) a mass transport treatment in which diffusion perpendicular to the interface dominates the process at high velocities [7] and (b) the free energy versus composition diagrams for the Ag-Cu system developed by Murray [20].

As the laser beam scans along the specimen's surface the isotherms advance at the same speed. Until the solid-liquid interface begins to move, its temperature, T, decreases in the following way:

$$T = T_m - \frac{\partial T}{\partial x} v_s t \tag{14}$$

where T_m is the melting temperature of the alloy, v_s is the scanning speed, t is time, and $\partial T/\partial x$ is the temperature gradient behind the interface. Using the temperature distribution model of Cline and Anthony [25], which neglects the effect of heat evolution at the advancing solid-liquid interface, $\partial T/\partial x$ was estimated to be approximately 1K μm^{-1} for the scanning velocity used in our experiment. In this case, $\partial T/\partial t \approx 3.4 \times 10^5$ K s^{-1}. Such a rapid drop in temperature may not allow sufficient time for the second eutectic phase to nucleate. It is proposed that this is why coupled growth has not been observed in the Ag-Cu system at velocities above 2.5 cm s^{-1}.

As the interface cools, its behavior can be predicted by the interface response functions. For small undercoolings single phase growth with $C^{L,\gamma} = C_0$, where $C_0 = 2.36 \times 10^{22}$ cm^{-3} Cu (38.3 at. pct. Cu) is the bulk concentration, is impossible, because single phase growth at any speed produces a solid phase with a much lower copper concentration than C_0. This would cause the liquid concentration at the interface to rise and the speed at which the interface can move to decrease almost instantly.

The analysis starts with $C^{S,\gamma} = C_0$ and $\Delta T_T = 70.8$ K (figure 8a). At this undercooling the speed of the interface is v = 93 cm s^{-1}, which is greater than the speed at which the isotherms are moving, so the interface will run into a slightly higher temperature region (figure 8b), causing its speed and the solid solute concentration to decrease. The model calculations show that the single phase will grow over a distance of approximately 2600 Å with gradually decreasing speed (figures 8c).

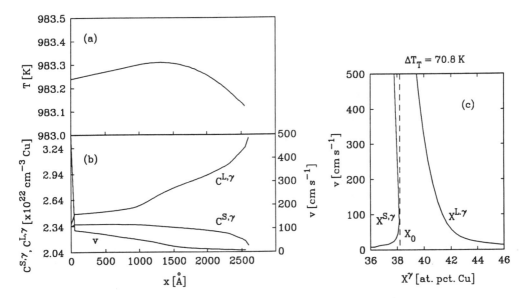

Figure 8:   Interface response functions for the single γ phase at $\Delta T_T = 70.8$ K (a). Model calculations: interface temperature (b) and concentrations at the interface and interface velocity (c) vs distance for the single phase growth during the initial cycle.

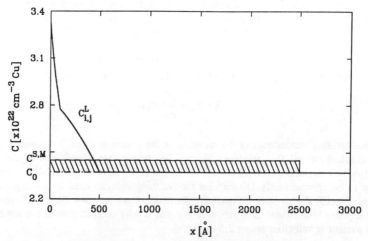

Figure 9:   Solute concentration profile, $C_{i,j}^L$, in the liquid at the end of the first cycle of the single phase growth. In order for the cross-hatched areas to be equal, $C^{S,M}$ for the coupled growth must be 2.44×10^{22} cm^{-3} Cu (39.3 at. pct. Cu).

Referring to the TTT diagram, the single phase growth causes a thermal arrest so eventually the cooling curve intersects the nucleation curve. As the experimental observations demonstrate that coupled growth occurs over approximately 2500 Å, it can be shown by mass balance that the single phase must stop when the interface velocity decreases below 5 cm s^{-1}. The exact velocity at which this occurs is not important in this model, due to the fact that at the end of the single phase growth the velocity drops very steeply with the distance. The distance travelled by the interface with $v < 5$ cm s^{-1} is less than 10 Å, which cannot affect the overall solute balance and the results of the model calculations. At this point the excess rejected copper is sufficient to form the coupled growth structure, assuming that nucleation occurs immediately. The concentration profile, $C_{i,j}^L$, in the liquid at the instant when the interface velocity reaches 5 cm s^{-1}, is given in figure 9. The total undercooling at this point is $\Delta T_T = 70.9$ K.

In order for the coupled growth structure to absorb the excess solute in the liquid over 2500 Å it must solidify with an average solute concentration $C^{S,M} = 2.44 \times 10^{22}$ cm^{-3} Cu (39.3 at. pct. Cu) as shown in figure 9. Applying our theory for far-from-equilibrium coupled growth [7], for $\Delta T_T = 70.9$ K and $C^{S,M} = 2.44 \times 10^{22}$ cm^{-3} Cu (39.3 at. pct. Cu), we can show that coupled growth must occur at a speed of 18 cm s^{-1} (figure 10). The coupled growth structure involves two phases: γ, with a concentration $C^{S,\gamma} = 1.86 \times 10^{22}$ cm^{-3} Cu (30.9 at. pct. Cu) and a lamellar width of 90 Å; and β', with concentration $C^{S,\beta'} = 7.06 \times 10^{22}$ cm^{-3} Cu (92 at. pct. Cu) and lamellar width of 10 Å. Taking into account the difference between the speed of the isotherms, which equals that of the beam, and the speed of the solid-liquid interface, we can calculate that the undercooling at the end of this cycle will be $\Delta T_T = 71.1$ K. This gives a temperature variation between the start and the end of the coupled growth of only 0.2 K, which validates the steady state approach (i.e. the speed will remain the same).

After 2500 Å of coupled growth the eutectic-like structure runs out of solute (figure 9) and it will be replaced by the single phase, which will start growing with $C^{S,\gamma} = 2.36 \times 10^{22}$ cm^{-3} Cu (38.3 at. pct. Cu) and a velocity greater than 93 cm s^{-1}, because of the greater undercooling (figure 11a). The single phase growth will proceed as described above, with the only difference being that the single phase now runs over a distance of approximately 7500 Å instead of 2600 Å (figures 11b,c).

In this way, after the first run of 2600 Å of the single phase we will have constant repetition of 2500 Å of steady state coupled growth of γ and β' with a mean solid concentration $C^{S,M} = 2.44 \times 10^{22}$ cm^{-3}

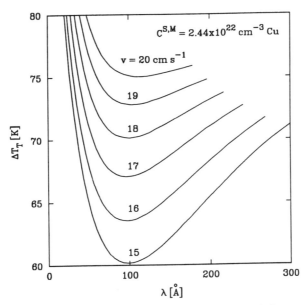

Figure 10: Results of the coupled growth theory. It is assumed that growth proceeds with a minimum undercooling at a given velocity.

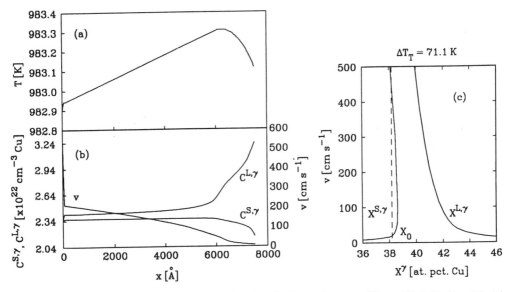

Figure 11: Interface response functions for the single γ phase at $\Delta T_T = 71.1$ K (a). Model calculations: interface temperature (b) and concentrations at the interface and interface velocity (c) *vs* distance for the single phase growth -subsequent cycles.

Cu (39.3 at. pct. Cu) and $v = 18$ cm s^{-1} followed by 7500 Å of single phase growth of γ occurring at a decreasing speed. Figure 12,a shows the position of the solid/liquid interface as a function of the time, validating that the average velocity of the process of solidification in the banded region is equal to the scanning velocity. The concentration of γ is initially $C^{S,\gamma} = 2.36 \times 10^{22}$ cm^{-3} Cu (38.3 at. pct. Cu). This value is maintained during most of the growth decreasing to 1.86×10^{22} cm^{-3} Cu (30.9 at. pct. Cu) in the last 1500 Å. Consequently, the mean solid concentration during the single phase growth is 2.34×10^{22} cm^{-3} Cu (38 at. pct. Cu). This verifies the model calculations, since $(C^{S,M} - C_0)$7500 Å for the single growth is equal to $(C^{S,M} - C_0)$2500 Å for coupled growth.

INTERFACE PRECIPITATION

In this section, banding in the Ag-Cu system is compared to interphase precipitation in vanadium steels. The recent analysis of this phenomenon by Todd et al. [1-5] provided the initial impetus for the present work.

The main features of interphase precipitation are illustrated in figure 12b, c. The velocity of the interphase boundary between austenite and the advancing ferrite is greatest immediately following the nucleation of the VC precipitates and then decreases with time. When the interphase boundary slows to a critical value, nucleation of the next sheet of precipitates occurs but does not pin the boundary.

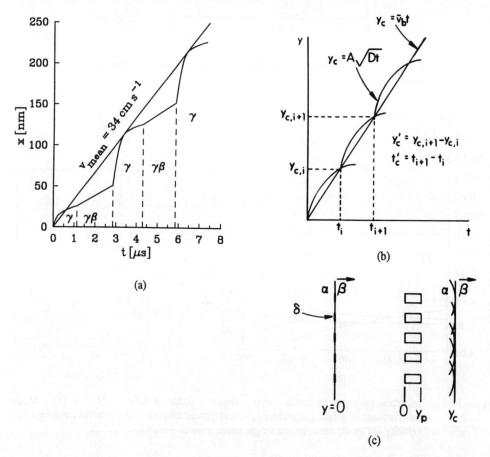

Figure 12: Plots of interphase boundary position *versus* time for banding in Ag-Cu alloys (a) and interphase precipitation in vanadium steels (b). Also, a schematic diagram showing nucleation and growth of VC during interphase precipitation in vanadium steels is shown in (c).

Immediately following, local equilibrium solute concentrations, which remain constant between nucleation events, are postulated at the boundary in the ferrite and austenite. Between the sheet of precipitates and the advancing interphase boundary is a layer of ferrite, which supplies vanadium and carbon to the growing sheet of precipitates. The interphase precipitation reaction in vanadium steels is controlled by the diffusivity of vanadium in ferrite.

Todd has proposed that banding in Ag-Cu and interphase precipitation in vanadium steels can be regarded as a generalized type of coupled growth closely related to conventional eutectic and eutectoid growth [26]. In all four types of growth one phase separates into two phases differing in solute content but with the same net content as the parent phase. As a consequence, no long range diffusion fields develop in front of the advancing interphase boundary during growth. In contrast to conventional eutectoid and eutectic transformations, which produce lamellae or rods aligned parallel to the growth direction, banding and interphase precipitation produce equally spaced sheets of precipitates with the sheets oriented perpendicular to the growth direction. In the case of eutectic and eutectoid growth, a steady state is possible in which the interphase boundary moves at constant velocity. In the case of banding and interphase precipitation, the velocity of the boundary varies in a periodic manner so growth gives the appearance of occurring at steady state.

REFERENCES

1) Todd, J.A. and Li, P.: Metall. Trans. A, 1986, 17A, 1191.
2) Todd, J.A., Li, P., and Copley, S.M.: Metall. Trans. A, 1988, 19A, 2133.
3) Li, P. and Todd, J.A.: Metall. Trans. A, 1988, 19A, 2139.
4) Todd, J.A. and Copley, S.M.: Scr. Metall., 1988, 22, 1771.
5) Todd, J.A. and Su, Y.J.: Metall. Trans. A, 1989, 20A, 1647.
6) Yankov, E.Y., Todd J.A., and Copley, S.M.: in "Proceedings of the Morris E. Fine Symposium", Liaw, P.K., Weertman, J.R., Marcus, H.L., and Santner, J.S., eds., TMS, Warrendale, PA, 1991, 29.
7) Yankov, E.Y., Copley, S.M., Yankova M.I., and Todd, J.A.: in "Proceedings of the Morris E. Fine Symposium", Liaw, P.K., Weertman, J.R., Marcus, H.L., and Santner, J.S., eds., TMS, Warrendale, PA, 1991, 33.
8) Yankov, E.Y., Copley, S.M., Todd J.A., and Yankova, M.I.: Mater. Res. Soc. Proc., 1991, 205, in press.
9) Beck, D.G., Copley, S.M., and Bass, M.: Metall. Trans. A, 1981, 12A, 1687.
10) Beck, D.G., Copley, S.M., and Bass, M.: Metall. Trans. A, 1982, 13A, 1879.
11) Boettinger, W.J., Shechtman, D., Schaefer, R.J., and Biancaniello, F.S.: Metall. Trans. A, 1984, 15A, 55.
12) Zimmermann, M. Carrard, M., and Kurz, W.: Acta Metall., 1989, 37, 3305.
13) Chattopadhyay, K., Swamy, V.T., and Agarwala, S.L.: Acta Metall., 1990, 38, 521.
14) Sastry, G.V.S. and Suryanarayana, C.: Mat. Sci. Engng., 1981, 47, 193.
15) Pandey, S.K., Gangopadhyay, D.K., and Suryanarayana, C.: Z. Metallk., 1986, 77, 13.
16) Chattopadhyay, K. and Mukhopadhyay, N.K.: J. Crystal Growth, 1990, 106, 387.
17) Aziz, M.J. and Kaplan, T.: Acta Metall., 1988, 36, 2335.
18) Kaplan, T., Aziz, M.J., and Gray, L.J.: J. Chem. Phys., 1989, 90, 1133.
19) Onsager, L.: Phys. Rev., 1931, 37, 405; 1931, 38, 2265.
20) Murray, J.L.: Metall. Trans. A, 1984, 15A, 261.
21), Smith, P.M., West, J.A., and Aziz, M.J.: Mater. Res. Soc. Proc., 1991, 205, in press.
22) Yamamura, T. and Ejima, T.: J. Japanese Inst. Met., 1973, 37, 901.
23) Cline, H.E. and Stein, D.F.: Trans. AIME, 1969, 245, 841.
24) Cline, H.E. and Lee, H.: Acta Metall., 1970, 18, 315.
25) Cline, H.E. and Anthony, T.R.: J. Appl. Phys., 1977, 48, 3895.
26) Todd, J. A.: J. Metals, 1991, 43, 45.

Materials Science Forum Vols. 102 - 104 (1992) pp. 433-442

CO$_2$ LASER SUBCRITICAL ANNEALING OF 2Cr-1Mo STEEL LASER WELDED JOINT

W. Cerri, G.P. Mor (a), M. Balbi and T. Zavanella (b)

(a) CISE S.p.a., Segrate, Italy
(b) Dipartimento di Meccanica, Politecnico di Milano, Italy

ABSTRACT

A study has been carried out regarding the feasibility of an integrated process of laser welding and a successive tempering treatment on a low alloy steel.
The heat treatment is applied at the end of the welding pass, modulating the power density transmitted to the piece in order to permit homogeneous distribution of the heat inside the welded joint.
The first part of the study was dedicated to the definition of the laser welding parameters while, in the second part, the whole process comprising welding and the post-weld treatment was studied.
For these types of steel, traditional welding techniques require a cycle of heat treatment which includes pre and post weld heating in order to prevent cracks and martensitic structures in melting and heat affected zones. The integrated laser process is much faster and can provide joints which are free of defects without the need for pre-heating. Post-weld tempering can reduce the hardness in the melting zones to a value of 270-290 HV (as-welded hardness is around 400 HV).

1. INTRODUCTION

The laser as an instrument for mechanical processing represents a valid alternative for all local heat treatment processes. The advantages of the laser are the high concentration of energy and the elevated speed of transfer of power from the source to the piece, thus permitting the operation to be carried out with a lower specific transfer of heat.
With the laser it is possible to obtain localised action on limited portions of material without involving the surrounding areas.
The use of the laser for surface treatments has been limited to

sporadic applications in industry, some of the most successful being metal gears and cam shafts in the automobile industry.
With the objective of expanding the potential field of application of laser heat treatments, a feasibility study of a post-weld heat treatment process integrated with a laser welding process was carried out.
A low alloy Cr-Mo steel was chosen as the material because the laser treatment drastically reduces the process times, a critical factor for this type of material, while at the same time guaranteeing that the welded joint has good mechanical characteristics. Thus, it becomes competitive compared to the traditional processes.
The welding process for this category of steel requires pre and post-heating with particular specifications in terms of speed of heating and cooling. The complete cycle of welding and subsequent heat treatment takes approximately four hours.
With reference to the particular applications of the material (high temperature) the reference standards for welding qualification are those of the ASME codes, section IX [1].

2. EXPERIMENTAL CONDITIONS

The ASTM A335 grade P22 steel utilised in the tests is a low alloy Cr-Mo steel. The best mechanical characteristics are obtained after a heat treatment which produces a compromise between the fineness of the dispersed phases and their thermal stability.
The welds were made on pipes with a diameter of 42 mm and a thickness of 3 mm.
The chemical composition and mechanical characteristics are provided in tab. 1.

Table 1 - Chemical composition and mechanical properties

%C	%Mn	%Si	%P	%S	%Mo		R	Rs	A
0.08	0.4	0.22	0.014	0.02	0.94		517 N/mm²	355 N/mm²	29%

The tests of welding and subcritical reheating were performed with a continuous cross-flow CO_2 laser (max. power 2.5 kW). The work station is equipped with a four axis transport system interfaced to a numerical control system, thus providing precise coupling and synchronisation of the laser source and the test piece.
Temperature control during annealing was ensured on-line by a control system comprising a two-color pyrometer and a feed-back control of incident laser power.
The vertical positioning of the laser beam over the workpiece was automated. At the end of the welding pass, the focussing head is raised above the surface of the pipe, projecting onto the pipe a spot of a size which is sufficient to provide ample coverage of the welded zone and the heat affected zone.
In figure 1 a layout of the control system is shown.
A preliminary analysis of different welding parameters sets have been carried out; in particular three different welded joints have been considered in order to evaluate the metallurgical transformations related to different parameters sets.
Welds have been then subjected to the full series of examinations laid down in the standards.
Post-weld heat treatment stage was studied later than the

preliminar approach to evaluate the potential of laser annealing technique; laser welded and annealed samples were therefore subjected to the qualification procedure already applied for the welded joints by the performance of mechanical tests (bending and

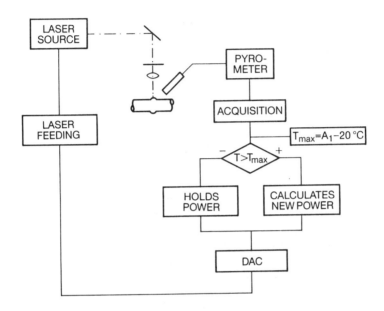

Figure 1 - Layout of the on-line temperature control system consisting of an optical two-colour pyrometer focused on the weld to be treated and a data acquisition card interfaced to a PC and to the power control of the laser source.

traction). In particular the results given by laser annealing have been compared with first to laser-welded and annealed by conventional methods and then to conventional TIG-welded joints. Since full penetration welding of the joints was required, the process parameters values giving complete penetration of the thickness have been chosen ; in particular different sets of parameters have been tested in order to reduce as much as possible the content of porosities in the fused zone (due to assist gas).
Three different optimized sets are reported in Tab.2.

Table 2 - Welding parameter values

	laser power W	welding speed m/min	protective gas flow rate coaxially l/min	laterally l/min
A	2100	0.85	25	40
B	1800	0.60	25	40
C	2500	1.50	25	40

Radiographic, macrographic and micrographic examinations have been performed, as well as mechanical tests (tensile and bending tests).
No cracks have been noted and mechanical test criteria given by ASME have been satisfied by all specimens.
In Fig.2 macrographic sections of laser welded joints realized with the three sets of parameters are shown.

Figure 2 - Macrophotographs of laser welding carried out with the three parameters: welded zone and heat affected zone (13x).

3. LASER SUBCRITICAL ANNEALING PROCESS

In a traditional welding processes for this type of material, the standards prescribe tempering or annealing of the welded joint

carried out locally by induction or resistence, in order to
attenuate excess hardening.
This research project consisted of a study of the preparation of
an original process of post-heating using a laser heat source.
The advantages of an innovative solution of this kind are the
fact that the post-heating can be executed in continuation after
the end of the welding pass and the fact that the treatment is so
rapid.
Processes of post-welding tempering effected with high density
energy fluxes have already been described in literature [2][3].
These processes are effected by utilising the beam as an energy
source which is no longer concentrated but distributed over an
area of an extent which is sufficient to include the welding bead
and the HAZ.
However, the effectiveness of this treatment is limited since the
heating is not sufficiently homogeneous throughout the thickness
of the welded joint.
To overcome this difficulty, the technique which has been
developed provides for temporal variation of the power density
applied, thus making it possible to maintain the surface
temperature below the critical point Ac1 while still continuing
to supply heat to the welded area.

4. TESTS ON PIPES

The tests carried out on the pieces used a technique whose
parameters were characterised by the lowest transfer of heat.
This choice was based on both metallurgical and economic factors.
As regards the structure produced in the three joints and their
extension, it must be noted that the test implemented with the
most heat showed a very extensive martensitic melted area. While,
from a purely metallurgical point of view, it could be
significant to test the laser heat tempering treatment on this
joint, from the economic point of view this would not be
advantageous. In fact, if we consider that the welding is done at
low speed, this would partially cancel out the "advantage" in
terms of productivity which could make a laser plant competitive
compared to a traditional welding process.
Laser tempering at 650°C was applied to welds carried out with
parameters "A".
The temperature was kept below the levels given in the post-weld
heat treatment specifications (700°C) to avoid quenching risks
and, given the measurement error deriving from the
instrumentation used, it was considered that it was opportune to
maintain the temperature 50°C lower than this limit.
The most striking aspects of the treatment which was tested are:
rather rapid heating (= 700°C/min) and very brief maintenance
time (. 3 min) compared to traditional post-heating treatments
(of the order of 15 min) [4].
No type of surfacing (krilon or graphite) was necessary in order
to permit immediate absorption of the incident radiation, since
the state of oxidation of the material after welding was
sufficient for this purpose.
The test piece was cooled at the end of the process by insulating
the pipe with quartz wool for no longer than 10 min.
A series of metallographic tests was then carried out on the
tempered test piece, including obervation under an optical
microscope and an electronic microscope and hardness tests.
The laser welded and tempered joints were finally subjected to
the same qualification procedures, carrying out the mechanical
tests (bend test and tensile test) already used for the laser
welded joints.

5. RESULTS

The joints welded and tempered by laser passed the bend and tension tests and were qualified to ASME standards. The melted zone is the most critical zone in a laser weld in terms of the structures obtained after welding. Since no type of weld material can be used during the process, the effect of dilution in the melted area which is characteristic of traditional welding methods cannot be exploited.

Thus putting a tempering method right for the central zone of the bead is sufficient to ensure that there are no hard phases in heat affected zones.

Metallographic examination demonstrated that there were perceptible variations between the non-tempered and tempered states in both the melted zones and the heat affected zones (figures 3 and 4).

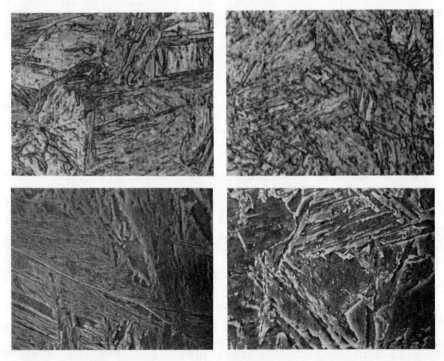

Figure 3 - Structure of the welded zone.   Left: as-weld (upper, 1000x, light; lower, 2500x, SEM); right: tempered (upper, 1000x, light; lower, 2500x, SEM).

In particular, in the melted zone, the morphology of the martensite of the non-tempered joint presented a clearly defined acicular structure while, in the tempered joint, the structure of the tempered martensite had a less-defined acicular structure surrounded by a finely dispersed carbide grain edge.

It is well known that one of the pecularities of Cr-Mo low alloy steels is the fact of carbide dispersion, thanks to which the material maintains its high mechanical properties even at high temperatures.

The results obtained in this first stage of the study were confirmed by the micro-hardness tests carried out on the tempered

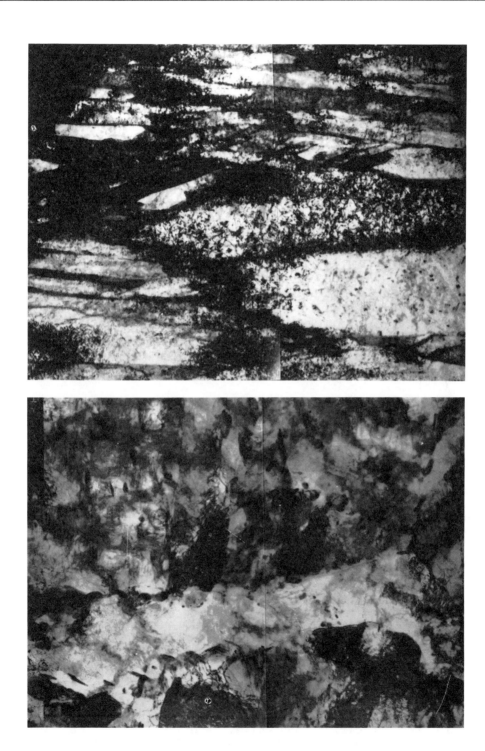

Figure 4 - Structure of the melted zone: upper) in the as-weld state; lower) after laser tempering (33000X, TEM).

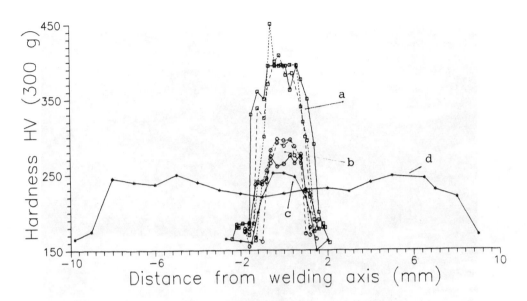

Figure 5 - Micro-hardness profiles. Comparison of as-laser welded (a), laser welded and laser tempered (b) (hardness curves carried out on three passes: upper, intermediate and lower), laser welded and oven tempered (c), TIG welded and oven tempered (d).

test pieces. The Vickers micro-hardness curves are given in figure 5 (300 g). Hardness in the melted zone decreased from around 400 HV to less than 300 HV; with an average hardness of 280 HV in the melted zone which indicated that the martensite had been transformed.

6. CONCLUSIONS

The comparison of joints which have been welded by laser and tempered by traditional methods and those both welded and tempered by laser demonstrated the intrinsic potential of the latter process: primarily, the rapidity of such a treatment which can be carried out immediately after the welding, raising the focussing head just enough to cover the seam with the focussing spot, and then, the duration of the treatment which, when applied to a laser-welded joint with a seam of limited dimensions, does not require those speed limits in the reheating stage which is necessary to relieve the stresses which are typical of traditional welding with high heat input.

The hardness analysis showed that traditionally tempered joints had lower values; with laser tempered joints having a hardness of 280 HV compared to the 250 HV of those tempered in an oven.

For this category of steel, the standards [4] specify that the hardness in joints which require post-weld heat treatment must be less than 270 HV while, in the case of those joints which are employed without treatment, the set limit is 300 HV.

Joints which are welded by laser and tempered in an oven fall within the specifications while, laser welded joints without treatment and those tempered by laser exceed this limit.

Thus, laser tempered joints require more detailed analysis, at least with regard to their resistence to creep and fatigue, since the limits laid down in the standards are based on technical tests of the type carried out on joints which are morphologically and structurally completely different from laser welded joints. Therefore, it can be maintained that, at this stage, the feasibility of an original laser tempering process has been verified at experimental level, while the applicability of this technique to real products requires a further study of their technical characteristics under operating conditions.

REFERENCES

[1] - ASME IX Welding and Brazing Qualification, Section IX, 1983
[2] - Fantini, V. and Incerti, G. - High performance 2,5kW
       industrial CO2 laser - Proc. SPIE, 1986, Vol.650.
[3] - Applicazioni laser in meccanica - C.N.R. Progetto
       finalizzato Laser di potenza, Giornate di lavoro 13-14
       dicembre 1983.
[4] - ANCC - raccolta S , 1978.

Materials Science Forum Vols. 102 - 104 (1992) pp. 443-458
Copyright Trans Tech Publications, Switzerland

CREATING TAILOR-MADE SURFACES WITH HIGH POWER CO_2-LASERS

C.F. Magnusson, G. Wiklund, E. Vuorinen, H. Engström (a) and
T.F. Pedersen (b)

(a) Lulea University of Technology, Sweden
(b) FORCE Institutes, Denmark

ABSTRACT

Laser surface modification offers a wide variety of possibilities to create surfaces with high wear- and corrosion resistance. In fact, the unique properties of the laser as a highly flexible and controllable heat source provides opportunities of tailor-made microstructures suitable for different industrial applications.

In this paper investigations in laser hardening, remelting and alloying of cast irons and its influence on microstructure, wear properties and resistance to tempering are reported. The results indicate possibilities of creating tailor-made microstructures, optimized for components servicing in different environments.

Also reported are investigations in microstructure control by cladding of Stellite No 6 on mild steel where the size and chemical analyses of the microstructure is modified by control of the energy input. The microstructure and results of wear tests are compared to TIG-cladding.

1. INTRODUCTION

In this work, a new group of precision methods has been used to modify the surfaces of cast irons and steels. Laserhardening, melting, alloying and cladding by laser respectively are methods that makes it possible to affect the surface structures of metals in a very accurate way.

The work is devided into two parts, part one in which different cast irons has been modified by hardening, melting and alloying techniques and part two, in which the surface of steels has been modified by cladding.

The purpose of this work has been to modify the structure of metallic surfaces by different laser techniques and to tailor-made the properties in this way.

2. SURFACE MODIFICATION OF CAST IRONS

2.1. Introduction

The surface of cast irons can be modified in different ways, for instance by hardening, melting or alloying of one or several overlapping tracks. Wear resistance is one property that can be increased by these techniques [1].

The most attractive advantage of surface hardening is the minimum shape distorsion of the part and hence the possibility to perform the hardening as the last step in the manufacturing process. Computer calculations can be used to predict the width of the hardened zones [2]. Surface melting has it´s main advantages in a more heat resistant structure than the martensitic achieved after hardening and in the possibility to create deeper layers with a high hardness. The properties of the alloyed zone after surface alloying will depend on the elements inserted to the melt.

Microstructures, hardness, wear- and tempering resistance for different cast irons have been studied in this work after hardening and melting respectively. Surface alloying with chromium of ductile cast iron has also been investigated.

2.2. Experimental set up

Kaleidoscope has been used to create a square distribution of the beam energy in hardening of single and overlapping tracks respectively. In the melting as well as in the alloying experiments a focused beam has been used. In the surface melting and alloying experiments, preheating and heating after the processes has been necessary to prevent cracking of the structures. The same experimental set up has been used in the alloying as in the cladding experiments (see figure 3.1).

The modified cast irons have been investigated by optical microscopy, by SEM and by hardness testing. EDX in the SEM has been used for chemical analysis. Wear testing with a pin on disc machine as well as tempering of cast irons has been performed. A 2.5 kW Spectra Physics 973 CO_2-laser has been used in the experiments. The process parameters are summarized in table 2.1.

Table 2.1. Process parameters.

	Hardening Kaleidoscope	Melting Focused beam	Alloying Focused beam
Beam size	8*8 mm	diameter = 2 mm	diameter = 2 mm
Power	1.57 kW	2.4 kW	2.4 kW
Coating	soot or "gun sight black"	none	none
Shielding gas	nitrogen	helium	argon
Preheating	none	450 °C	450 °C
Stabilizing	none	450 °C	450 °C
Velocity	0.7 m/min	2.2 m/min	1.0, 1.5 m/min
Powder	none	none	Cr < 100 mm
Powder feed	none	none	8.5-27.4 gram/min

Grey- (Swedish Standard; SS 140219), ductile- (SS 0717, SS 0727 and SS 0737) and malleable- (SS 0856) cast irons have been studied in the experiments. In the alloying experiments only the pearlitic ductile cast iron has been studied.

2.3. Results

2.3.1. Microstructure and hardness

The microstructures after the different processes are summarized in table 2.2. The structures consists of ferrite, pearlite and graphite before the heating process. The hardness after the different processes are dependent on the achieved microstructures and the amounts of the different phases.

Table 2.2. Microstructures of cast irons after different processes.

Hardening	Melting	Melting. Preheated and stabilized above Ms-temperature	Melting.Alloyed with Cr. Preheated and stabilized above Ms-temperature
Martensite, retained austenite and graphite	Martensite, retained austenite and cementite	Ferrite-carbide mixture and cementite	Ferrite-carbide mixture and cementite

The hardness values for the unprocessed materials and for the materials after hardening and melting respectively are given in figure 2.1.

The large differences in macrohardness after laser hardening (500-720 HV) depends mainly on the base structure of the cast iron, the presence of retained austenite and the amount of graphite.The differences between the microhardness values in the martensite are small (800 to 850 $HV_{0.3}$). A likely explanation is that the carbon content of the martensite in the different cast irons are almost the same.
The microstructure of the pearlitic ductile cast iron before and after hardening is shown in figure 2.2. In hardening of overlapping tracks a narrow zone will temper in adjacent tracks.

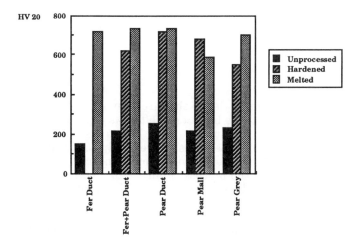

Figure 2.1. Hardness for unprocessed, hardened and melted materials of ferritic, ferritic-pearlitic and pearlitic ductile and of pearlitic malleable and grey cast irons.

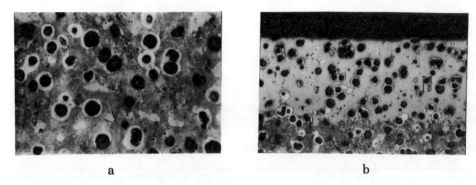

a   b

Figure 2.2. Pearlitic ductile cast iron. (a) Unprocessed structure, magnification 100 X. (b) Hardened structure, 50 X.

Melting gives a hardness of 600 HV for the malleable and of about 700 HV for the grey and the ductile cast irons (preheated and stabilized).

Hardness profiles after melting of overlapping tracks with a focused beam are shown in figure 2.3. The hardness profile for the grey cast iron is uniform. The melted structure consists of dendrites surrounded by eutectic. The microstructure consists of cementite and a mixture of ferrite and cementite.The hardness for the malleable cast iron is lower than those of the ductile- and grey cast irons, because of a higher content of the ferrite-cementite mixture.

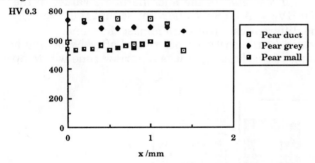

Figure 2.3. The surface hardness, at right angles to the direction of the tracks for pearlitic cast irons after melting. The distance between the tracks is 1.0 mm.

The hardness will decrease to a low value in the areas between adjacent tracks for the ductile cast irons after remelting. The material will temper and regraphitize in these areas. An increase of the velocity in the process (from 2.2 to >3.0 m/min) will prevent this tempering. Figure 2.4 shows the structure of ferritic ductile cast iron after melting.

The chromium content varied from 3.6 wt% up to 12 wt% after alloying, depending on the parameters used. It varied also as a function of the depth from the surface down to the bottom of the melted zone due to uncomplete mixing of the alloying elements with the base material. The chromium content could be increased up to 24.5 wt% if the distance between the melted surface and the powder nozzle was decreased. This is caused by the divergence of the powder flow from the nozzle.

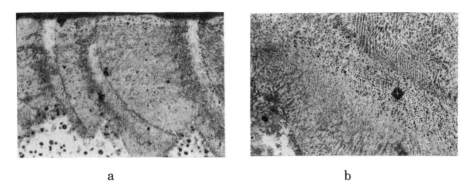

a   b

Figure 2.4. Ferritic ductile cast iron. Microstructure after surface melting. (a) Cross section, magnification 25 X. (b) Tempered zone, magnification 100 X.

The results shows that it is easy to achieve different amounts of chromium in the alloyed zone by changing some laser or powder parameter. Some results are presented in figure 2.5.

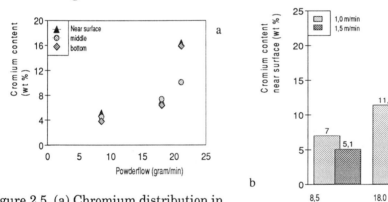

Figure 2.5. (a) Chromium distribution in alloyed zones for different powder flows. (b) Chromium content near the surface in alloyed zones for different velocities and powder flows.

The alloyed zones consists of primary precipitated dendrites surrounded by eutectic, figure 2.6. For Cr < 10 wt% the dendrites are dark and surrounded by a light matrix (after etching in nital). For Cr > 10wt% the dendrites are light and they are surrounded by dark areas. This indicates that there could be an eutectic point somewhere around 10 wt% Cr.

2.3.2 Tempering resistance

Tempering of melted structures (not preheated or stabilized) shows that the hardness value for the malleable pearlitic cast iron decreases to about 550 HV already after 5 hours at 400 °C. This could be explained by the high amount of martensite in the melted structure because of the undereutectic composition of this type of cast irons. Grey and ductile cast irons respectively has a nearly eutectic

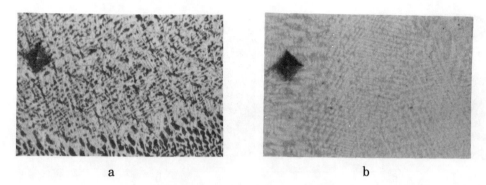

a   b

Figure 2.6. Surface alloyed pearlitic ductile cast iron with (a) Cr < 10 wt%, magnification 250 X. (b) Cr > 10 wt%, magnification 250 X.

composition, why their content of cementite will be higher than that of the malleable cast irons.

The ductile pearlitic cast iron will soften after 24 hours at 500 °C to the same level as for the malleable cast iron (400 HV). Softening of ductile cast irons is influenced by a secondary graphitization at approximately 500 °C. This secondary graphitization is strongly influenced by the silicon content of the cast iron [3]. The pearlitic grey iron has good tempering resistance at 500 °C but it will soften at 600 °C already after 24 hours to a level of about 300 HV. Figure 2.7 shows the hardness values for surfacemelted pearlitic structures after tempering at 500 °C.

The microhardness after laseralloying varied between 720 and 800 HV, as a function of the velocity and the amount of chromium in the alloyed zone. The alloyed specimens showed good resistance against tempering for 24 hours at 520 °C independent of the chromium content in the alloyed zone. The hardness decreased to approximately 710 HV after 24 hours as can be seen in figure 2.8.

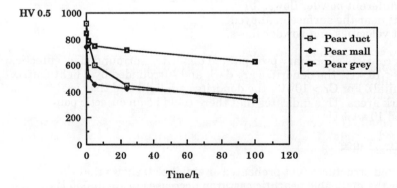

Figure 2.7. Hardness of surface melted cast irons after tempering at 500 °C.

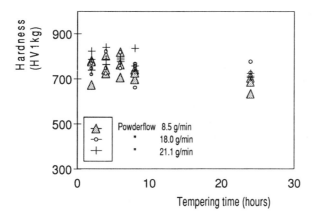

Figure 2.8. Hardness of alloyed pearlitic ductile cast iron after tempering at 520 °C.

2.3.3 Wear resistance

Untreated samples and samples with laserhardened single circular tracks were
exposed to dry sliding wear in a conventional pin-on disc machine. Additionally
samples, treated with overlapping (hardened as well as melted) tracks on a larger
area were also tested. In the set-up the disc was the sample, and the pin was
spherically tipped with a diameter of 4.7 mm.The material was tested to a total wear
length of 1000 m, at a rotational speed of 100 r.p.m. Test conditions are summarized
in table 2.3 The wear tracks were also examined in SEM to evaluate the active wear
mechanism.

Table 2.3 Conditions used in the dry wear tests

Type of sample	Pin type	Load Kg	Surface preparation	
as cast	steel*	2.5	as received (ground)	*Hardness = 927 HV
hardened	Si_3N_4	5.0	ground on SiC paper,	grid 1000
melted	Si_3N_4	5.0	as received (ground)	

With the present test parameters, hardening or surface melting lead to a decrease in
the wear rate of approximately 10 times. For laser treated samples, the wear rates
are in all cases small, but appear to be generally lower for surface melted than for
surface hardened materials. There is also a tendency for continuously hardened
surfaces to have better wear resistance than for the single tracks although the
difference is very small. The results of the wear tests are summarized in figure 2.9.
For the laser treated samples, the wear process leads to an annealing of the surface,
which gives a decrease in hardness of 100 HV.

The wear rate as well as the wear mechanism depend on what type of cast iron is being used : grey cast iron has the lowest wear rate in both the surface hardened and surface melted condition. On surface hardened grey cast iron, the wear loss appears to take place by brittle fracture in the material at the graphite flakes. This is also the case for the ferritic/pearlitic ductile iron that was hardened in overlapping tracks. In all other cases, the wear track shows a more or less oxidized, smeared structure.

The load, frictional force and rotational speed were monitored continuously throughout the test. After the test , the amount of wear was determined by calculation of weight loss from profilometry measurements of the wear track.

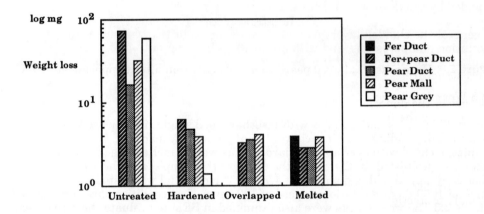

Figure 2.9. Wear test results for different cast irons. The untreated samples were tested with a 2.5 kg load on a steel pin, the other with a 5 kg load on a Si_3N_4 pin.

2.4. Conclusions

The surface hardness of cast irons can be increased from about 200 HV to 500-800 HV by the three methods used in the experiments, hardening, melting and alloying of the surface respectively by a CO-2-laser.

The tempering resistance of the treated materials increases in order hardening - melting - surface alloying by Cr.

The wear rate of different cast irons will decrease by a factor of approx. 10 after surface hardening or melting of the material.

Finally, these experiments have shown that it´s possible to combine wear resistance and tempering resistance by an appropriate selection of cast iron and laser process.

3. TAILOR-MADE CLADDINGS OF STELLITE No6 ON MILD STEEL

3.1 Introduction

Laser cladding is one of the new methods for surface modification introduced in industry. The process offers possibilities to achieve thin layers with low dilution of corrosion- and wear resistant alloys [4, 5].

Cladding with laser offers a new possibility of precision work on components which have so far not been possible to clad because of the thermal distortions caused by conventional methods. Mainly, it is the low and easily controllable energy input provided by the laser which gives this opportunity.

In general the cladded layers exhibit a very fine dendritic structure with low amount of dilution and porosities, and with an excellent bonding to the substrate. By regulating the energy input it is possible to control the grain size of the microstructure.

The size of the dendrite arms is controlled by the cooling rate, which in turn is controlled by the heat input from the laser and interaction time. In the present paper the microstructure of laser claddings of Stellite™ No 6[1], (28.3% Cr, 4.5 % W, 2.2% Fe, 2.0 % Ni, 1.2 % C, 0.1 % Si, bal Co), powder grade W (150/45 μm) on mild steel, SS1312 (AISI 1020), processed by two different lasers of 2.5 and 10 kW, is being compared to conventional TIG-welded cladding. The wear properties of the different cladding types have been compared from pin-on-disc wear tests in artificial sea water under cathodic protection.

3.2 Experimental set up

The samples of dimensions 50*50*10 mm were ground and blasted before cladding. The Stellite powder was applied to the surface by powder injection as shown in figure 3.1. The powder feeder used was a Tech Flow 5102, which is a commercially available standard feeder originally designed for plasma spraying applications.

Figure 3.1 Laser cladding by powder injection

---

[1] Stellite is a trademark of Cabot Corp., USA

The lasers used were two DC-exited fast transversal flow CO_2-lasers, Spectra Physics 973 (2.5 kW) and United Technology Research Centre (10 kW).

For the processing with the United Technology-laser, an integrating mirror (Spawr) giving a flat energy distribution with a beam size of 12*12 mm was used. All other samples were processed by the use of lenses with a focal length of 131 mm.

Table 3.1 summarizes the laser- and TIG parameters used in processing.

Table 3.1 Process parameters

PROCESS PARAMETERS		LASER		TIG
		RS 1700	UTRC	
Beam intensity distribution		Ring mode	Flat*	
Power density	[kw/cm^2]	23	6	
Energy input**	[J/mm]	230	864	1000
Interaction time	[s]	0.37	1.2	
Beam overlap	[%]	62	33	
Shied gas		N_2	Ar	
Shield gas flow	[l/min]	23	16	
Powder feed rate	[g/s]	1.8	2	
Current	[A]	-	-	180
Voltage	[V]			10
Pre-heating	[^{0}C]	-	-	150-300

* Spawr integrator used
** Reflection losses excluded

After processing the TIG-samples where slowly cooled while the laser samples where cooled in ambient air.

The equipment used for wear testing was a conventional pin-on-disc wear testing machine in which the sample is a flat rotating disc. On the disc at a radius of 10 mm rides a loaded stationary pin with a spherical Si_3N_4 tip of Ø 4,7 mm .

The instrument has been modified by the addition of an electrochemical cell [6]. This consists of a cylinder which is clamped directly onto the disc, a platinum counter electrode and a capillary tube connected to a saturated calomel reference electrode, thus allowing the electrochemical potential to be controlled. The electrolyte used was synthetic sea water which was circulated through a pump to create agitation in the cell.

The independent variables are load, rotational speed, cell temperature and electrochemical potential (or corrosion current). The dependent variables are friction force and corrosion current (or electrochemical potential). All variables are registered continuously.

In order to study the effects of mechanical wear alone without the effects of corrosion, the disc was held at a cathodic potential of -800 mV vs. SCE. Two loads, 4 and 8 kg, have been used in the experiment. The result of combined corrosion and wear has been reported earlier [6].

3.3 Results and discussion

3.3.1 Microstructure

The main factors influencing the microstructure of the three different cladding types is the heat input and the cooling rate. TIG-cladding is characterized by a large energy input, high utilization of the electric energy providing a high deposition rate, thus giving slower cooling rates than laser cladding. This results in a very coarse microstructure as shown in figure 3.2 with dendrite diameters of typically ≈15 µm. Suitable parameters for the 2.5 kW lasers produces a microstructure, which is much finer as shown in figure 3.3. The dendrites are here typically 7-10µm in diameter.

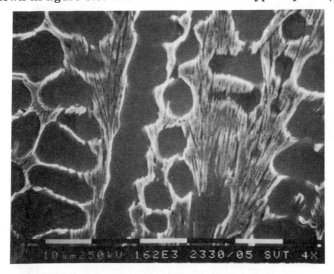

Figure 3.2. Microstructure of TIG-cladded Stellite No6. Energy input 1000 J/mm.

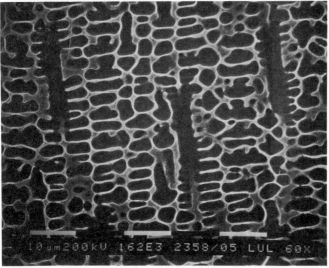

Figure 3.3. Microstructure of laser cladded Stellite No6. Energy input 230 J/mm.

With the integrating mirror the 10 kW laser produces a microstructure, which is an intermediate between the other two, as shown in figure 3.4.

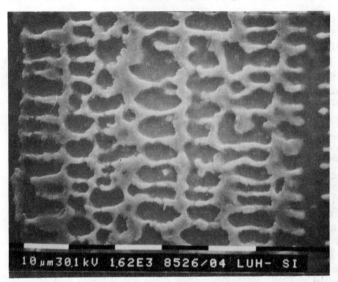

Figure 3.4. Microstructure of laser cladded Stellite No6. Energy input 864 J/mm.

The difference in heat input also causes the iron content of the cladding to vary as seen in table 3.2

Table 3.2. Properties of cladded Stellite No6.

Cladding method	Iron content [%]	Hardness [HV$_{0.3}$]	Carbide [vol%]
TIG-welding	10-11	425	54
10 kW laser	15	450	59
2.5 kW laser	6-7	502	37

A stereologic analysis of the micrographs in Figs 3.2-3.4 gives the approximate volume fractions of the interdendritic carbides also shown in table 3.2. The smaller volume fraction of carbides in the 2.5 kW laser cladding, is due to incomplete precipitation of the carbides caused by the rapid cooling. Furthermore, there is an obvious difference in the structure of the carbides in the TIG-welded and the laser welded claddings.The carbides in the laser claddings are more coherent than the "spiky" carbides in the TIG-cladding. This probably has a large influence on the strength of the carbides, which is decisive for wear strength. There is usually a clear correlation between the iron content and the hardness of Stellite claddings, but the larger hardness of the 10 kW laser-cladding compared to the TIG-cladding must be due to the smaller grain size of the former.

3.3.2 Wear tests

The results of the wear tests in terms of weight loss after 1000 m of wear are shown in table 3.3. The weight loss was calculated from the dimensions of the wear track, as determined by profilometry.

Table 3.3 Pin-on-disc wear of Stellite No6 claddings

Cladding method	Weight loss [mg]	
	4 kg load	8 kg load
TIG-welding	1.7	12.3
10 kW laser	1.0	4.5
2.5 kW laser	6.2	12.1

At the low load (4 kg) it is clear that the TIG-welded and 10 kW laser welded claddings are far superior to the 2.5 kW laser cladding. This is not entirely unexpected because sliding wear resistance is determined by the hardest phase present. It must be concluded that the coarse carbides provide a higher resistance to sliding wear. It should be noted that, in contrast, erosion resistance is controlled by the softest phase present, usually the matrix, and it would be expected, that the laser-clad samples would exhibit better erosion resistance than the TIG samples, simply because of the smaller grain size.

At the high load (8 kg) the TIG cladding and the 2.5 kW laser cladding show similar resistance, both inferior to the 10 kW laser cladding. This could be due to the different structure of the carbides in the TIG cladding. The "spiky" appearance of the carbides in the TIG-cladding could reduce it's fracture toughness and thus reduce the wear resistance. The combination of coherent carbides, small grain size and complete precipitation of the carbides that has been obtained with the 10 kW laser has also resulted in the best wear resistance.

3.4 Conclusions

The results from this work demonstrate the possibilities to create surfaces of different microstructures by varying the heat input. This has been accomplished by using different lasers, but it is also possible to vary the heat input using only one laser. In the case of cladding using the 2.5 kW laser the energy input has been varied between 185-300 J/mm, all giving high quality claddings, but with smaller variation of the size of the microstructure.

Wear resistance is influenced not only by the size of microstructure but probably also by the geometrical shape of the carbides, indicating increasing wear resistance advantages for laser cladding with increasing load.

4. ACKNOWLEDGEMENTS

The authors gratefully acknowledges the financial support for this project provided by the Nordic Fund for Technology and Industrial Development, The Swedish National Board for Technical Development, and The Danish Technical Research Council.

5. REFERENCES

1. Vuorinen, E.,et al. Surface Hardening and Melting of Cast Irons by CO_2-laser. Final Report Project No P86051 Laser Surface Treated Materials, Nordic Fund for Technology and Industrial Development, 1990, Oslo.

2. Li, W., Laser Transformation Hardening of Steel Surfaces. Luleå University of Technology, 1984:35 D

3. Elliot, A., Cast Iron Technology, Butterworth & Co, 1988.

4. Engström, H., et al. Laser Cladding of Stellite No6 on Mild Steel. Final Report Project No P86051 Laser Surface Treated Materials, Nordic Fund for Technology and Industrial Development, Oslo, 1990.

5. Sörensen, B., et al. A Comparison of Claddings Prepared by Laser Melting of Powder Layers and by TIG welding. 1st Nordic Laser Materials Processing Conference, Sept.30- Oct. 2, 1988, Oslo.

6. Sörensen, B., et al. The Combined Corrosion and Wear Resistance of Laser-Clad Stellite 6. 2nd European Conf. on Laser Treatment of Materials, Oct. 1988, Bad Neuheim, Germany